Economic Dynamics

Economic Dynamics

Theory and Computation

John Stachurski

The MIT Press
Cambridge, Massachusetts
London, England

MIT Press books may be purchased at special quantity discounts for business or sales promotional use. For information, please email special_sales@mitpress.mit.edu or write to Special Sales Department, The MIT Press, 55 Hayward Street, Cambridge, MA 02142.

This book was typeset in LATEX by the author and was printed and bound in the United States of America.

Library of Congress Cataloging-in-Publication Data

Stachurski, John, 1969–
Economic dynamics : theory and computation / John Stachurski.
    p. cm.
Includes bibliographical references and index.
ISBN 978-0-262-01277-5 (hbk. : alk. paper) 1. Statics and dynamics (Social sciences)—Mathematical models. 2.
Economics—Mathematical models. I. Title.
HB145.S73 2009
330.1′519—dc22   2008035973

10 9 8 7 6 5 4 3

To Cleo

Contents

Preface   xiii

Common Symbols   xvii

1 Introduction   1

I Introduction to Dynamics   9

2 Introduction to Programming   11
   2.1 Basic Techniques   11
        2.1.1   Algorithms   11
        2.1.2   Coding: First Steps   14
        2.1.3   Modules and Scripts   19
        2.1.4   Flow Control   21
   2.2 Program Design   25
        2.2.1   User-Defined Functions   25
        2.2.2   More Data Types   27
        2.2.3   Object-Oriented Programming   29
   2.3 Commentary   33

3 Analysis in Metric Space   35
   3.1 A First Look at Metric Space   35
        3.1.1   Distances and Norms   36
        3.1.2   Sequences   38
        3.1.3   Open Sets, Closed Sets   41
   3.2 Further Properties   44
        3.2.1   Completeness   44
        3.2.2   Compactness   46

        3.2.3   Optimization, Equivalence   48
        3.2.4   Fixed Points   51
    3.3   Commentary   54

4   Introduction to Dynamics   55
    4.1   Deterministic Dynamical Systems   55
        4.1.1   The Basic Model   55
        4.1.2   Global Stability   59
        4.1.3   Chaotic Dynamic Systems   62
        4.1.4   Equivalent Dynamics and Linearization   66
    4.2   Finite State Markov Chains   68
        4.2.1   Definition   68
        4.2.2   Marginal Distributions   72
        4.2.3   Other Identities   76
        4.2.4   Constructing Joint Distributions   80
    4.3   Stability of Finite State MCs   83
        4.3.1   Stationary Distributions   83
        4.3.2   The Dobrushin Coefficient   88
        4.3.3   Stability   90
        4.3.4   The Law of Large Numbers   93
    4.4   Commentary   96

5   Further Topics for Finite MCs   99
    5.1   Optimization   99
        5.1.1   Outline of the Problem   99
        5.1.2   Value Iteration   102
        5.1.3   Policy Iteration   105
    5.2   MCs and SRSs   107
        5.2.1   From MCs to SRSs   107
        5.2.2   Application: Equilibrium Selection   110
        5.2.3   The Coupling Method   112
    5.3   Commentary   116

6   Infinite State Space   117
    6.1   First Steps   117
        6.1.1   Basic Models and Simulation   117
        6.1.2   Distribution Dynamics   122
        6.1.3   Density Dynamics   125
        6.1.4   Stationary Densities: First Pass   129
    6.2   Optimal Growth, Infinite State   133

	6.2.1	Optimization	133
	6.2.2	Fitted Value Iteration	135
	6.2.3	Policy Iteration	142
6.3	Stochastic Speculative Price		145
	6.3.1	The Model	145
	6.3.2	Numerical Solution	150
	6.3.3	Equilibria and Optima	153
6.4	Commentary		155

II Advanced Techniques   157

7 Integration   **159**
7.1	Measure Theory		159
	7.1.1	Lebesgue Measure	159
	7.1.2	Measurable Spaces	163
	7.1.3	General Measures and Probabilities	166
	7.1.4	Existence of Measures	168
7.2	Definition of the Integral		171
	7.2.1	Integrating Simple Functions	171
	7.2.2	Measurable Functions	173
	7.2.3	Integrating Measurable Functions	177
7.3	Properties of the Integral		178
	7.3.1	Basic Properties	179
	7.3.2	Finishing Touches	180
	7.3.3	The Space L_1	183
7.4	Commentary		186

8 Density Markov Chains   **187**
8.1	Outline		187
	8.1.1	Stochastic Density Kernels	187
	8.1.2	Connection with SRSs	189
	8.1.3	The Markov Operator	195
8.2	Stability		197
	8.2.1	The Big Picture	197
	8.2.2	Dobrushin Revisited	201
	8.2.3	Drift Conditions	204
	8.2.4	Applications	207
8.3	Commentary		210

9   Measure-Theoretic Probability   **211**
  9.1   Random Variables   211
        9.1.1   Basic Definitions   211
        9.1.2   Independence   215
        9.1.3   Back to Densities   216
  9.2   General State Markov Chains   218
        9.2.1   Stochastic Kernels   218
        9.2.2   The Fundamental Recursion, Again   223
        9.2.3   Expectations   225
  9.3   Commentary   227

10 Stochastic Dynamic Programming   **229**
  10.1 Theory   229
        10.1.1 Statement of the Problem   229
        10.1.2 Optimality   231
        10.1.3 Proofs   235
  10.2 Numerical Methods   238
        10.2.1 Value Iteration   238
        10.2.2 Policy Iteration   241
        10.2.3 Fitted Value Iteration   244
  10.3 Commentary   246

11 Stochastic Dynamics   **247**
  11.1 Notions of Convergence   247
        11.1.1 Convergence of Sample Paths   247
        11.1.2 Strong Convergence of Measures   252
        11.1.3 Weak Convergence of Measures   254
  11.2 Stability: Analytical Methods   257
        11.2.1 Stationary Distributions   257
        11.2.2 Testing for Existence   260
        11.2.3 The Dobrushin Coefficient, Measure Case   263
        11.2.4 Application: Credit-Constrained Growth   266
  11.3 Stability: Probabilistic Methods   271
        11.3.1 Coupling with Regeneration   272
        11.3.2 Coupling and the Dobrushin Coefficient   276
        11.3.3 Stability via Monotonicity   279
        11.3.4 More on Monotonicity   283
        11.3.5 Further Stability Theory   288
  11.4 Commentary   293

12 More Stochastic Dynamic Programming   **295**
 12.1 Monotonicity and Concavity   295
  12.1.1 Monotonicity   295
  12.1.2 Concavity and Differentiability   299
  12.1.3 Optimal Growth Dynamics   302
 12.2 Unbounded Rewards   306
  12.2.1 Weighted Supremum Norms   306
  12.2.2 Results and Applications   308
  12.2.3 Proofs   311
 12.3 Commentary   312

III   Appendixes   **315**

A Real Analysis   **317**
 A.1 The Nuts and Bolts   317
  A.1.1 Sets and Logic   317
  A.1.2 Functions   320
  A.1.3 Basic Probability   324
 A.2 The Real Numbers   327
  A.2.1 Real Sequences   327
  A.2.2 Max, Min, Sup, and Inf   331
  A.2.3 Functions of a Real Variable   334

B Chapter Appendixes   **339**
 B.1 Appendix to Chapter 3   339
 B.2 Appendix to Chapter 4   342
 B.3 Appendix to Chapter 6   344
 B.4 Appendix to Chapter 8   345
 B.5 Appendix to Chapter 10   347
 B.6 Appendix to Chapter 11   349
 B.7 Appendix to Chapter 12   350

Bibliography   **357**

Index   **367**

Preface

The aim of this book is to teach topics in economic dynamics such as simulation, stability theory, and dynamic programming. The focus is primarily on stochastic systems in discrete time. Most of the models we meet will be nonlinear, and the emphasis is on getting to grips with nonlinear systems in their original form, rather than using crude approximation techniques such as linearization. As we travel down this path, we will delve into a variety of related fields, including fixed point theory, laws of large numbers, function approximation, and coupling.

In writing the book I had two main goals. First, the material would present the modern theory of economic dynamics in a rigorous way. I wished to show that sound understanding of the mathematical concepts leads to effective algorithms for solving real world problems. The other goal was that the book should be easy and enjoyable to read, with an emphasis on building intuition. Hence the material is driven by examples—I believe the fastest way to grasp a new concept is through studying examples—and makes extensive use of programming to illustrate ideas. Running simulations and computing equilibria helps bring abstract concepts to life.

The primary intended audience is graduate students in economics. However, the techniques discussed in the book add some shiny new toys to the standard tool kit used for economic modeling, and as such they should be of interest to researchers as well as graduate students. The book is as self-contained as possible, providing background in computing and analysis for the benefit of those without programming experience or university-level mathematics.

Part I of the book covers material that all well-rounded graduate students should know. The style is relatively mathematical, and those who find the going hard might start by working through the exercises in appendix A. Part II is significantly more challenging. In designing the text it was not my intention that all of those who read part I should go on to read part II. Rather, part II is written for researchers and graduate students with a particular aptitude for technical problems. Those who do read the majority of part II will gain a very strong understanding of infinite-horizon dynamic programming and (nonlinear) stochastic models.

How does this book differ from related texts? There are several books on computational macroeconomics and macrodynamics that have some similarity. In comparison, this book is not specific to macroeconomics. It should be of interest to (at least some) people working in microeconomics, operations research, and finance. Second, it is more focused on analysis and techniques than on applications. Even when numerical methods are discussed, I have tried to emphasize mathematical analysis of the algorithms, and acquiring a strong knowledge of the probabilistic and function-analytic framework that underlies proposed solutions.

The flip-side of the focus on analysis is that the models are more simplistic than in applied texts. This is not so much a book from which to learn about economics as it is a book to learn about techniques that are useful for economic modeling. The models we do study in detail, such as the optimal growth model and the commodity pricing model, are stripped back to reveal their basic structure and their links with one another.

The text contains a large amount of Python code, as well as an introduction to Python in chapter 2. Python is rapidly maturing into one of the major programming languages, and is a favorite of many high technology companies. The fact that Python is open source (i.e., free to download and distribute) and the availability of excellent numerical libraries has led to a surge in popularity among the scientific community. All of the Python code in the text can be downloaded from the book's homepage, at http://johnstachurski.net/book.html.

MATLAB aficionados who have no wish to learn Python can still read the book. All of the code listings have MATLAB counterparts. They can be found alongside the Python code on the text homepage.

A word on notation. Like any text containing a significant amount of mathematics, the notation piles up thick and fast. To aid readers I have worked hard to keep notation minimal and consistent. Uppercase symbols such as A and B usually refer to sets, while lowercase symbols such as x and y are elements of these sets. Functions use uppercase and lowercase symbols such as f, g, F, and G. Calligraphic letters such as $\mathcal{A}$ and $\mathcal{B}$ represent sets of sets or, occasionally, sets of functions. Python keywords are typeset in **boldface**. Proofs end with the symbol $\square$.

I provide a table of common symbols on page xvii. Furthermore, the index begins with an extensive list of symbols, along with the number of the page on which they are defined.

In the process of writing this book I received invaluable comments and suggestions from my colleagues. In particular, I wish to thank Robert Becker, Toni Braun, Roger Farmer, Onésimo Hernández-Lerma, Timothy Kam, Takashi Kamihigashi, Noritaka Kudoh, Vance Martin, Sean Meyn, Len Mirman, Tomoyuki Nakajima, Kevin Reffett, Ricard Torres, and Yiannis Vailakis.

Many graduate students also contributed to the end product, including Real Arai,

Rosanna Chan, Katsuhiko Hori, Murali Neettukattil, Jenö Pál, and Katsunori Yamada. I owe particular thanks to Dimitris Mavridis, who made a large number of thoughtful suggestions on both computational and theoretical content; and to Yu Awaya, who—in a remarkable feat of brain-power and human endurance—read part II of the book in a matter of weeks, solving every exercise, and checking every proof.

I have benefited greatly from the generous help of my intellectual elders and betters. Special thanks go to John Creedy, Cuong Le Van, Kazuo Nishimura, and Rabee Tourky. Extra special thanks go to Kazuo, who has helped me at every step of my personal random walk.

The editorial team at MIT Press has been first rate. I deeply appreciate their enthusiasm and their professionalism, as well as their gentle and thoughtful criticism: Nothing is more valuable to the author who believes he is always right—especially when he isn't.

I am grateful to the Department of Economics at Melbourne University, the Center for Operations Research and Econometrics at Université Catholique de Louvain, and the Institute for Economic Research at Kyoto University for providing me with the time, space, and facilities to complete this text.

I thank my parents for their love and support, and Andrij, Nic, and Roman for the same. Thanks also go to my extended family of Aaron, Kirdan, Merric, Murdoch, and Simon, who helped me weed out all those unproductive brain cells according to the principles set down by Charles Darwin. Dad gets an extra thanks for his long-running interest in this project, and for providing the gentle push that is always necessary for a task of this size.

Finally, I thank my beautiful wife Cleo, for suffering with patience and good humor the absent-minded husband, the midnight tapping at the computer, the highs and the lows, and then some more lows, until the job was finally done. This book is dedicated to you.

Common Symbols

$\overset{\text{IID}}{\sim} F$	independent and identically distributed according to F
$N(\mu, \sigma^2)$	the normal distribution with mean μ and variance σ^2
$\sim F$	distributed according to F
$\mathfrak{P}(A)$	the set of all subsets of A
$\|x\|_p$	the norm $(\sum_{i=1}^k x_i^p)^{1/p}$ on $\mathbb{R}^k$
$d_p(x, y)$	the distance $\|x - y\|_p$ on $\mathbb{R}^k$
bS	the set of bounded functions mapping S into $\mathbb{R}$
$\|f\|_\infty$	the norm $\sup_{x \in S} |f(x)|$ on bS
$d_\infty(f, g)$	the distance $\|f - g\|_\infty$ on bS
bcS	the continuous functions in bS
ibS	the increasing (i.e., nondecreasing) functions in bS
$ibcS$	the continuous functions in ibS
$B(\epsilon; x)$	the ϵ-ball centered on x
$\mathbb{1}_B$	the indicator function of set B
$\mathscr{B}(S)$	the Borel subsets of S
$\mathscr{P}(S)$	the distributions on S
δ_x	the probability measure concentrated on x
$s\mathscr{S}$	the simple functions on measure space $(S, \mathscr{S})$
$m\mathscr{S}$	the measurable real-valued functions on $(S, \mathscr{S})$
$b\mathscr{S}$	the bounded functions in $m\mathscr{S}$
$\mathscr{L}_1(S, \mathscr{S}, \mu)$	the μ-integrable functions on $(S, \mathscr{S})$
$L_1(S, \mathscr{S}, \mu)$	the metric space generated by $\mathscr{L}_1(S, \mathscr{S}, \mu)$
$\|f\|_1$	the norm $\mu(|f|)$ on $L_1(S, \mathscr{S}, \mu)$
$d_1(f, g)$	the distance $\|f - g\|_1$ on $L_1(S, \mathscr{S}, \mu)$
$D(S)$	the densities on S
$b\mathscr{M}(S)$	the finite signed measures on $(S, \mathscr{B}(S))$
$b\ell S$	the bounded, Lipschitz functions on metric space S
d_{FM}	the Fortet–Mourier distance on $\mathscr{P}(S)$

Chapter 1

Introduction

The teaching philosophy of this book is that the best way to learn is by example. In that spirit, consider the following benchmark modeling problem from economic dynamics: At time t an economic agent receives income y_t. This income is split into consumption c_t and savings k_t. Savings is used for production, with input k_t yielding output

$$y_{t+1} = f(k_t, W_{t+1}), \qquad t = 0, 1, 2, \ldots \tag{1.1}$$

where $(W_t)_{t \geq 1}$ is a sequence of independent and identically distributed shocks. The process now repeats, as shown in figure 1.1. The agent gains utility $U(c_t)$ from consumption $c_t = y_t - k_t$, and discounts future utility at rate $\rho \in (0, 1)$. Savings behavior is modeled as the solution to the expression

$$\max_{(k_t)_{t \geq 0}} \mathbb{E} \left[\sum_{t=0}^{\infty} \rho^t U(y_t - k_t) \right]$$

$$\text{subject to } y_{t+1} = f(k_t, W_{t+1}) \text{ for all } t \geq 0, \text{ with } y_0 \text{ given}$$

This problem statement raises many questions. For example, from what set of possible paths is $(k_t)_{t \geq 0}$ to be chosen? And how do we choose a path at the start of time such that the resource constraint $0 \leq k_t \leq y_t$ holds at each t, given that output is random? Surely the agent cannot choose k_t until he learns what y_t is. Finally, how does one go about computing the expectation implied by the symbol $\mathbb{E}$?

A good first step is to rephrase the problem by saying that the agent seeks a savings *policy*. In the present context this is a map σ that takes a value y and returns a number $\sigma(y)$ satisfying $0 \leq \sigma(y) \leq y$. The interpretation is that upon observing y_t, the agent's response is $k_t = \sigma(y_t)$. Next period output is then $y_{t+1} = f(\sigma(y_t), W_{t+1})$, next period

1

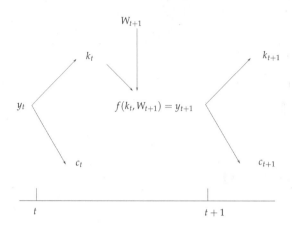

Figure 1.1 Timing

savings is $\sigma(y_{t+1})$, and so on. We can evaluate total reward as

$$\mathbb{E}\left[\sum_{t=0}^{\infty}\rho^{t}U(y_{t}-\sigma(y_{t}))\right]\quad\text{where}\quad y_{t+1}=f(\sigma(y_{t}),W_{t+1}),\text{ with }y_{0}\text{ given}\qquad(1.2)$$

The equation $y_{t+1}=f(\sigma(y_{t}),W_{t+1})$ is called a stochastic recursive sequence (SRS), or stochastic difference equation. As we will see, it completely defines $(y_{t})_{t\geq 0}$ as a sequence of random variables for each policy σ. The expression on the left evaluates the utility of this income stream.

Regarding the expectation $\mathbb{E}$, we know that in general expectation is calculated by integration. But how are we to understand this integral? It seems to be an expectation over the set of nonnegative sequences (i.e., possible values of the path $(y_{t})_{t\geq 0}$). High school calculus tells us how to take an integral over an interval, or perhaps over a subset of n-dimensional space $\mathbb{R}^{n}$. But how does one take an integral over an (infinite-dimensional) space of sequences?

Expectation and other related issues can be addressed in a very satisfactory way, but to do so, we will need to know at least the basics of a branch of mathematics called measure theory. Almost all of modern probability theory is defined and analyzed in terms of measure theory, so grasping its basics is a highly profitable investment. Only with the tools that measure theory provides can we pin down the meaning of (1.2).[1]

Once the meaning of the problem is clarified, the next step is considering how to solve it. As we have seen, the solution to the problem is a policy *function*. This is

---

[1]The golden rule of research is to carefully define your question before you start searching for answers.

rather different from undergraduate economics, where solutions are usually *numbers*, or perhaps *vectors* of numbers—often found by differentiating objective functions. In higher level applied mathematics, many problems have functions as their solutions.

The branch of mathematics that deals with problems having functions as solutions is called "functional analysis." Functional analysis is a powerful tool for solving real-world problems. Starting with basic functional analysis and a dash of measure theory, this book provides the tools necessary to optimize (1.2), including algorithms and numerical methods.

Once the problem is solved and an optimal policy is obtained, the income path $(y_t)_{t \geq 0}$ is determined as a sequence of random variables. The next objective is to study the dynamics of the economy. What statements can we make about what will "happen" in an economy with this kind of policy? Might it settle down into some sort of equilibrium? This is the ideal case because we can then make firm predictions. And predictions are the ultimate goal of modeling, partly because they are useful in their own right and partly because they allow us to test theory against data.

To illustrate analysis of dynamics, let's specialize our model so as to dig a little further. Suppose now that $U(c) = \ln c$ and $f(k, W) = k^\alpha W$. For this very special case, no computation is necessary: pencil and paper can be used to show (e.g., Stokey and Lucas 1989, §2.2) that the optimal policy is given by $\sigma(y) = \theta y$, where $\theta := \alpha \rho$. From (1.2) the law of motion for the "state" variable y_t is then

$$y_{t+1} = (\theta y_t)^\alpha W_{t+1}, \qquad t = 0, 1, 2, \ldots \tag{1.3}$$

To make life simple, let's assume that $\ln W_t \sim N(0,1)$. Here $N(\mu, v)$ represents the normal distribution with mean μ and variance v, and the notation $X \sim F$ means that X has distribution F.

If we take the log of (1.3), it is transformed into the linear system

$$x_{t+1} = b + \alpha x_t + w_{t+1}, \text{ where } x_t := \ln y_t, \ w_{t+1} \sim N(0,1), \text{ and } b := \alpha \ln \theta \tag{1.4}$$

This system is easy to analyze. In fact every x_t is normally distributed because x_1 is normally distributed (x_0 is constant and constant plus normal equals normal), and moreover x_{t+1} is normally distributed whenever x_t is normally distributed.[2]

One of the many nice things about normal distributions is that they are determined by only two parameters, the mean and the variance. If we can find these parameters, then we know the distribution. So suppose that $x_t \sim N(\mu_t, v_t)$, where the constants μ_t and v_t are given. If you are familiar with manipulating means and variances, you will be able to deduce from (1.4) that $x_{t+1} \sim N(\mu_{t+1}, v_{t+1})$, where

$$\mu_{t+1} = b + \alpha \mu_t \quad \text{and} \quad v_{t+1} = \alpha^2 v_t + 1 \tag{1.5}$$

---

[2]Recall that linear combinations of normal random variables are themselves normal.

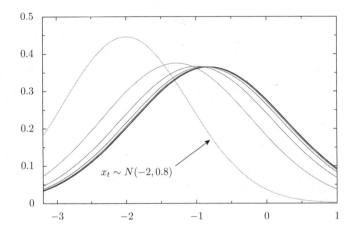

Figure 1.2 Sequence of marginal distributions

Paired with initial conditions μ_0 and v_0, these laws of motion pin down the sequences $(\mu_t)_{t\geq 0}$ and $(v_t)_{t\geq 0}$, and hence the distribution $N(\mu_t, v_t)$ of x_t at each point in time. A sequence of distributions starting from $x_t \sim N(-2, 0.8)$ is shown in figure 1.2. The parameters are $\alpha = 0.4$ and $\rho = 0.75$.

In the figure it appears that the distributions are converging to some kind of limiting distribution. This is due to the fact that $\alpha < 1$ (i.e., returns to capital are diminishing), which implies that the sequences in (1.5) are convergent (don't be concerned if you aren't sure how to prove this yet). The limits are

$$\mu^* := \lim_{t\to\infty} \mu_t = \frac{b}{1-\alpha} \quad \text{and} \quad v^* := \lim_{t\to\infty} v_t = \frac{1}{1-\alpha^2} \tag{1.6}$$

Hence the distribution $N(\mu_t, v_t)$ of x_t converges to $N(\mu^*, v^*)$.[3] Note that this "equilibrium" is a distribution rather than a single point.

All this analysis depends, of course, on the law of motion (1.4) being linear, and the shocks being normally distributed. How important are these two assumptions in facilitating the simple techniques we employed? The answer is that they are both critical, and without either one we must start again from scratch.

---

[3]What do we really mean by "convergence" here? We are talking about convergence of a sequence of *functions* to a given function. But how to define this? There are many possible ways, leading to different notions of equilibria, and we will need to develop some understanding of the definitions and the differences.

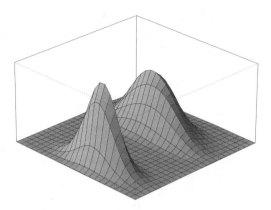

Figure 1.3 Stationary distribution

To illustrate this point, let's briefly consider the threshold autoregression model

$$X_{t+1} = \begin{cases} A_1 X_t + b_1 + W_{t+1} & \text{if } X_t \in B \subset \mathbb{R}^n \\ A_2 X_t + b_2 + W_{t+1} & \text{otherwise} \end{cases} \tag{1.7}$$

Here X_t is $n \times 1$, A_i is $n \times n$, b_i is $n \times 1$, and $(W_t)_{t \geq 1}$ is an IID sequence of normally distributed random $n \times 1$ vectors. Although for this system the departure from linearity is relatively small (in the sense that the law of motion is at least piecewise linear), analysis of dynamics is far more complex. Through the text we will build a set of tools that permit us to analyze nonlinear systems such as (1.7), including conditions used to test whether the distributions of $(X_t)_{t \geq 0}$ converge to some stationary (i.e., limiting) distribution. We also discuss how one should go about *computing* the stationary distributions of nonlinear stochastic models. Figure 1.3 shows the stationary distribution of (1.7) for a given set of parameters, based on such a computation.

Now let's return to the linear model (1.4) and investigate its sample paths. Figure 1.4 shows a simulated time series over 250 periods. The initial condition is $x_0 = 4$, and the parameters are as before. The horizontal line is the mean μ^* of the stationary distribution. The sequence is obviously correlated, and not surprisingly, shows no tendency to settle down to a constant value. On the other hand, the sample mean $\bar{x}_t := \frac{1}{t} \sum_{i=1}^{t} x_i$ seems to converge to μ^* (see figure 1.5).

That this convergence should occur is not obvious. Certainly it does not follow from the classical law of large numbers, since $(x_t)_{t \geq 0}$ is neither independent nor identically distributed. Nevertheless, the question of whether sample moments converge to the corresponding moments of the stationary distribution is an important one, with implications for both theory and econometrics.

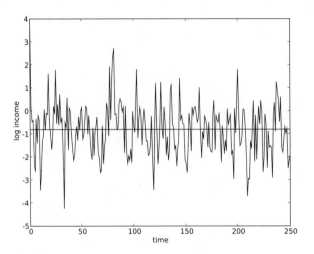

Figure 1.4 Time series

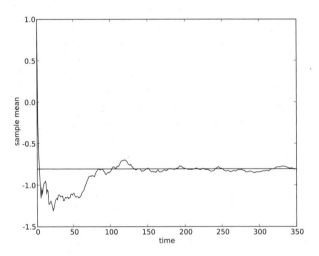

Figure 1.5 Sample mean of time series

For example, suppose that our simple model is being used to represent a given economy over a given period of time. Suppose further that the precise values of the underlying parameters α and ρ are unknown, and that we wish to estimate them from the data.[4] The method of moments technique proposes that we do this by identifying the first and second moments with their sample counterparts. That is, we set

$$\text{first moment} = \mu^*(\alpha, \rho) = \frac{1}{t} \sum_{i=1}^{t} x_t$$

$$\text{second moment} = v^*(\alpha, \rho) + \mu^*(\alpha, \rho)^2 = \frac{1}{t} \sum_{i=1}^{t} x_t^2$$

The right-hand side components $\frac{1}{t} \sum_{i=1}^{t} x_t$ and $\frac{1}{t} \sum_{i=1}^{t} x_t^2$ are collected from data, and the two equalities are solved simultaneously to calculate values for α and ρ.

The underlying assumption that underpins this whole technique is that sample means converge to their population counterparts. Figure 1.5 is not sufficient proof that such convergence does occur. We will need to think more about how to establish these results. More importantly, our linear normal-shock model is very special. Does convergence of sample moments occur for other related economies? The question is a deep one, and we will have to build up some knowledge of probability theory before we can tackle it.

To end our introductory comments, note that as well as studying theory, we will be developing computer code to tackle the problems outlined above. All the major code listings in the text can be downloaded from the text homepage, which can be found at `http://johnstachurski.net/book.html`. The homepage also collects other resources and links related to our topic.

---

[4]We have also used parameters b and θ, but $b = \alpha \ln \theta$, and $\theta = \alpha \rho$.

Part I

Introduction to Dynamics

Chapter 2

Introduction to Programming

Some readers may disagree, but to me computers and mathematics are like beer and potato chips: two fine tastes that are best enjoyed together. Mathematics provides the foundations of our models and of the algorithms we use to solve them. Computers are the engines that run these algorithms. They are also invaluable for simulation and visualization. Simulation and visualization build intuition, and intuition completes the loop by feeding into better mathematics.

This chapter provides a brief introduction to scientific computing, with special emphasis on the programming language Python. Python is one of several outstanding languages that have come of age in recent years. It features excellent design, elegant syntax, and powerful numerical libraries. It is also free and open source, with a friendly and active community of developers and users.

2.1 Basic Techniques

This section provides a short introduction to the fundamentals of programming: algorithms, control flow, conditionals, and loops.

2.1.1 Algorithms

Many of the problems we study in this text reduce to a search for algorithms. The language we will use to describe algorithms is called *pseudocode*. Pseudocode is an informal way of presenting algorithms for the benefit of human readers, without getting tied down in the syntax of any particular programming language. It's a good habit to begin every program by sketching it first in pseudocode.

Our pseudocode rests on the following four constructs:

if–then–else,   while,   repeat–until,   and   for

The general syntax for the **if–then–else** construct is

if *condition* **then**
| first sequence of actions
else
| second sequence of actions
end

The condition is evaluated as either true or false. If found true, the first sequence of actions is executed. If false, the second is executed. Note that the **else** statement and alternative actions can be omitted: If the condition fails, then no actions are performed. A simple example of the **if–then–else** construct is

if *there are cookies in the jar* **then**
| eat them
else
| go to the shops and buy more
| eat them
end

The **while** construct is used to create a loop with a test condition at the beginning:

while *condition* **do**
| sequence of actions
end

The sequence of actions is performed only if the condition is true. Once they are completed, the condition is evaluated again. If it is still true, the actions in the loop are performed again. When the condition becomes false the loop terminates. Here's an example:

while *there are cookies in the jar* **do**
| eat one
end

The algorithm terminates when there are no cookies left. If the jar is empty to begin with, the action is never attempted.

The **repeat–until** construct is similar:

repeat
| sequence of actions
until *condition*

Here the sequence of actions is always performed once. Next the condition is checked. If it is true, the algorithm terminates. If not, the sequence is performed again, the condition is checked, and so on.

The **for** construct is sometimes called a definite loop because the number of repetitions is predetermined:

for *element in sequence* **do**
| do something
end

For example, the following algorithm computes the maximum of a function f over a finite set S using a **for** loop and prints it to the screen.[1]

set $c = -\infty$
for *x in S* **do**
| set $c = \max\{c, f(x)\}$
end
print c

In the **for** loop, x is set equal to the first element of S and the statement "set $c = \max\{c, f(x)\}$" is executed. Next x is set equal to the second element of S, the statement is executed again, and so on. The statement "set $c = \max\{c, f(x)\}$" should be understood to mean that $\max\{c, f(x)\}$ is first evaluated, and the resulting value is assigned to the variable c.

Exercise 2.1.1 Modify this algorithm so that it prints the maximizer rather than the maximum. Explain why it is more useful to know the maximizer.

Let's consider another example. Suppose that we have two arrays A and B stored in memory, and we wish to know whether the elements of A are a subset of the elements of B. Here's a prototype algorithm that will tell us whether this is the case:

set subset = True
for *a in A* **do**
| **if** $a \notin B$ **then** set subset = False
end
print subset

---

[1]That is, it displays the value to the user. The term "print" dates from the days when sending output to the programmer required generating hardcopy on a printing device.

Exercise 2.1.2 The statement "$a \notin B$" may require coding at a lower level.[2] Rewrite the algorithm with an inner loop that steps through each b in B, testing whether $a = b$, and setting `subset` = False if no matches occur.

Finally, suppose we wish to model flipping a biased coin with probability p of heads, and have access to a random number generator that yields uniformly distributed variates on $[0, 1]$. The next algorithm uses these random variables to generate and print the outcome (either "heads" or "tails") of ten flips of the coin, as well as the total number of heads.[3]

```
set H = 0
for i in 1 to 10 do
    draw U from the uniform distribution on [0, 1]
    if U < p then                          // With probability p
        print "heads"
        H = H + 1
    else                                   // With probability 1 − p
        print "tails"
    end
end
print H
```

Note the use of indentation, which helps maintaining readability of our code.

Exercise 2.1.3 Consider a game that pays $1 if three consecutive heads occur in ten flips and zero otherwise. Modify the previous algorithm to generate a round of the game and print the payoff.

Exercise 2.1.4 Let b be a vector of zeros and ones. The vector corresponds to the employment history of one individual, where 1 means employed at the associated point in time, and 0 means unemployed. Write an algorithm to compute the longest (consecutive) period of employment.

2.1.2 Coding: First Steps

When it comes to programming, which languages are suitable for scientific work? Since the time it takes to complete a programming project is the sum of the time spent writing the code and the time that a machine spends running it, an ideal language would minimize both these terms. Unfortunately, designing such a language is not

---

[2]Actually, in many high-level languages you will have an operator that tests whether a variable is a member of a list. For the sake of the exercise, suppose this is not the case.

[3]What is the probability distribution of this total?

easy. There is an inherent trade-off between human time and computer time, due to the fact that humans and computers "think" differently: Languages that cater more to the human brain are usually less optimal for the computer and vice versa.

Using this trade-off, we can divide languages into (1) robust, lower level languages such as Fortran, C/C++, and Java, which execute quickly but can be a chore when it comes to coding up your program, and (2) the more nimble, higher level "scripting" languages, such as Python, Perl, Ruby, Octave, R, and MATLAB. By design, these languages are easy to write with and debug, but their execution can be orders of magnitude slower.

To give an example of these different paradigms, consider writing a program that prints out "Hello world." In C, which is representative of the first class of languages, such a program might look like this:

```
#include <stdio.h>

int main(int argc, char *argv[]) {
    printf("Hello world\n");
    return 0;
}
```

Let's save this text file as hello.c, and compile it at the shell prompt[4] using the gcc (GNU project C) compiler:

```
gcc -o hello hello.c
```

This creates an executable file called hello that can be run at the prompt by typing its name.

For comparison, let's look at a "Hello world" program in Python—which is representative of the second class of languages. This is simply

```
print("Hello world")
```

and can be run from my shell prompt as follows:

```
python hello.py
```

What differences do we observe with these two programs? One obvious difference is that the C code contains more boilerplate. In general, C will be more verbose, requiring us to provide instructions to the computer that don't seem directly relevant to our task. Even for experienced programmers, writing boilerplate code can be tedious and error prone. The Python program is much shorter, more direct, and more intuitive.

Second, executing the C program involves a two-step procedure: First compile, then run. The compiler turns our program into machine code specific to our operating system. By viewing the program as a whole prior to execution, the compiler can

---

[4]Don't worry if you aren't familiar with the notion of a shell or other the details of the program. We are painting with broad strokes at the moment.

optimize this machine code for speed. In contrast, the Python interpreter sends individual instructions to the CPU for processing as they are encountered. While slower, the second approach is more interactive, which is helpful for testing and debugging. We can run parts of the program separately, and then interact with the results via the interpreter in order to evaluate their performance.

In summary, the first class of languages (C, Fortran, etc.) put more burden on the programmer to specify exactly what they want to happen, working with the operating system on a more fundamental level. The second class of languages (Python, MATLAB, etc.) shield the programmer from such details and are more interactive, at the cost of slower execution. As computers have become cheaper and more powerful, these "scripting" languages have naturally become more popular. Why not ask the computer to do most of the heavy lifting?

In this text we will work exclusively with an interpreted language, leaving the first class of languages for those who wish to dig deeper. However, we note in passing that one can often obtain the best of both worlds in the speed versus ease-of-use trade-off. This is achieved by *mixed language programming*. The idea here is that in a typical program there are only a few lines of code that are executed many times, and it is these bottlenecks that should be optimized for speed. Thus the modern method of numerical programming is to write the entire program in an interpreted language such as Python, *profile* the code to find bottlenecks, and outsource those (and only those) parts to fast, compiled languages such as C or Fortran.[5]

The interpreted language we work with in the text is Python. However, MATLAB code is provided on the text homepage for those who prefer it. MATLAB has a gentler learning curve, and its numerical libraries are better documented. *Readers who are comfortable with MATLAB and have no interest in Python can skip the rest of this chapter.*

Python is a modern and highly regarded object-oriented programming language used in academia and the private sector for a wide variety of tasks. In addition to being powerful enough to write large-scale applications, Python is known for its minimalist style and clean syntax—designed from the start with the human reader in mind. Among the common general-purpose interpreted languages (Perl, Visual Basic, etc.), Python is perhaps the most suitable for scientific and numerical applications, with a large collection of MATLAB-style libraries organized primarily by the SciPy project (scipy.org).[6] The power of Python combined with these scientific libraries makes it an excellent choice for numerical programming.

Python is open source, which means that it can be downloaded free of charge,[7] and that we can peruse the source code of the various libraries to see how they work

---

[5] Another promising option is Cython, which is similar to Python but generates highly optimized C code.

[6] See also Sage, which is a mathematical tool kit built on top of Python.

[7] The main Python repositories are at python.org. Recently several distributions of Python that come bundled with various scientific tools have appeared. The book home page contains suggestions and links.

(and to make changes if necessary).

Rather than working directly with the Python interpreter, perhaps the best way to begin interacting with Python is by using IDLE.[8] IDLE is a free, cross-platform development environment for Python that comes bundled with most Python distributions. After you start IDLE, you will meet an interface to the interpreter that is friendlier than the standard one, providing color syntax highlighting, tab completion of commands, and more.[9] At the IDLE prompt you can start typing in commands and viewing the results:

```
>>> 10 * 10
100
>>> 10**2   # exponentiation
100
```

The result of the calculations is printed when you hit the return key. Notice that with the second calculation we added a *comment*, which is the hash symbol followed by text. Anything to the right of a hash is ignored by the interpreter. Comments are only for the benefit of human readers.

Continuing on with our tour of Python, next let's try assigning values to *variables*. Variables are names for values stored in memory. Here is an example:

```
>>> x = 3           # Bind x to integer 3
>>> y = 2.5         # Bind y to floating point number 2.5
>>> x               # Query the value of x
3
>>> z = x * y       # Bind z to x * y = 7.5
>>> x = x + 1       # Rebind x to integer 4
>>> a, b = 1, 2     # Bind a to 1 and b to 2
```

Names (x, y, etc.) are also called *identifiers*, and the values assigned to them are called *objects*. You can think of the identifiers as "pointers" to the location in memory where their values are stored. Identifiers are case sensitive (X is different from x), must start with a letter or an underscore, and cannot be one of Python's keywords.[10]

Observe that assignment of names to values (identifiers to objects) is accomplished via the "=" symbol. Assignment is also called *binding*: setting x = 3 *binds* the identifier x to the integer object 3. Passed the statement z = x * y, the interpreter evaluates

---

[8]See http://johnstachurski.net/book.html for more information on how to get started with Python.

[9]IDLE is not the best development environment for Python, but it is the easiest to get started with. A more powerful alternative is the IPython shell combined with a good text editor (such as Emacs or Vim). The following discussion assumes that you are using IDLE—if you are using something else, then you probably know what you are doing and hence need less instruction.

[10]These are: **and, del, from, not, while, as, elif, global, or, with, assert, else, if, pass, yield, break, except, import, class, exec, in, raise, continue, finally, is, return, def, for, lambda, try, True, False,** and **None**.

the expression x * y on the right-hand side of the = sign to obtain 7.5, stores this result in memory, and binds the identifier specified on the left (i.e., z) to that value. Passed x = x + 1, the statement executes in a similar way (right to left), creating a new integer object 4 and *rebinding* x to that value.

Objects stored in memory have different *types*. The type of our objects can be queried using the built-in function type():

```
>>> type(x)          # x = 4
<type 'int'>
>>> type(y)          # y = 2.5
<type 'float'>
```

The identifier x is bound to integer 4, while y is bound to floating point number 2.5. A floating point number is a number with a decimal point. Like most programming languages, Python distinguishes between floats and integers because integers are more efficient in terms of operations such as addition and multiplication.

Another common type of object is a *string*:

```
>>> s = "godzilla"          # Single or double quotes
>>> s.count("g")            # How many g's?  Returns 1
>>> s.startswith("god")     # Returns True
>>> s.upper()               # Returns "GODZILLA"
>>> s.replace("l", "m")     # Returns "godzimma"
>>> s2 = """
... Triple quotes can be used to create multi-line
... strings, which is useful when you want to
... record a lot of text."""
```

We are using some of Python's *string methods* (e.g., count(), upper()) to manipulate the string "godzilla". Before discussing methods, let's introduce a fourth data type, called *lists*. Lists are *containers* used to store a collection of objects.

```
>>> X = [20, 30, 40]     # Bind X to a list of integers
>>> sum(X)               # Returns 90
>>> max(X)               # Returns 40
```

We can extract elements of the list using square brackets notation. Like most programming languages, the first index is 0 rather than 1, so X[0] references the first element of the list (which is 20), X[1] references the second element, and so on. In this context the integers 0, 1, 2 are called the *indexes* of the list. The list can be modified using indexes as follows:

```
>>> X[0] = "godzilla"   # Now X = ["godzilla", 30, 40]
>>> del X[1]            # Now X = ["godzilla", 40]
```

Lists can be *unpacked* into variables containing their elements:

```
>>> x, y = ["a", "b"]   # Now x = "a" and y = "b"
```

We saw that Python has methods for operations on strings, as in s.count("g") above. Here s is the variable name (the identifier), bound to string "godzilla", and count() is the name of a string method, which can be called on any string. Lists also have methods. Method calls on strings, lists and other objects follow the general syntax

```
identifier.methodName(arguments)   # e.g., s.count("g")
```

For example, X.append(3) appends 3 to the end of X, while X.count(3) counts the number of times that 3 occurs in X. In IDLE you can enter X. at the prompt and then press the TAB key to get a list of methods that can be applied to lists (more generally, to objects of type type(X)).

2.1.3   Modules and Scripts

There are several ways to interact with the Python interpreter. One is to type commands directly into the prompt as above. A more common way is to write the commands in a text file and then run that file through the interpreter. There are many ways to do this, and in time you will find one that best suits your needs. The easiest is to use the editor found in IDLE: Open up a new window under the 'File' menu. Type in a command such as print("Hello world") and save the file in the current directory as hello.py. You can now run the file by pressing F5 or selecting 'Run Module'. The output Hello world should appear at the command prompt.

A text file such as hello.py that is run through an interpreter is known as a *script*. In Python such files are also known as *modules*, a module being any file with Python functions and other definitions. Modules can be run through the interpreter using the keyword **import**. Thus, as well as executing hello.py using IDLE's 'Run Module' command as above, we can type the following at the Python prompt:

```
>>> import hello   # Load the file hello.py and run
'Hello world'
```

When you first import the module hello, Python creates a file called hello.pyc, which is a byte-compiled file containing the instructions in hello.py. Note that if you now change hello.py, resave, and import again, the changes will not be noticed because hello.pyc is not altered. To affect the changes in hello.py, use reload(hello), which rewrites hello.pyc.

There are vast libraries of Python modules available,[11] some of which are bundled with every Python distribution. A useful example is math:

```
>>> import math              # Module math
```

---

[11]See, for example, http://pypi.python.org/pypi.

```
>>> math.pi                    # Returns 3.1415926535897931
>>> math.sqrt(4)               # Returns 2.0
```

Here pi is a float object supplied by math and sqrt() is a function object. Collectively these objects are called *attributes* of the module math.

Another handy module in the standard library is random.

```
>>> import random
>>> X = ["a", "b", "c"]
>>> random.choice(X)           # Returned "b"
>>> random.shuffle(X)          # X is now shuffled
>>> random.gammavariate(2, 2)  # Returned 3.433472
```

Notice how module attributes are accessed using moduleName.identifier notation. Each module has it's own *namespace*, which is a mapping from identifiers to objects in memory. For example, pi is defined in the namespace of math, and bound to the float $3.14\cdots 1$. Identifiers in different namespaces are independent, so modules mod1 and mod2 can both have distinct attribute a. No confusion arises because one is accessed as mod1.a and the other is accessed as mod2.a.[12]

When x = 1 is entered at the command line (i.e., Python prompt), the identifier x is registered in the interactive namespace.[13] If we import a module such as math, only the module name is registered in the interactive namespace. The attributes of math need to be accessed as described above (i.e, math.pi references the float object pi registered in the math namespace).[14] Should we wish to, we can however import attributes directly into the interactive namespace as follows:

```
>>> from math import sqrt, pi
>>> pi * sqrt(4)               # Returns 6.28...
```

Note that when a module mod is run from within IDLE using 'Run Module', commands are executed within the interactive namespace. As a result, attributes of mod can be accessed directly, without the prefix mod. The same effect can be obtained at the prompt by entering

```
>>> from mod import *    # Import everything
```

In general, it is better to be selective, importing only necessary attributes into the interactive namespace. The reason is that our namespace may become flooded with variable names, possibly "shadowing" names that are already in use.

---

[12]Think of the idea of a namespace as like a street address, with street name being the namespace and street number being the attribute. There is no confusion if two houses have street number 10, as long as we supply the street names of the two houses.

[13]Internally, the interactive namespace belongs to a top-level module called __main__.

[14]To view the contents of the interactive namespace type vars() at the prompt. You will see some stuff we have not discussed yet (__doc__, etc.), plus any variables you have defined or modules you have imported. If you import math and then type vars(math), you will see the attributes of this module.

Modules such as math, sys, and os come bundled with any Python distribution. Others will need to be installed. Installation is usually straightforward, and documented for each module. Once installed, these modules can be imported just like standard library modules. For us the most important third-party module[15] is the scientific computation package SciPy, which in turn depends on the fast array processing module NumPy. The latter is indispensable for serious number crunching, and SciPy provides many functions that take advantage of the facilities for array processing in NumPy, based on efficient C and Fortran libraries.[16]

Documentation for SciPy and NumPy can be found at the SciPy web site and the text home page. These examples should give the basic idea:

```
>>> from scipy import *
>>> integral, error = integrate.quad(sin, -1, 1)
>>> minimizer = optimize.fminbound(cos, 0, 2 * pi)
>>> A = array([[1, 2], [3, 4]])
>>> determinant = linalg.det(A)
>>> eigenvalues = linalg.eigvals(A)
```

SciPy functions such as sin() and cos() are called *vectorized* (or *universal*) functions, which means that they accept either numbers or sequences (lists, NumPy arrays, etc.) as arguments. When acting on a sequence, the function returns an array obtained by applying the function elementwise on the sequence. For example:

```
>>> cos([0, pi, 2 * pi]) # Returns array([ 1., -1.,  1.])
```

There are also modules for plotting and visualization under active development. At the time of writing, Matplotlib and PyX are popular and interact well with SciPy. A bit of searching will reveal many alternatives.

2.1.4 Flow Control

Conditionals and loops can be used to control which commands are executed and the order in which they are processed. Let's start with the **if/else** construct, the general syntax for which is

```
if <expression>:      # If <expression> is true, then
    <statements>      # this block of code is executed
else:
    <statements>      # Otherwise, this one
```

The **else** block is optional. An *expression* is any code phrase that yields a value when executed, and conditionals like **if** may be followed by any valid expression. Expressions are regarded as false if they evaluate to the boolean value **False** (e.g., 2 < 1),

---

[15] Actually a package (i.e., collection of modules) rather than a module.

[16] For symbolic (as opposed to numerical) algebra see SymPy or Sage.

to zero, to an empty list [], and one or two other cases. All other expressions are regarded as true:

```
>>> if 42 and 99: print("foo")    # Both True, prints foo
>>> if [] or 0.0: print("bar")    # Both False
```

As discussed above, a single = is used for *assignment* (i.e., binding an identifier to an object), rather than *comparison* (i.e., testing equality). To test the equality of objects, two equal signs are used:

```
>>> x = y = 1                     # Bind x and y to 1
>>> if x == y: print("foobar")    # Prints foobar
```

To test whether a list contains a given element, we can use the Python keyword **in**:

```
>>> 1 in [1, 2, 3]       # Evaluates as True
>>> 1 not in [1, 2, 3]   # Evaluates as False
```

To repeat execution of a block of code until a condition fails, Python provides the **while** loop, the syntax of which is

```
while <expression>:
    <statements>
```

Here a **while** loop is used to create the list X = [1, ..., 10]:[17]

```
X = []               # Start with an empty list
i = 1                # Bind i to integer object 1
while len(X) < 10:   # While length of list X is < 10
    X.append(i)      # Append i to end of list X
    i += 1           # Equivalent to i = i + 1
print("Loop completed")
```

At first X is empty and i = 1. Since len(X) is zero, the expression following the **while** keyword is true. As a result we enter the **while** loop, setting X = [1] and i = 2. The expression len(X) < 10 is now evaluated again and, since it remains true, the two lines in the loop are again executed, with X becoming [1, 2] and i taking the value 3. This continues until len(X) is equal to 10, at which time the loop terminates and the last line is executed.

Take note of the syntax. The two lines of code following the colon, which make up the body of the **while** loop, are indented the same number of spaces. This is not just to enhance readability. In fact the Python interpreter *determines the start and end of code blocks using indentation.* An increase in indentation signifies the start of a code block,

---

[17] In the following code, the absence of a Python prompt at the start of each line means that the code is written as a script (module) and then run.

whereas a decrease signifies its end. In Python the convention is to use four spaces to indent each block, and I recommend you follow it.[18]

Here's another example that uses the **break** keyword. We wish to simulate the random variable $T := \min\{t \geq 1 : W_t > 3\}$, where $(W_t)_{t \geq 1}$ is an IID sequence of standard normal random variables:

```
from random import normalvariate
T = 1
while 1:                              # Always true
    X = normalvariate(0, 1)           # Draw X from N(0,1)
    if X > 3:                         # If X > 3,
        print(T)                      # print the value of T,
        break                         # and terminate while loop.
    T += 1                            # Else T = T + 1, repeat
```

The program returns the first point in time t such that $W_t > 3$.

Another style of loop is the **for** loop. Often **for** loops are used to carry out operations on lists. Suppose, for example, that we have a list X, and we wish to create a second list Y containing the squares of all elements in X that are strictly less than zero. Here is a first pass:

```
Y = []
for i in range(len(X)):     # For all indexes of X
    if X[i] < 0:
        Y.append(X[i]**2)
```

This is a traditional C-style **for** loop, iterating over the indexes 0 to len(X)-1 of the list X. In Python, **for** loops can iterate over *any* list, rather than just sequences of integers (i.e., indexes), which means that the code above can be simplified to

```
Y = []
for x in X:     # For all x in X, starting with X[0]
    if x < 0:
        Y.append(x**2)
```

In fact, Python provides a very useful construct, called a *list comprehension*, that allows us to achieve the same thing in one line:

```
Y = [x**2 for x in X if x < 0]
```

A **for** loop can be used to code the algorithm discussed in exercise 2.1.2 on page 14:

```
subset = True
for a in A:
```

---

[18]In text files, tabs are different to spaces. If you are working with a text editor other than IDLE, you should configure the tab key to insert four spaces. Most decent text editors have this functionality.

```
    if a not in B:
        subset = False
print(subset)
```

Here A and B are expected to be lists.[19]

We can also code the algorithm on page 14 along the following lines:

```
from random import uniform
H, p = 0, 0.5                      # H = 0, p = 0.5
for i in range(10):                # Iterate 10 times
    U = uniform(0, 1)              # U is uniform on (0, 1)
    if U < p:
        print("heads")
        H += 1                     # H = H + 1
    else:
        print("tails")
print(H)
```

Exercise 2.1.5 Turn the pseudocode from exercise 2.1.3 into Python code.

Python **for** loops can step through any object that is *iterable*. For example:

```
from urllib import urlopen
webpage = urlopen("http://johnstachurski.net")
for line in webpage:
    print(line)
```

Here the loop acts on a "file-like" object created by the call to urlopen(). Consult the Python documentation on iterators for more information.

Finally, it should be noted that in scripting languages **for** loops are inherently slow. Here's a comparison of summing an array with a **for** loop and summing with NumPy's sum() function:

```
import numpy, time
Y = numpy.ones(100000)             # NumPy array of 100,000 ones
t1 = time.time()                   # Record time
s = 0
for y in Y:                        # Sum elements with for loop
    s += y
t2 = time.time()                   # Record time
s = numpy.sum(Y)                   # NumPy's sum() function
t3 = time.time()                   # Record time
print((t2-t1)/(t3-t2))
```

---

[19] Actually they are required to be iterable. Also note that Python has a set data type, which will perform this test for you. The details are omitted.

On my computer the output is about 200, meaning that, at least for this array of numbers, NumPy's sum() function is roughly 200 times faster than using a **for** loop. The reason is that NumPy's sum() function passes the operation to efficient C code.

2.2 Program Design

The material we have covered so far is already sufficient to solve useful programming problems. The issue we turn to next is that of design: How to construct programs so as to retain clarity and readability as our projects grow. We begin with the idea of functions, which are labeled blocks of code written to perform a specific operation.

2.2.1 User-Defined Functions

The first step along the road to good program design is learning to break your program up into functions. Functions are a key tool through which programmers implement the time-honored strategy of divide and conquer: problems are broken up into smaller subproblems, which are then coded up as functions. The main program then coordinates these functions, calling on them to do their jobs at the appropriate time.

Now the details. When we pass the instruction x = 3 to the interpreter, an integer "object" with value 3 is stored in the memory and assigned the identifier x. In a similar way we can also create a set of instructions for accomplishing a given task, store the instructions in memory, and bind an identifier (name) that can be used to *call* (i.e., run) the instructions. The set of instructions is called a *function*. Python supplies a number of built-in functions, such as max() and sum() above, as well as permitting users to define their own functions. Here's a fairly useless example of the latter:

```
def f(x, y):        # Bind f to a function that
    print(x + y)    # prints the value of x + y
```

After typing this into a new window in IDLE, saving it and running it, we can then call f at the command prompt:

```
>>> f(2,3)                  # Prints 5
>>> f("code ", "monkey")    # Prints "code monkey"
```

Take note of the syntax used to define f. We start with **def**, which is a Python keyword used for creating functions. Next follows the name and a sequence of arguments in parentheses. After the closing bracket a colon is required. Following the colon we have a code block consisting of the function body. As before, indentation is used to delimit the code block.

Notice that the order in which arguments are passed to functions is important. The calls f("a", "b") and f("b", "a") produce different output. When there are many arguments, it can become difficult to remember which argument should be passed first, which should be passed second, and so on. In this case one possibility is to use *keyword* arguments:

```
def g(word1="Charlie ", word2="don't ", word3="surf."):
    print(word1 + word2 + word3)
```

The values supplied to the parameters are defaults. If no value is passed to a given parameter when the function is called, then the parameter name is bound to its default value:

```
>>> g()                    # Prints "Charlie don't surf"
>>> g(word3="swim")    # Prints "Charlie don't swim"
```

If we do not wish to specify any particular default value, then the convention is to use **None** instead, as in x=**None**.

Often one wishes to create functions that *return* an object as the result of their internal computations. To do so, we use the Python keyword **return**. Here is an example that computes the norm distance between two lists of numbers:

```
def normdist(X, Y):
    Z = [(x - y)**2 for x, y in zip(X, Y)]
    return sum(Z)**0.5
```

We are using the built-in function zip(), which allows us to step through the x, y pairs, as well as a list comprehension to construct the list Z. A call such as

```
>>> p = normdist(X, Y)    # X, Y are lists of equal length
```

binds identifier p to the value returned by the function.

It's good practice to include a *doc string* in your functions. A doc string is a string that documents the function, and comes at the start of the function code block. For example,

```
def normdist(X, Y):
    "Computes euclidean distance between two vectors."
    Z = [(x - y)**2 for x, y in zip(X, Y)]
    return sum(Z)**0.5
```

Of course, we could just use the standard comment notation, but doc strings have certain advantages that we won't go into here. In the code in this text, doc strings are used or omitted depending on space constraints.

Python provides a second way to define functions, using the **lambda** keyword. Typically **lambda** is used to create small, in-line functions such as

```
f = lambda x, y: x + y    # E.g. f(1,2) returns 3
```

Note that functions can return any Python object, including functions. For example, suppose that we want to be able to create the Cobb–Douglas production function $f(k) = Ak^{\alpha}$ for any parameters A and α. The following function takes parameters (A, α) as arguments and creates and returns the corresponding function f:

```
def cobbdoug(A, alpha):
    return lambda k: A * k**alpha
```

After saving and running this, we can call cobbdoug at the prompt:

```
>>> g = cobbdoug(1, 0.5)    # Now g(k) returns 1 * k**0.5
```

2.2.2 More Data Types

We have already met several native Python data types, such as integers, lists and strings. Another native Python data type is the *tuple*:

```
>>> X = (20, 30, 40)    # Parentheses are optional
```

Tuples behave like lists, in that one can access elements of the tuple using indexes. Thus X[0] returns 20, X[1] returns 30, and so on. There is, however, one crucial difference: Lists are a *mutable* data type, whereas tuples (like strings) are *immutable*. In essence, mutable data types such as lists can be changed (i.e., their contents can be altered without creating a new object), whereas immutable types such as tuples cannot. For example, X[0] = 3 raises an exception (error) if X is a tuple. If X is a list, then the same statement changes the first element of X to 3.

Tuples, lists, and strings are collectively referred to as *sequence* types, and they support a number of common operations (on top of the ability to access individual elements via indexes starting at zero). For example, adding two sequences *concatenates* them. Thus (1, 2) + (3, 4) creates the tuple (1, 2, 3, 4), while "ab" + "cd" creates "abcd". Multiplication of a sequence by an integer n concatenates with n copies of itself, so [1] * 3 creates [1, 1, 1]. Sequences can also be *unpacked*: x, y = (1, 2) binds x to 1 and y to 2, while x, y = "ab" binds x to "a" and y to "b".

Another useful data type is a *dictionary*, also known as a mapping, or an associative array. Mathematically, a dictionary is just a function on a finite set, where the programmer supplies the domain of the function plus the function values on elements of the domain. The points in the domain of a dictionary are referred to as its *keys*. A dictionary d is created by specifying the key/value pairs. Here's an example:

```
>>> d = {"Band": "AC/DC", "Track": "Jailbreak"}
>>> d["Track"]
"Jailbreak"
```

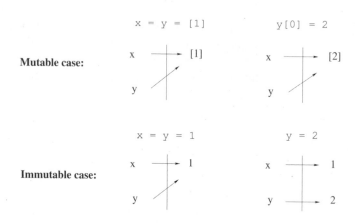

Figure 2.1 Mutable and immutable types

Just as lists and strings have methods that act on them, dictionaries have dictionary methods. For example, d.keys() returns the list of keys, and d.values() returns the list of values. Try the following piece of code, which assumes d is as defined above:

```
>>> for key in d.keys(): print(key, d[key])
```

Values of a dictionary can be any object, and keys can be any immutable object.

This brings us back to the topic of mutable versus immutable. A good understanding of the difference between mutable and immutable data types is helpful when trying to keep track of your variables. At the same time the following discussion is relatively technical, and can probably be skipped if you have little experience with programming.

To begin, consider figure 2.1. The statement x = y = [1] binds the identifiers x and y to the (mutable) list object [1]. Next y[0] = 2 modifies that same list object, so its first (and only) element is 2. Note that *the value of x has now changed,* since the object it is bound to has changed. On the other hand, x = y = 1 binds x and y to an immutable integer object. Since this object cannot be altered, the assignment y = 2 *rebinds* y to a *new* integer object, *and x remains unchanged.*[20]

A related way that mutable and immutable data types lead to different outcomes is when passing arguments to functions. Consider, for example, the following code segment:

---

[20]You can check that x and y point to different objects by typing id(x) and id(y) at the prompt. Their unique identifier (which happens to be their location in memory) should be different.

```
def f(x):
    x = x + 1
    return x
x = 1
print(f(x), x)
```

This prints 2 as the value of f(x) and 1 as the value of x, which works as follows: After the function definition, x = 1 creates a *global variable* x and binds it to the integer object 1. When f is called with argument x, a *local namespace* for its variables is allocated in memory, and the x inside the function is created as a *local variable* in that namespace and bound to the same integer object 1. Since the integer object is immutable, the statement x = x + 1 creates a *new* integer object 2 and rebinds the local x to it. This reference is now passed back to the calling code, and hence f(x) references 2. Next, the local namespace is destroyed and the local x disappears. Throughout, the global x remains bound to 1.

The story is different when we use a mutable data type such as a list:

```
def f(x):
    x[0] = x[0] + 1
    return x
x = [1]
print(f(x), x)
```

This prints [2] for both f(x) and x. Here the global x is bound to the list object [1]. When f is called with argument x, a local x is created and bound to the same list object. Since [1] is mutable, x[0] = x[0] + 1 modifies this object without changing its location in memory, so both the local x and the global x are now bound to [2]. Thus the global variable x is modified by the function, in contrast to the immutable case.

2.2.3   Object-Oriented Programming

Python supports both *procedural* and *object-oriented* programming (OOP). While any programming task can be accomplished using the traditional procedural style, OOP has become a central part of modern programming design, and will reward even the small investment undertaken here. It succeeds because its design pattern fits well with the human brain and its natural logic, facilitating clean, efficient code. It fits well with mathematics because it encourages *abstraction*. Just as abstraction in mathematics allows us to develop general ideas that apply in many contexts, abstraction in programming lets us build structures that can be used and reused in different programming problems.

Procedural programming is based around functions (procedures). The program has a *state*, which is the values of its variables, and functions are called to act on these data according to the task. Data are passed to functions via function calls. Functions return output that modifies the state. With OOP, however, data and functions are bundled together into logical entities called *abstract data types* (ADTs). A *class* definition is a blueprint for such an ADT, describing what kind of data it stores, and what functions it possesses for acting on these data. An *object* is an *instance* of the ADT; an individual realization of the blueprint, typically with its own unique data. Functions defined within classes are referred to as *methods*.

We have already met objects and methods. Recall that when the Python interpreter receives the instruction X = [1, 2], it stores the data [1, 2] in memory, recording it as an *object of type list*. The identifier X is bound to this object, and we can use it to call methods that act on the data. For example, X.reverse() changes the data to [2, 1]. This method is one of several list methods. There are also string methods, dictionary methods, and so on.

What we haven't done so far is create our own ADTs using class definitions. You will probably find the class definition syntax a little fiddly at first, but it does become more intuitive as you go along. To illustrate the syntax we will build a simple class to represent and manipulate polynomial functions. The data in this case are the coefficients $(a_0, \ldots, a_N)$, which define a unique polynomial

$$p(x) = a_0 + a_1 x + a_2 x^2 + \cdots a_N x^N = \sum_{n=0}^{N} a_n x^n \qquad (x \in \mathbb{R})$$

To manipulate these data we will create two methods, one to evaluate the polynomial from its coefficients, returning the value $p(x)$ for any x, and another to differentiate the polynomial, replacing the original coefficients $(a_0, \ldots, a_N)$ with the coefficients of p'.

Consider, first, listing 2.1, which sketches a class definition in pseudo-Python. This is *not* real Python code—it is intended to give the feeling of how the class definition might look, while omitting some boilerplate. The name of the class is Polynomial, as specified after the keyword **class**. The class definition consists of three methods. Let's discuss them in the order they appear.

The first method is called initialize(), and represents a *constructor*, which is a special method most languages provide to build (construct an instance of) an object from a class definition. Constructor methods usually take as arguments the data needed to set up a specific instance, which in this case is the vector of coefficients $(a_0, \ldots, a_N)$. The function should be passed a list or tuple, to which the identifier coef is then bound. Here coef[i] represents a_i.

The second method evaluate() evaluates $p(x)$ from x and the coefficients. We are using the built-in function enumerate(), which allows us to step through the i, X[i] pairs of any list X. The third method is differentiate(), which modifies the data

Listing 2.1 (`polyclass0.py`) A polynomial class in pseudo-Python

```
class Polynomial:

    def initialize(coef):
        """Creates an instance p of the Polynomial class,
        where p(x) = coef[0] x^0 + ... + coef[N] x^N."""

    def evaluate(x):
        y = sum(a*x**i for i, a in enumerate(coef))
        return y

    def differentiate():
        new_coef = [i*a for i, a in enumerate(coef)]
        # Remove the first element, which is zero
        del new_coef[0]
        # And reset coefficients data to new values
        coef = new_coef
```

of a Polynomial instance, rebinding coef from $(a_0, \ldots, a_N)$ to $(a_1, 2a_2, \ldots, Na_N)$. The modified instance represents p'.

Now that we have written up an outline of a class definition in pseudo-Python, let's rewrite it in proper Python syntax. The modified code is given in listing 2.2. Before working through the additional syntax, let's look at an example of how to use the class, which is saved in a file called polyclass.py in the current working directory:

```
>>> from polyclass import Polynomial
>>> data = [2, 1, 3]
>>> p = Polynomial(data)    # Creates instance of Polynomial class
>>> p.evaluate(1)           # Returns 6
>>> p.coef                  # Returns [2, 1, 3]
>>> p.differentiate()       # Modifies coefficients of p
>>> p.coef                  # Returns [1, 6]
>>> p.evaluate(1)           # Returns 7
```

The filename polyclass.py becomes the name of the module (with the ".py" extension omitted), and from it we import our class `Polynomial`. An instance p is created by a call of the form p = Polynomial(data). Behind the scenes this generates a call to the constructor method, which realizes the instance as an object stored in memory, and binds the name p to this instance. As part of this process a namespace for the object is created, and the name coef is registered in that namespace and bound to the

Listing 2.2 (`polyclass.py`) A polynomial class, correct syntax

```
class Polynomial:

    def __init__(self, coef):
        """Creates an instance p of the Polynomial class,
        where p(x) = coef[0] x^0 + ... + coef[N] x^N."""
        self.coef = coef

    def evaluate(self, x):
        y = sum(a*x**i for i, a in enumerate(self.coef))
        return y

    def differentiate(self):
        new_coef = [i*a for i, a in enumerate(self.coef)]
        # Remove the first element, which is zero
        del new_coef[0]
        # And reset coefficients data to new values
        self.coef = new_coef
```

data [2, 1, 3].[21] The attributes of p can be accessed using p.attribute notation, where the attributes are the methods (in this case evaluate() and differentiate()) and instance variables (in this case coef).

Let's now walk through the new syntax in listing 2.2. First, the constructor method is given its correct name, which is __init__. The double underscore notation reminds us that this is a special Python method—we will meet another example in a moment. Second, every method has self as its first argument, and attributes referred to within the class definition are also preceded by self (e.g., self.coef).

The idea with the self references is that *they stand in for the name of any instance that is subsequently created.* As one illustration of this, note that calling p.evaluate(1) is equivalent to calling

```
>>> Polynomial.evaluate(p, 1)
```

This alternate syntax is more cumbersome and not generally used, but we can see how p does in fact replace self, passed in as the first argument to the evaluate() method. And if we imagine how the evaluate() method would look with p instead of self, our code starts to appear more natural:

---

[21]To view the contents of this namespace type p.__dict__ at the prompt.

```
def evaluate(p, x):
    y = sum(a*x**i for i, a in enumerate(p.coef))
    return y
```

Before finishing, let's briefly discuss another useful special method. One rather ungainly aspect of the `Polynomial` class is that for a given instance p corresponding to a polynomial p, the value $p(x)$ is obtained via the call `p.evaluate(x)`. It would be nicer—and closer to the mathematical notation—if we could replace this with the syntax `p(x)`. Actually this is easy: we simply replace the word `evaluate` in listing 2.2 with `__call__`. Objects of this class are now said to be *callable*, and `p(x)` is equivalent to `p.__call__(x)`.

2.3 Commentary

Python was developed by Guido van Rossum, with the first release in 1991. It is now one of the major success stories of the open source model, with a vibrant community of users and developers. van Rossum continues to direct the development of the language under the title of BDFL (Benevolent Dictator For Life). The use of Python has increased rapidly in recent years.

There are many good books on Python programming. A gentle introduction is provided by Zelle (2003). A more advanced book focusing on numerical methods is Langtangen (2008). However, the best place to start is on the Internet. The Python homepage (`python.org`) has links to the official Python documentation and various tutorials. Links, lectures, MATLAB code, and other information relevant to this chapter can be found on the text home page, at `http://johnstachurski.net/book.html`.

At the time of writing, Python is undergoing a transition from the 2.x versions to the newer Python 3.x. In order to correct some of the design flaws of Python 2.x, full backward compatibility has not been maintained, and some 2.x code will not run under Python 3.x. At the same time, the transition to 3.x is anticipated to be slow, since the major libraries (including NumPy and SciPy) will take time to be upgraded. I've tried to write code which is compatible with both 2.x and 3.x. As a result some of the syntax might look strange to a seasoned 2.x programmer.

Computational economics is a rapidly growing field. For a sample of the literature, consult Amman et al. (1996), Heer and Maussner (2005), Kendrick et al. (2005), or Tesfatsion and Judd (2006).

Chapter 3

Analysis in Metric Space

Metric spaces are sets (spaces) with a notion of distance between points in the space that satisfies certain axioms. From these axioms we can deduce many properties relating to convergence, continuity, boundedness, and other concepts needed for the study of dynamics. Metric space theory provides both an elegant and powerful framework for analyzing the kinds of problems we wish to consider, and a great sandpit for playing with analytical ideas: A careful read of this chapter should strengthen your ability to read and write proofs.

The chapter supposes that you have at least some exposure to introductory real analysis or advanced calculus. A review of this material is given in appendix A. On the other hand, if you are already familiar with the fundamentals of metric spaces, then the best approach is to skim through this chapter quickly and return as necessary.

3.1 A First Look at Metric Space

Consider the set $\mathbb{R}^k$, a typical element of which is $x = (x_1, \ldots, x_k)$, where $x_i \in \mathbb{R}$. These elements are also called *vectors*. There are a number of important "topological" notions we need to introduce for $\mathbb{R}^k$. These notions concern sets and functions on or into such space. In order to introduce them, it is convenient to begin with the concept of euclidean distance between vectors: Define $d_2 \colon \mathbb{R}^k \times \mathbb{R}^k \to \mathbb{R}$ by

$$d_2(x, y) =: \|x - y\|_2 := \left[\sum_{i=1}^{k} (x_i - y_i)^2 \right]^{1/2} \tag{3.1}$$

Doubtless you have met with this notion of distance before. You might know that it satisfies the following three conditions:

1. $d_2(x,y) = 0$ if and only if $x = y$,

2. $d_2(x,y) = d_2(y,x)$, and

3. $d_2(x,y) \leq d_2(x,v) + d_2(v,y)$.

for any $x,y,v \in \mathbb{R}^k$. The first property says that a point is at zero distance from itself, and also that distinct points always have positive distance. The second property is symmetry, and the third—the only one that is not immediately apparent—is the triangle inequality.

These three properties are fundamental to our understanding of distance. In fact if you look at the proofs of many important results—for example, the proof that every continuous function f from a closed bounded subset of $\mathbb{R}^k$ to $\mathbb{R}$ has a maximizer and a minimizer—you will notice that no other properties of d_2 are actually used.

Now it turns out that there are many other "distance" functions we can impose on $\mathbb{R}^k$ that also satisfy properties 1–3. Any proof for the euclidean (i.e., d_2) case that only uses properties 1–3 continues to hold for other distances, and in certain problems alternative notions of distance are easier to work with. This motivates us to generalize the concept of distance in $\mathbb{R}^k$.

While we are generalizing the notion of distance between vectors in $\mathbb{R}^k$, it is worth thinking about distance between other kinds of objects. If we could define the distance between two (infinite) sequences, or between a pair of functions, or two probability distributions, we could then give a definition for things like the "convergence" of distributions discussed informally in chapter 1.

3.1.1   Distances and Norms

Here is the key definition:

Definition 3.1.1 A *metric space* is a nonempty set S and a *metric* or *distance* $\rho \colon S \times S \to \mathbb{R}$ such that, for any $x,y,v \in S$,

1. $\rho(x,y) = 0$ if and only if $x = y$,

2. $\rho(x,y) = \rho(y,x)$, and

3. $\rho(x,y) \leq \rho(x,v) + \rho(v,y)$.

Apart from being nonempty, the set S is completely arbitrary. In the context of a metric space the elements of the set are usually called points. As in the case of euclidean distance, the third axiom is called the triangle inequality.

An immediate consequence of the axioms in definition 3.1.1 (which are sometimes referred to as the Hausdorff postulates) is that $\rho(x,y) \geq 0$ for any $x,y \in S$. To see this, note that if x and y are any two points in S, then $0 = \rho(x,x) \leq \rho(x,y) + \rho(y,x) = \rho(x,y) + \rho(x,y) = 2\rho(x,y)$. Hence $\rho(x,y) \geq 0$ as claimed.

The space $(\mathbb{R}^k, d_2)$ is a metric space, as discussed above. The most important case is $k = 1$, when $d_2(x,y)$ reduces to $|x - y|$ for $x, y \in \mathbb{R}$. *The notation $(\mathbb{R}, |\cdot|)$ will be used to denote this one-dimensional space.*

Many additional metric spaces on $\mathbb{R}^k$ are generated by what is known as a norm:

Definition 3.1.2 A *norm* on $\mathbb{R}^k$ is a mapping $\mathbb{R}^k \ni x \mapsto \|x\| \in \mathbb{R}$ such that, for any $x, y \in \mathbb{R}^k$ and any $\gamma \in \mathbb{R}$,

1. $\|x\| = 0$ if and only if $x = 0$,

2. $\|\gamma x\| = |\gamma| \|x\|$, and

3. $\|x + y\| \leq \|x\| + \|y\|$.

Each norm $\|\cdot\|$ on $\mathbb{R}^k$ generates a metric ρ on $\mathbb{R}^k$ via $\rho(x,y) := \|x - y\|$.

Exercise 3.1.3 Show that ρ is indeed a metric, in the sense that it satisfies the three axioms in definition 3.1.1.

Exercise 3.1.4 Let $\|\cdot\|$ be a norm on $\mathbb{R}^k$. Show that for any $x, y \in \mathbb{R}^k$ we have $|\|x\| - \|y\|| \leq \|x - y\|$.

The most important norm on $\mathbb{R}^k$ is $\|x\|_2 := (\sum_{i=1}^k x_i^2)^{1/2}$, which generates the euclidean distance d_2. A class of norms that includes $\|\cdot\|_2$ as a special case is the family $\|\cdot\|_p$ defined by

$$\|x\|_p := \left(\sum_{i=1}^k |x_i|^p\right)^{1/p} \qquad (x \in \mathbb{R}^k) \tag{3.2}$$

where $p \geq 1$. It is standard to admit $p = \infty$ in this family, with $\|x\|_\infty := \max_{1 \leq i \leq k} |x_i|$.

Proving that $\|\cdot\|_p$ is indeed a norm on $\mathbb{R}^k$ for arbitrary $p \geq 1$ is not difficult, but neither is it entirely trivial. In particular, establishing the triangle inequality requires the services of Minkowski's inequality. The latter is found in any text covering norms and is omitted.

Exercise 3.1.5 Prove that $\|\cdot\|_p$ is a norm on $\mathbb{R}^k$ for $p = 1$ and $p = \infty$.

The class of norms $\|\cdot\|_p$ gives rise to the class of metric spaces $(\mathbb{R}^k, d_p)$, where $d_p(x,y) := \|x - y\|_p$ for all $x, y \in \mathbb{R}^k$.

So far all our spaces have involved different metrics on finite-dimensional vector space. Next let's consider an example of a "function space." Let U be any set, let bU be the collection of all bounded functions $f: U \to \mathbb{R}$ (i.e., $\sup_{x \in U} |f(x)| < \infty$), and let

$$d_\infty(f,g) :=: \|f - g\|_\infty := \sup_{x \in U} |f(x) - g(x)| \tag{3.3}$$

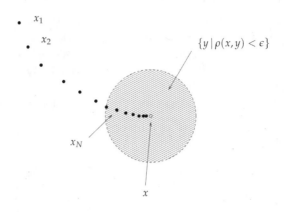

Figure 3.1 Limit of a sequence

The space (bU, d_∞) is a metric space. The reader can verify the first two properties of the definition of a metric space. The triangle inequality is verified as follows. Let $f, g, h \in bU$. Pick any $x \in U$. We have

$$|f(x) - g(x)| \leq |f(x) - h(x)| + |h(x) - g(x)| \leq d_\infty(f, h) + d_\infty(h, g)$$

Since x is arbitrary, we obtain $d_\infty(f, g) \leq d_\infty(f, h) + d_\infty(h, g)$.[1]

3.1.2 Sequences

Let $S = (S, \rho)$ be a metric space. A sequence $(x_n) \subset S$ is said to *converge* to $x \in S$ if, for all $\epsilon > 0$, there exists an $N \in \mathbb{N}$ such that $n \geq N$ implies $\rho(x_n, x) < \epsilon$. In other words (x_n) converges to x if and only if the *real* sequence $\rho(x_n, x) \to 0$ in $\mathbb{R}$ as $n \to \infty$ (see §A.2 for more on real sequences). If this condition is satisfied, we write $\lim_{n \to \infty} x_n = x$, or $x_n \to x$. The point x is referred to as the *limit* of the sequence. Figure 3.1 gives an illustration for the case of two-dimensional euclidean space.

Theorem 3.1.6 *A sequence in (S, ρ) can have at most one limit.*

Proof. You might like to try a proof by contradiction as an exercise. Here is a direct proof. Let (x_n) be an arbitrary sequence in S, and let x and x' be two limit points. We have

$$0 \leq \rho(x, x') \leq \rho(x, x_n) + \rho(x_n, x') \qquad \forall\, n \in \mathbb{N}$$

---

[1]As an aside, you may have noticed that the metric space (bU, d_∞) seems to be defined by a "norm" $\|f\|_\infty := \sup_{x \in U} |f(x)|$. This is not a norm in the sense of definition 3.1.2, as that definition requires that the underlying space is $\mathbb{R}^k$, rather than bU. However, more general norms can be defined for abstract "vector space," and $\|\cdot\|_\infty$ is a prime example. The details are omitted: See any text on functional analysis.

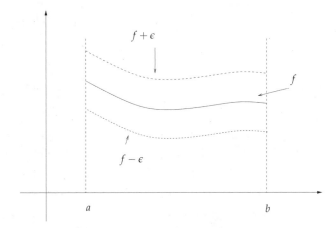

Figure 3.2 An ϵ-ball for d_∞

From theorems A.2.12 and A.2.13 (page 330) we have $\rho(x, x') = 0$. Therefore $x = x'$. (Why?)   □

Exercise 3.1.7 Let (x_n) and (y_n) be sequences in S. Show that if $x_n \to x \in S$ and $\rho(x_n, y_n) \to 0$, then $y_n \to x$.

One of the most important creatures defined from the distance function is the humble open ball. The *open ball* or ϵ-ball $B(\epsilon; x)$ centered on $x \in S$ with *radius* $\epsilon > 0$ is the set

$$B(\epsilon; x) := \{z \in S : \rho(z, x) < \epsilon\}$$

In the plane with $\rho = d_2$ the ϵ-ball is a circle; in $\mathbb{R}^3$ it is a sphere. Figure 3.2 gives a visualization of the ϵ-ball around $f \in (b[a, b], d_\infty)$.

Exercise 3.1.8 Let $(x_n) \subset S$ and $x \in S$. Show that $x_n \to x$ if and only if for all $\epsilon > 0$, the ball $B(\epsilon; x)$ contains all but finitely many terms of (x_n).

A subset E of S is called *bounded* if $E \subset B(n; x)$ for some $x \in S$ and some (suitably large) $n \in \mathbb{N}$. A sequence (x_n) in S is called bounded if its range $\{x_n : n \in \mathbb{N}\}$ is a bounded set.

Exercise 3.1.9 Show that every convergent sequence in S is also bounded.

Given sequence $(x_n) \subset S$, a subsequence is defined analogously to the case of real sequences: (y_n) is called a *subsequence* of (x_n) if there is a strictly increasing function $f \colon \mathbb{N} \to \mathbb{N}$ such that $y_n = x_{f(n)}$ for all $n \in \mathbb{N}$. It is common to use the notation (x_{n_k}) to denote a subsequence of (x_n).

Exercise 3.1.10 Show that for $(x_n) \subset S$, $x_n \to x$ for some $x \in S$ if and only if every subsequence of (x_n) converges to x.

For the euclidean space $(\mathbb{R}^k, d_2)$ we have the following result:

Lemma 3.1.11 *A sequence* $(x_n) = (x_n^1, \ldots, x_n^k)$ *in* $(\mathbb{R}^k, d_2)$ *converges to* $x = (x^1, \ldots, x^k) \in \mathbb{R}^k$ *if and only if* $x_n^j \to x^j$ *in* $\mathbb{R} = (\mathbb{R}, |\cdot|)$ *for all* j *in* $1, \ldots, k$.

Proof. For j in $1, \ldots, k$ we have $|x_n^j - x^j| \leq d_2(x_n, x)$. (Why?) Hence if $d_2(x_n, x) \to 0$, then $|x_n^j - x^j| \to 0$ for each j. For the converse, fix $\epsilon > 0$ and choose for each j in $1, \ldots, k$ an $N^j \in \mathbb{N}$ such that $n \geq N^j$ implies $|x_n^j - x^j| < \epsilon/\sqrt{k}$. Now $n \geq \max_j N^j$ implies $d_2(x_n, x) \leq \epsilon$. (Why?)   □

Lemma 3.1.11 is important, and you should try sketching it for the case $k = 2$ to build intuition. We will see that in fact *the same result holds not just for d_2, but for the metric induced by any norm on $\mathbb{R}^k$.*

Let S and Y be two metric spaces. Parallel to §A.2.3, define $f\colon S \supset A \to Y$ to be *continuous* at $a \in A$ if for every sequence (x_n) in A converging to a we have $f(x_n) \to f(a)$ in Y, and continuous on A whenever it is continuous at every $a \in A$. For the same $f\colon A \to Y$ and for $a \in A$, we say that $y = \lim_{x \to a} f(x)$ if $f(x_n) \to y$ for every sequence $(x_n) \subset A$ with $x_n \to a$. Clearly, f is continuous at a if and only if $\lim_{x \to a} f(x) = f(a)$.

Example 3.1.12 Let S be a metric space, and let $\bar{x}$ be any given point in S. The map $S \ni x \mapsto \rho(x, \bar{x}) \in \mathbb{R}$ is continuous on all of S. To see this, pick any $x \in S$, and any $(x_n) \subset S$ with $x_n \to x$. Two applications of the triangle inequality yield

$$\rho(x, \bar{x}) - \rho(x_n, x) \leq \rho(x_n, \bar{x}) \leq \rho(x_n, x) + \rho(x, \bar{x}) \qquad \forall n \in \mathbb{N}$$

Now take the limit (i.e., apply theorem A.2.12 on page 330).

Exercise 3.1.13 Let $f(x, y) = x^2 + y^2$. Show that f is a continuous function from $(\mathbb{R}^2, d_2)$ into $(\mathbb{R}, |\cdot|)$.[2]

Throughout the text, if S is some set, $f\colon S \to \mathbb{R}$, and $g\colon S \to \mathbb{R}$, then $f + g$ denotes the function $x \mapsto f(x) + g(x)$ on S, while fg denotes the function $x \mapsto f(x)g(x)$ on S.

Exercise 3.1.14 Let f and g be as above, and let S be a metric space. Show that if f and g are continuous, then so are $f + g$ and fg.

Exercise 3.1.15 A function $f\colon S \to \mathbb{R}$ is called upper-semicontinuous (usc) at $x \in S$ if, for every $x_n \to x$, we have $\limsup_n f(x_n) \leq f(x)$; and lower-semicontinuous (lsc) if, for every $x_n \to x$, we have $\liminf_n f(x_n) \geq f(x)$. Show that f is usc at x if and only if $-f$ is lsc at x. Show that f is continuous at x if and only if it is both usc and lsc at x.

---

[2]Hint: Use lemma 3.1.11.

3.1.3   Open Sets, Closed Sets

Arbitrary subsets of arbitrary spaces can be quite unruly. It is useful to identify classes of sets that are well-behaved, interacting nicely with common functions, and co-operating with attempts to measure them, or to represent them in terms of simpler elements. In this section we investigate a class of sets called the open sets, as well as their complements the closed sets.

Let's say that $x \in S$ *adheres* to $E \subset S$ if, for each $\epsilon > 0$, the ball $B(\epsilon; x)$ contains at least one point of E;[3] and that x is *interior* to E if $B(\epsilon; x) \subset E$ for some $\epsilon > 0$.[4] A set $E \subset S$ is called *open* if all points in E are interior to E, and *closed* if E contains all points that adhere to E. In the familiar metric space $(\mathbb{R}, |\cdot|)$, canonical examples are the intervals (a, b) and $[a, b]$, which are open and closed respectively.[5] The concepts of open and closed sets turn out to be some of the most fruitful ideas in all of mathematics.

Exercise 3.1.16 Show that a point in S adheres to $E \subset S$ if and only if it is the limit of a sequence contained in E.

Theorem 3.1.17 *A set $F \subset S$ is closed if and only if for every convergent sequence entirely contained in F, the limit of the sequence is also in F.*

Proof. Do the proof as an exercise if you can. If not, here goes. Suppose that F is closed, and take a sequence in F converging to some point $x \in S$. Then x adheres to F by exercise 3.1.16, and is therefore in F by definition. Suppose, on the other hand, that the limit of every convergent sequence in F belongs to F. Take any $x \in S$ that adheres to F. By exercise 3.1.16, there is a sequence in F converging to it. Therefore $x \in F$, and F is closed.   □

Open sets and closed sets are closely related. In fact we have the following fundamental theorem:

Theorem 3.1.18 *A subset of an arbitrary metric space S is open if and only if its complement is closed, and closed if and only if its complement is open.*

Proof. The proof is a good exercise. If you need a start, here is a proof that G open implies $F := G^c$ closed. Take $(x_n) \subset F$ with $x_n \to x \in S$. We wish to show that $x \in F$. In fact this must be the case because, if $x \notin F$, then $x \in G$, in which case there is an $\epsilon > 0$ such that $B(\epsilon, x) \subset G$. (Why?) Such a situation is not possible when $(x_n) \subset F$ and $x_n \to x$. (Why?)   □

---

[3] In some texts, x is said to be a *contact point* of E.

[4] Try sketching some examples for the case of $(\mathbb{R}^2, d_2)$.

[5] If you find it hard to verify this now, you won't by the end of the chapter.

We call $D(\epsilon; x) := \{z \in S : \rho(z, x) \leq \epsilon\}$ the *closed ϵ-ball* centered on x. Every $D(\epsilon; x) \subset S$ is a closed set, as anticipated by the notation. To see this, take $(a_n) \subset D(\epsilon; x)$ converging to $a \in S$. We need to show that $a \in D(\epsilon; x)$ or, equivalently, that $\rho(a, x) \leq \epsilon$. Since $\rho(a_n, x) \leq \epsilon$ for all $n \in \mathbb{N}$, since limits preserve orders and since $y \mapsto \rho(y, x)$ is continuous, we have $\rho(a, x) = \lim \rho(a_n, x) \leq \epsilon$.

Exercise 3.1.19 Likewise every open ball $B(\epsilon; x)$ in S is an open set. Prove this directly, or repeat the steps of the previous example applied to $B(\epsilon; x)^c$.

You will not find it difficult to convince yourself that if (S, ρ) is any metric space, then the whole set S is itself both open and closed. (Just check the definitions carefully.) This can lead to some confusion. For example, suppose that we consider the metric space $(S, |\cdot|)$, where $S = (0, 1)$. Since $(0, 1)$ is the whole space, it is closed. At the same time, $(0, 1)$ is open as a subset of $(\mathbb{R}, |\cdot|)$. The properties of openness and closedness are relative rather than absolute.

Exercise 3.1.20 Argue that for any metric space (S, ρ), the empty set $\varnothing$ is both open and closed.

Exercise 3.1.21 Show that if (S, ρ) is an arbitrary metric space, and if $x \in S$, then the set $\{x\}$ is always closed.

Theorem 3.1.22 *If F is a closed, bounded subset of $(\mathbb{R}, |\cdot|)$, then $\sup F \in F$.*

Proof. Let $s := \sup F$. Since F is closed it is sufficient to show there exists a sequence $(x_n) \subset F$ with $x_n \to s$. (Why?) By lemma A.2.19 (page 332) such a sequence exists. $\square$

Exercise 3.1.23 Prove that a sequence converges to a point x if and only if the sequence is eventually in every open set containing x.

Exercise 3.1.24 Prove: If $\{G_\alpha\}_{\alpha \in A}$ are all open, then so is $\cup_{\alpha \in A} G_\alpha$.

Exercise 3.1.25 Show that if A is finite and $\{G_\alpha\}_{\alpha \in A}$ is a collection of open sets, then $\cap_{\alpha \in A} G_\alpha$ is also open.

In other words, arbitrary unions and finite intersections of open sets are open. But be careful: An infinite intersection of open sets is not necessarily open. For example, consider the metric space $(\mathbb{R}, |\cdot|)$. If $G_n = (-1/n, 1/n)$, then $\cap_{n \in \mathbb{N}} G_n = \{0\}$ because

$$x \in \cap_n G_n \iff -\frac{1}{n} < x < \frac{1}{n} \quad \forall n \in \mathbb{N} \iff x = 0$$

Exercise 3.1.26 Show that $\cap_{n \in \mathbb{N}}(a - 1/n, b + 1/n) = [a, b]$.

Exercise 3.1.27 Prove that if $\{F_\alpha\}_{\alpha \in A}$ are all closed, then so is $\cap_{\alpha \in A} F_\alpha$.

Exercise 3.1.28 Show that if A is finite and $\{F_\alpha\}_{\alpha \in A}$ is a collection of closed sets, then $\cup_{\alpha \in A} F_\alpha$ is closed. On the other hand, show that the union $\cup_{n \in \mathbb{N}}[a + 1/n, b - 1/n] = (a, b)$. (Why is this not a contradiction?)

Exercise 3.1.29 Show that $G \subset S$ is open if and only if it can be formed as the union of an arbitrary number of open balls.

The *closure* of E is the set of all points that adhere to E, and is written cl E. In view of exercise 3.1.16, $x \in$ cl E if and only if there exists a sequence $(x_n) \subset E$ with $x_n \to x$. The *interior* of E is the set of its interior points, and is written int E.

Exercise 3.1.30 Show that cl E is always closed. Show in addition that for all closed sets F such that $F \supset E$, cl $E \subset F$. Using this result, show that cl E is equal to the intersection of all closed sets containing E.

The last exercise tells us that the closure of a set is the smallest closed set that contains that particular set. The next one shows us that the interior of a set is the largest open set contained in that set.

Exercise 3.1.31 Show that int E is always open. Show also that for all open sets G such that $G \subset E$, int $E \supset G$. Using this result, show that int E is equal to the union of all open sets contained in E.

Exercise 3.1.32 Show that $E = $ cl E if and only if E is closed. Show that $E = $ int E if and only if E is open.

Open sets and continuous functions interact very nicely. For example, we have the following fundamental theorem.

Theorem 3.1.33 *A function $f \colon S \to Y$ is continuous if and only if the preimage $f^{-1}(G)$ of every open set $G \subset Y$ is open in S.*

Proof. Suppose that f is continuous, and let G be any open subset of Y. If $x \in f^{-1}(G)$, then x must be interior, for if it is not, then there is a sequence $x_n \to x$ where $x_n \notin f^{-1}(G)$ for all n. But, by continuity, $f(x_n) \to f(x)$, implying that $f(x) \in G$ is not interior to G. (Why?) Contradiction.

Conversely, suppose that the preimage of every open set is open, and take any $\{x_n\}_{n \geq 1} \cup \{x\} \subset S$ with $x_n \to x$. Pick any ϵ-ball B around $f(x)$. The preimage $f^{-1}(B)$ is open, so for N sufficiently large we have $x_n \in f^{-1}(B)$ for all $n \geq N$, in which case $f(x_n) \in B$ for all $n \geq N$. $\qquad \square$

Exercise 3.1.34 Let S, Y, and Z be metric spaces, and let $f \colon S \to Y$ and $g \colon Y \to Z$. Show that if f and g are continuous, then so is $h := g \circ f$.

Exercise 3.1.35 Let $S = \mathbb{R}^k$, and let $\rho^*(x, y) = 0$ if $x = y$ and 1 otherwise. Prove that ρ^* is a metric on $\mathbb{R}^k$. Which subsets of this space are open? Which subsets are closed? What kind of functions $f: S \to \mathbb{R}$ are continuous? What kinds of sequences are convergent?

3.2 Further Properties

Having covered the fundamental ideas of convergence, continuity, open sets and closed sets, we now turn to two key concepts in metric space theory: completeness and compactness. After stating the definitions and covering basic properties, we will see how completeness and compactness relate to existence of optima and to the theory of fixed points.

3.2.1 Completeness

A sequence (x_n) in metric space (S, ρ) is said to be a *Cauchy* sequence if, for all $\epsilon > 0$, there exists an $N \in \mathbb{N}$ such that $\rho(x_j, x_k) < \epsilon$ whenever $j \geq N$ and $k \geq N$. A subset A of a metric space S is called *complete* if every Cauchy sequence in A converges to some point in A. Often the set A of interest is the whole space S, in which case we say that S is a complete metric space. As discussed in §A.2, the set of reals $(\mathbb{R}, |\cdot|)$ has this property. Many other metric spaces do not.

Notice that completeness is intrinsic to a given set A and a metric ρ on A. Either every Cauchy sequence in (A, ρ) converges or there exists a Cauchy sequence that does not. On the other hand, openness and closedness are *relative* properties. The set $A := [0, 1)$ is not open as a subset of $(\mathbb{R}, |\cdot|)$, but it is open as a subset of $(\mathbb{R}_+, |\cdot|)$.

The significance of completeness is that when searching for the solution to a problem, it sometimes happens that we are able to generate a Cauchy sequence whose limit would be a solution if it does in fact exist. In a complete space we can rest assured that our solution does exist as a well-defined element of the space.

Exercise 3.2.1 Show that a sequence (x_n) in metric space (S, ρ) is Cauchy if and only if $\lim_{n \to \infty} \sup_{k \geq n} \rho(x_n, x_k) = 0$.

Exercise 3.2.2 Show that if a sequence (x_n) in metric space (S, ρ) is convergent, then it is Cauchy. Show that if (x_n) is Cauchy, then it is bounded.

Which metric spaces are complete? Observe that while $\mathbb{R} = (\mathbb{R}, |\cdot|)$ is complete, subsets of $\mathbb{R}$ may not be. For example, consider the metric space $(S, \rho) = ((0, \infty), |\cdot|)$. Some manipulation proves that while $(x_n) = (1/n)$ is Cauchy in S, it converges to no point in S. On the other hand, for $(S, \rho) = (\mathbb{R}_+, |\cdot|)$ the limit point of the sequence $(1/n)$ is in S. Indeed this space is complete, as is any closed subset of $\mathbb{R}$. More generally,

Theorem 3.2.3 *Let S be a complete metric space. Subset $A \subset S$ is complete if and only if it is closed as a subset of S.*

Proof. Let A be complete. To see that A is closed, let $(x_n) \subset A$ with $x_n \to x \in S$. Since (x_n) is convergent it must be Cauchy (exercise 3.2.2). Because A is complete we have $x \in A$. Thus A contains its limit points, and is therefore closed. Conversely, suppose that A is closed. Let (x_n) be a Cauchy sequence in A. Since S is complete, (x_n) converges to some $x \in S$. As A is closed, the limit point x must be in A. Hence A is complete. $\square$

The euclidean space $(\mathbb{R}^k, d_2)$ is complete. To see this, observe first that

Lemma 3.2.4 *A sequence $(x_n) = (x_n^1, \ldots, x_n^k)$ in $(\mathbb{R}^k, d_2)$ is Cauchy if and only if each component sequence x_n^j is Cauchy in $\mathbb{R} = (\mathbb{R}, |\cdot|)$.*

The proof of lemma 3.2.4 is an exercise.[6] The lemma is important because it implies that $(\mathbb{R}^k, d_2)$ inherits the completeness of $\mathbb{R}$ (axiom A.2.4, page 328):

Theorem 3.2.5 *The euclidean space $(\mathbb{R}^k, d_2)$ is complete.*

Proof. If (x_n) is Cauchy in $(\mathbb{R}^k, d_2)$, then each component is Cauchy in $\mathbb{R} = (\mathbb{R}, |\cdot|)$, and, by completeness of $\mathbb{R}$, converges to some limit in $\mathbb{R}$. It follows from lemma 3.1.11 that (x_n) is convergent in $(\mathbb{R}^k, d_2)$. $\square$

Recall that (bU, d_∞) is the bounded real-valued functions $f: U \to \mathbb{R}$, endowed with the distance d_∞ defined on page 37. This space also inherits completeness from $\mathbb{R}$:

Theorem 3.2.6 *Let U be any set. The metric space (bU, d_∞) is complete.*

Proof. Let $(f_n) \subset bU$ be Cauchy. We claim the existence of a $f \in bU$ such that $d_\infty(f_n, f) \to 0$. To see this, observe that for each $x \in U$ we have $\sup_{k \geq n} |f_n(x) - f_k(x)| \leq \sup_{k \geq n} d_\infty(f_n, f_k) \to 0$, and hence $(f_n(x))$ is Cauchy (see exercise 3.2.1). By the completeness of $\mathbb{R}$, $(f_n(x))$ is convergent, and we define a new function $f \in bU$ by $f(x) = \lim_{n \to \infty} f_n(x)$.[7]

To show that $d_\infty(f_n, f) \to 0$, fix $\epsilon > 0$, and choose $N \in \mathbb{N}$ such that $d_\infty(f_n, f_m) < \epsilon/2$ whenever $n, m \geq N$. Pick any $n \geq N$. For arbitrary $x \in U$ we have $|f_n(x) - f_m(x)| < \epsilon/2$ for all $m \geq n$, and hence, taking limits with respect to m, we have $|f_n(x) - f(x)| \leq \epsilon/2$. Since x was arbitrary, $d_\infty(f_n, f) \leq \epsilon/2 < \epsilon$. $\square$

---

[6]Hint: You might like to begin by rereading the proof of lemma 3.1.11.
[7]Why is $f \in bU$ (i.e., why is f bounded on U)? Consult exercise 3.2.2.

This is a good opportunity to briefly discuss convergence of functions. A sequence of functions (f_n) from arbitrary set U into $\mathbb{R}$ converges *pointwise* to $f\colon U \to \mathbb{R}$ if $|f_n(x) - f(x)| \to 0$ as $n \to \infty$ for every $x \in U$; and *uniformly* if $d_\infty(f_n, f) \to 0$. Pointwise convergence is certainly important, but it is also significantly weaker than convergence in d_∞. For example, suppose that U is a metric space, that $f_n \to f$, and that all f_n are continuous. It might then be hoped that the limit f inherits continuity from the approximating sequence. For pointwise convergence this is not generally true,[8] while for uniform convergence it is:

Theorem 3.2.7 *Let (f_n) and f be real-valued functions on metric space U. If f_n is continuous on U for all n and $d_\infty(f_n, f) \to 0$, then f is also continuous on U.*

Proof. Take $(x_k) \subset U$ with $x_k \to \bar{x} \in U$. Fix $\epsilon > 0$. Choose $n \in \mathbb{N}$ such that $|f_n(x) - f(x)| < \epsilon/2$ for all $x \in U$. For any given $k \in \mathbb{N}$ the triangle inequality gives

$$|f(x_k) - f(\bar{x})| \leq |f(x_k) - f_n(x_k)| + |f_n(x_k) - f_n(\bar{x})| + |f_n(\bar{x}) - f(\bar{x})|$$

$$\therefore \quad |f(x_k) - f(\bar{x})| \leq |f_n(x_k) - f_n(\bar{x})| + \epsilon \qquad (k \in \mathbb{N})$$

From exercise A.2.28 (page 333) we have $0 \leq \limsup_k |f(x_k) - f(\bar{x})| \leq \epsilon$. Since ϵ is arbitrary, $\limsup_k |f(x_k) - f(\bar{x})| = \lim_k |f(x_k) - f(\bar{x})| = 0$. $\square$

Now let's introduce another important metric space.

Definition 3.2.8 Given any metric space U, let (bcU, d_∞) be the continuous functions in bU endowed with the same metric d_∞.

Theorem 3.2.9 *The space (bcU, d_∞) is always complete.*

Proof. This follows from theorem 3.2.3 (closed subsets of complete spaces are complete), theorem 3.2.6 (the space (bU, d_∞) is complete) and theorem 3.2.7 (which implies that the space bcU is closed as a subset of bU). $\square$

3.2.2 Compactness

Now we turn to the notion of compactness. A subset K of $S = (S, \rho)$ is called *precompact* in S if every sequence contained in K has a subsequence that converges to a point of S. The set K is called *compact* if every sequence contained in K has a subsequence that converges to a point of K. (Thus every compact subset of S is precompact in S, and every closed precompact set is compact.) Compactness will play a major role in our analysis. As we will see, the existence of a converging subsequence often allows us to track down the solution to a difficult problem.

---

[8]A counterexample is $U = [0, 1]$, $f_n(x) = x^n$, $f(x) = 0$ on $[0, 1)$ and $f(1) = 1$.

As a first step, note that there is another important characterization of compactness, which at first sight bears little resemblance to the sequential definition above. To state the theorem, recall that for a set K in S, an *open cover* is a collection $\{G_\alpha\}$ of open subsets of S such that $K \subset \cup_\alpha G_\alpha$. The cover is called finite if it consists of only finitely many sets.

Theorem 3.2.10 *A subset K of an arbitrary metric space S is compact if and only if every open cover of K can be reduced to a finite cover.*

In other words, a set K is compact if and only if, given any open cover, we can discard all but a finite number of elements and still cover K. The proof of theorem 3.2.10 can be found in any text on real analysis.

Exercise 3.2.11 Exhibit an open cover of $\mathbb{R}^k$ that cannot be reduced to a finite subcover. Construct a sequence in $\mathbb{R}^k$ with no convergent subsequence.

Exercise 3.2.12 Use theorem 3.2.10 to prove that every compact subset of a metric space S is bounded (i.e., can be contained in an open ball $B(n; x)$ for some $x \in S$ and some (suitably large) $n \in \mathbb{N}$).

Exercise 3.2.13 Prove that every compact subset of a metric space is closed.

On the other hand, closed and bounded subsets of metric spaces are not always compact.

Exercise 3.2.14 Let $(S, \rho) = ((0, \infty), |\cdot|)$, and let $K = (0, 1]$. Show that although K is a closed, bounded subset of S, it is not precompact in S.

Exercise 3.2.15 Show that every subset of a compact set is precompact, and every closed subset of a compact set is compact.

Exercise 3.2.16 Show that in any metric space the intersection of an arbitrary number of compact sets and the union of a finite number of compact sets are again compact.

Exercise 3.2.17 For a more advanced exercise, you might like to try to show that the closure of a precompact set is compact. It follows that every precompact set is bounded. (Why?)

When it comes to precompactness and compactness, the space $(\mathbb{R}^k, d_2)$ is rather special. For example, the Bolzano–Weierstrass theorem states that

Theorem 3.2.18 *Every bounded sequence in euclidean space $(\mathbb{R}^k, d_2)$ has at least one convergent subsequence.*

Proof. Let's check the case $k = 2$. Let $(x_n) = (x_n^1, x_n^2) \subset (\mathbb{R}^2, d_2)$ be bounded. Since (x_n^1) is itself bounded in $\mathbb{R}$ (why?), we can find a sequence $n_1, n_2, \ldots =: (n_j)$ such that $(x_{n_j}^1)$ converges in $\mathbb{R}$ (theorem A.2.10, page 330). Now consider $(x_{n_j}^2)$. This sequence is also bounded, and must itself have a convergent subsequence, so if we discard more terms from $n_1, n_2, \ldots =: (n_j)$ we can obtain a subsubsequence $(n_i) \subset (n_j)$ such that $(x_{n_i}^2)$ converges. Since $(n_i) \subset (n_j)$, the sequence $(x_{n_i}^1)$ also converges. It follows from lemma 3.1.11 (page 40) that (x_{n_i}) converges in $(\mathbb{R}^k, d_2)$. $\qquad\square$

The next result (called the Heine–Borel theorem) follows directly.

Theorem 3.2.19 *A subset of $(\mathbb{R}^k, d_2)$ is precompact in $(\mathbb{R}^k, d_2)$ if and only if it is bounded, and compact if and only if it is closed and bounded.*

As we have seen, some properties of $(\mathbb{R}^k, d_2)$ carry over to arbitrary metric spaces, while others do not. For example, we saw that in an arbitrary metric space, closed and bounded sets are not necessarily compact. (This has important implications for the theory of Markov chains developed below.) However, we will see in §3.2.3 that any metric d on $\mathbb{R}^k$ induced by a norm (see definition 3.1.2 on page 37) is "equivalent" to d_2, and that, as a result, subsets of $(\mathbb{R}^k, d)$ are compact if and only if they are closed and bounded.

3.2.3 Optimization, Equivalence

Optimization is important not only to economics, but also to statistics, numerical computation, engineering, and many other fields of science. In economics, rationality is the benchmark assumption for agent behavior, and is usually imposed by requiring agents solve optimization problems. In statistics, optimization is used for maximum likelihood and other related procedures, which search for the "best" estimator in some class. For numerical methods and approximation theory, one often seeks a simple representation f_n of a given function f that is the "closest" to f in some suitable metric sense.

In any given optimization problem the first issue we must confront is whether or not optima exist. For example, a demand function is usually defined as the solution to a consumer optimization problem. It would be awkward then if no solution to the problem exists. The same can be said for supply functions, or for policy functions, which return the optimal action of a "controller" faced with a given state of the world. Discussions of existence typically begin with the following theorem:

Theorem 3.2.20 *Let $f: S \to Y$, where S and Y are metric spaces and f is continuous. If $K \subset S$ is compact, then so is $f(K)$, the image of K under f.*

Proof. Take an open cover of $f(K)$. The preimage of this cover under f is an open cover of K (recall theorem 3.1.33 on page 43). Since K is compact we can reduce this to a finite cover (theorem 3.2.10). The image of this finite cover under f contains $f(K)$, and hence $f(K)$ is compact. $\qquad\square$

Exercise 3.2.21 Give another proof of theorem 3.2.20 using the sequential definitions of compactness and continuity.

The following theorem is one of the most fundamental results in optimization theory. It says that in the case of continuous functions on compact domains, optima always exist.

Theorem 3.2.22 (Weierstrass) *Let $f\colon K \to \mathbb{R}$, where K is a subset of arbitrary metric space (S, ρ). If f is continuous and K is compact, then f attains its supremum and infimum on K.*

In other words, $\alpha := \sup f(K)$ exists, and, moreover, there is an $x \in K$ such that $f(x) = \alpha$. A corresponding result holds for the infimum.

Proof. Regarding suprema, the result follows directly from theorem 3.2.20 combined with theorem 3.1.22 (page 42). By these theorems you should be able to show that $\alpha := \sup f(K)$ exists, and, moreover, that $\alpha \in f(K)$. By the definition of $f(K)$, there is an $x \in K$ such that $f(x) = \alpha$. This proves the assertion regarding suprema. The proof of the assertion regarding infima is similar. $\qquad\square$

In general, for $f\colon S \to \mathbb{R}$, a value $y \in \mathbb{R}$ is called the *maximum* of f on $A \subset S$ if $f(x) \le y$ for all $x \in A$ and $f(\bar{x}) = y$ for some $\bar{x} \in A$. At most one maximum exists. The *maximizers* of f on A are the points

$$\operatorname*{argmax}_{x \in A} f(x) := \{x \in A : f(x) = y\} = \{x \in A : f(z) \le f(x) \text{ for all } z \in A\}$$

Minima and minimizers are defined in a similar way.

With this notation, we can restate theorem 3.2.22 as follows: If K is compact and $f\colon K \to \mathbb{R}$ is continuous, then K contains at least one maximizer and one minimizer of f on K. (Convince yourself that this is so.)

Exercise 3.2.23 Let $f\colon K \to \mathbb{R}$, where K is compact and f is continuous. Show that if f is strictly positive on K, then $\inf f(K)$ is strictly positive.

As an application of theorem 3.2.22, let's show that all norms on $\mathbb{R}^k$ induce essentially the same metric space. We begin with a definition: Let S be a nonempty set, and let ρ and ρ' be two metrics on S. We say that ρ and ρ' are *equivalent* if there exist constants K and J such that

$$\rho(x, y) \le K\rho'(x, y) \text{ and } \rho'(x, y) \le J\rho(x, y) \qquad \text{for any } x, y \in S \qquad (3.4)$$

The notion of equivalence is important because equivalent metrics share the same convergent sequences and Cauchy sequences, and the metric spaces (S, ρ) and (S, ρ') share the same open sets, closed sets, compact sets and bounded sets:

Lemma 3.2.24 *Let ρ and ρ' be equivalent on S, and let $(x_n) \subset S$. The sequence (x_n) ρ-converges to $x \in S$ if and only if it ρ'-converges to x.*[9]

Proof. If $\rho(x_n, x) \to 0$, then $\rho'(x_n, x) \le J\rho(x_n, x) \to 0$, and so forth.   □

Exercise 3.2.25 Let ρ and ρ' be equivalent on S, and let $(x_n) \subset S$. Show that (x_n) is ρ-Cauchy if and only if it is ρ'-Cauchy.[10]

Exercise 3.2.26 Let ρ and ρ' be equivalent on S, and let $A \subset S$. Show that A is ρ-complete if and only if it is ρ'-complete.

Exercise 3.2.27 Let ρ and ρ' be equivalent on S. Show that (S, ρ) and (S, ρ') share the same closed sets, open sets, bounded sets and compact sets.

Exercise 3.2.28 Let ρ and ρ' be equivalent on S, and let $f \colon S \to \mathbb{R} = (\mathbb{R}, |\cdot|)$. Show that f is ρ-continuous if and only if it is ρ'-continuous.

Exercise 3.2.29 Let S be any nonempty set, and let ρ, ρ', and ρ'' be metrics on S. Show that equivalence is transitive, in the sense that if ρ is equivalent to ρ' and ρ' is equivalent to ρ'', then ρ is equivalent to ρ''.

Theorem 3.2.30 *All metrics on $\mathbb{R}^k$ induced by a norm are equivalent.*

Proof. The claim is that if $\|\cdot\|$ and $\|\cdot\|'$ are any two norms on $\mathbb{R}^k$ (see definition 3.1.2 on page 37), and ρ and ρ' are defined by $\rho(x, y) := \|x - y\|$ and $\rho'(x, y) := \|x - y\|'$, then ρ and ρ' are equivalent. In view of exercise 3.2.29, it is sufficient to show that any one of these metrics is equivalent to d_1. To check this, it is sufficient (why?) to show that if $\|\cdot\|$ is any norm on $\mathbb{R}^k$, then there exist constants K and J such that

$$\|x\| \le K\|x\|_1 \text{ and } \|x\|_1 \le J\|x\| \qquad \text{for any } x \in \mathbb{R}^k \tag{3.5}$$

To check the first inequality, let e_j be the j-th basis vector in $\mathbb{R}^k$ (i.e., the j-th component of vector e_j is 1 and all other components are zero). Let $K := \max_j \|e_j\|$. Then for any $x \in \mathbb{R}^k$ we have

$$\|x\| = \|x_1 e_1 + \cdots x_k e_k\| \le \sum_{j=1}^{k} \|x_j e_j\| = \sum_{j=1}^{k} |x_j| \|e_j\| \le K\|x\|_1$$

---

[9] Here ρ-convergence means convergence in (S, ρ), etc., etc.
[10] Hint: Try a proof using exercise 3.2.1 (page 44).

To check the second inequality in (3.5), observe that $x \mapsto \|x\|$ is continuous on $(\mathbb{R}^k, d_1)$ because if $x_n \to x$ in d_1, then

$$| \|x_n\| - \|x\| | \leq \|x_n - x\| \leq K\|x_n - x\|_1 \to 0 \quad (n \to \infty)$$

Now consider the set $E := \{x \in \mathbb{R}^k : \|x\|_1 = 1\}$. Some simple alterations to theorem 3.2.19 (page 48) and the results that lead to it show that, just as for the case of $(\mathbb{R}^k, d_2)$, closed and bounded subsets of $(\mathbb{R}^k, d_1)$ are compact.[11] Hence E is d_1-compact. It now follows from theorem 3.2.22 that $x \mapsto \|x\|$ attains its minimum on E, in the sense that there is an $x^* \in E$ with $\|x^*\| \leq \|x\|$ for all $x \in E$. Clearly, $\|x^*\| \neq 0$. (Why?) Now observe that for any $x \in \mathbb{R}^k$ we have

$$\|x\| = \left\| \frac{x}{\|x\|_1} \right\| \|x\|_1 \geq \|x^*\| \|x\|_1$$

Setting $J := 1/\|x^*\|$ gives the desired inequality. $\qquad \square$

3.2.4 Fixed Points

Next we turn to fixed points. Fixed point theory tells us how to find an x that solves $Tx = x$ for some given $T: S \to S$.[12] Like optimization it has great practical importance. Very often the solutions of problems we study will turn out to be fixed points of some appropriately constructed function. Of the theorems we treat in this section, one uses convexity and is due to L. E. J. Brouwer while the other two are contraction mapping arguments: a famous one due to Stefan Banach and a variation of the latter.

Incidentally, fixed point and optimization problems are closely related. When we study dynamic programming, an optimization problem will be converted into a fixed point problem—in the process yielding an efficient means of computation. On the other hand, if $T: S \to S$ has a unique fixed point in metric space (S, ρ), then finding that fixed point is equivalent to finding the minimizer of $g(x) := \rho(Tx, x)$.

So let $T: S \to S$, where S is any set. An $x^* \in S$ is called a *fixed point* of T on S if $Tx^* = x^*$. If S is a subset of $\mathbb{R}$, then fixed points of T are those points in S where T meets the 45 degree line, as illustrated in figure 3.3.

Exercise 3.2.31 Show that if $S = \mathbb{R}$ and $T: S \to S$ is decreasing ($x \leq y$ implies $Tx \geq Ty$), then T has at most one fixed point.

---

[11] Alternatively, you can show directly that $(\mathbb{R}^k, d_2)$ and $(\mathbb{R}^k, d_1)$ are equivalent by establishing (3.5) for $\|\cdot\| = \|\cdot\|_2$. The first inequality is already done, and the second follows from the Cauchy–Schwartz inequality (look it up).

[12] It is common in fixed point theory to use upper case symbols like T for the function, and no brackets around its argument. One reason is that S is often a space of functions, and standard symbols like f and g are reserved for the elements of S.

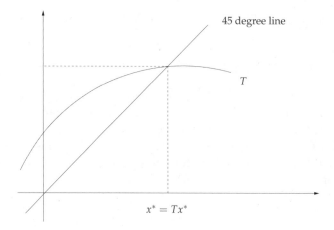

Figure 3.3 Fixed points in one dimension

A set $S \subset \mathbb{R}^k$ is called *convex* if for all $\lambda \in [0,1]$ and $a, a' \in S$ we have $\lambda a + (1 - \lambda)a' \in S$. Here is Brouwer's fixed point theorem:

Theorem 3.2.32 *Consider the space* $(\mathbb{R}^k, d)$, *where d is the metric induced by any norm.*[13] *Let* $S \subset \mathbb{R}^k$, *and let* $T: S \to S$. *If T is continuous and S is both compact and convex, then T has at least one fixed point in S.*

The proof is neither easy nor central to these notes, so we omit it[14] and move on to contraction mappings. Let (S, ρ) be a metric space, and let $T: S \to S$. The map T is called *nonexpansive* on S if

$$\rho(Tx, Ty) \leq \rho(x, y) \qquad \forall\, x, y \in S \tag{3.6}$$

It is called *contracting* on S if

$$\rho(Tx, Ty) < \rho(x, y) \qquad \forall\, x, y \in S \text{ with } x \neq y \tag{3.7}$$

It is called *uniformly contracting* on S with modulus λ if $0 \leq \lambda < 1$ and

$$\rho(Tx, Ty) \leq \lambda\rho(x, y) \qquad \forall\, x, y \in S \tag{3.8}$$

Exercise 3.2.33 Show that if T is nonexpansive on S then it is also continuous on S (with respect to the same metric ρ).

---

[13] All such metrics are equivalent. See theorem 3.2.30.

[14] You might like to sketch the case $S = [0,1]$ to gain some intuition.

Exercise 3.2.34 Show that if T is a contraction on S, then T has at most one fixed point in S.

For $n \in \mathbb{N}$ the notation T^n refers to the n-th composition of T with itself, so $T^n x$ means apply T to x, apply T to the result, and so on for n times. By convention, T^0 is the identity map $x \mapsto x$.[15]

Exercise 3.2.35 Let T be uniformly contracting on S with modulus λ, and let $x_0 \in S$. Define $x_n := T^n x_0$ for $n \in \mathbb{N}$. Use induction to show that $\rho(x_{n+1}, x_n) \leq \lambda^n \rho(x_1, x_0)$ for all $n \in \mathbb{N}$.

The next theorem is one of the cornerstones of functional analysis:

Theorem 3.2.36 (Banach) *Let* $T \colon S \to S$, *where* (S, ρ) *is a complete metric space. If T is a uniform contraction on S with modulus λ, then T has a unique fixed point $x^* \in S$. Moreover for every $x \in S$ and $n \in \mathbb{N}$ we have $\rho(T^n x, x^*) \leq \lambda^n \rho(x, x^*)$, and hence $T^n x \to x^*$ as $n \to \infty$.*

Proof. Let λ be as in (3.8). Let $x_n := T^n x_0$, where x_0 is some point in S. From exercise 3.2.35 we have $\rho(x_n, x_{n+1}) \leq \lambda^n \rho(x_0, x_1)$ for all $n \in \mathbb{N}$, suggesting that the sequence is ρ-Cauchy. In fact with a bit of extra work one can show that if $n, k \in \mathbb{N}$ and $n < k$, then $\rho(x_n, x_k) \leq \sum_{i=n}^{k-1} \lambda^i \rho(x_0, x_1)$.

$$\therefore \quad \rho(x_n, x_k) < \frac{\lambda^n}{1 - \lambda} \rho(x_0, x_1) \qquad (n, k \in \mathbb{N} \text{ with } n < k)$$

Since (x_n) is ρ-Cauchy, this sequence has a limit $x^* \in S$. That is, $T^n x_0 \to x^* \in S$. Next we show that x^* is a fixed point of T. Since T is continuous, we have $T(T^n x_0) \to Tx^*$. But $T(T^n x_0) \to x^*$ clearly also holds. (Why?) Since sequences in a metric space have at most one limit, it must be that $Tx^* \doteq x^*$.

Regarding uniqueness, let x and x' be fixed points of T in S. Then

$$\rho(x, x') = \rho(Tx, Tx') \leq \lambda \rho(x, x')$$

$$\therefore \quad \rho(x, x') = 0, \text{ and hence } x = x'$$

The estimate $\rho(T^n x, x^*) \leq \lambda^n \rho(x, x^*)$ in the statement of the theorem is left as an exercise. $\qquad \square$

If we take away uniformity and just have a contraction, then Banach's proof of stability does not work, and indeed a fixed point may fail to exist. Under the action of a *uniformly* contracting map T, the motion induced by iterating T slows down at a geometric rate. The limit of this process is a fixed point. On the other hand, with a

---

[15]In other words, $T^0 := \{x \mapsto x\}$ and $T^n := T \circ T^{n-1}$ for $n \in \mathbb{N}$.

contraction we know only that the process slows down at each step, and this is not enough to guarantee convergence. Imagine a particle that travels at speed $1 + 1/t$ at time t. Its motion slows down at each step, but the particle's speed is bounded away from zero.

Exercise 3.2.37 Let $S := \mathbb{R}_+$ with distance $|\cdot|$, and let $T\colon x \mapsto x + e^{-x}$. Show that T is a contraction on S, and that T has no fixed point in S.

However, if we add compactness of S to the contractiveness of T the problem is rectified. Now our particle cannot diverge, as that would violate the existence of a convergent subsequence.

Theorem 3.2.38 *If (S, ρ) is compact and $T\colon S \to S$ is contracting, then T has a unique fixed point $x^* \in S$. Moreover $T^n x \to x^*$ for all $x \in S$.*

The proof is consigned to the appendix in order to maintain continuity.

3.3   Commentary

The French mathematician Maurice Fréchet (1878–1973) introduced the notion of metric space in his dissertation of 1906. The name "metric space" is due to Felix Hausdorff (1868–1942). Other important spaces related to metric spaces are topological spaces (a generalization of metric space) and normed linear spaces (metric spaces with additional algebraic structure). Good references on metric space theory—sorted from elementary to advanced—include Light (1990), Kolmogorov and Fomin (1970), Aliprantis and Burkinshaw (1998), and Aliprantis and Border (1999). For a treatment with economic applications, see Ok (2007).

This chapter's discussion of fixed points and optimization only touched the surface of these topics. For a nice treatment of optimization theory, see Sundaram (1996). Various extensions of Brouwer's fixed point theorem are available, including Kakutani's theorem (for correspondences, see McLennan and Tourky 2005 for an interesting proof) and Schauder's theorem (for infinite-dimensional spaces). Aliprantis and Border (1999) is a good place to learn more.

Chapter 4

Introduction to Dynamics

4.1 Deterministic Dynamical Systems

Having covered programming and metric spaces in some depth, we now possess ample tools for analysis of dynamics. After starting with deterministic dynamical systems, setting up the basic theory and the notion of stability, we turn to stochastic models, where evolution of the state variable is affected by noise. While deterministic systems are clearly a kind of stochastic system (with zero-variance noise), we will see that the converse is also true: Stochastic models can be embedded in the deterministic framework. Through this embedding we can study the dynamic properties of stochastic systems using our knowledge of the deterministic model.

4.1.1 The Basic Model

Suppose that we are observing the time path of some variable x in a metric space S. At t, the current *state* of the system is denoted by x_t. Assume that from the current state x_t we can compute the time $t+1$ value x_{t+1} by applying a map h. That is, $x_{t+1} = h(x_t)$. The two primitives that make up this system are S and h:

Definition 4.1.1 A *dynamical system* is a pair (S, h), where $S = (S, \rho)$ is an arbitrary metric space and h is a map from S into itself.

By the *n-th iterate* of $x \in S$ under h we mean $h^n(x)$. It is conventional to set $h^0(x) := x$. The *trajectory* of $x \in S$ under h is the sequence $(h^t(x))_{t \geq 0}$. As before, $x^* \in S$ is a fixed point of h in S if $h(x^*) = x^*$. Fixed points are also said to be *stationary* or *invariant* under h.[1] Figure 4.1 illustrates the dynamics of one particular map h on $\mathbb{R}^2$

---

[1]Similar terminology applies to sets. For example, if $h(A) \subset A$, then A is said to be *invariant* under h.

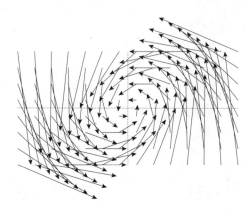

Figure 4.1 A dynamical system

by showing an arrow from x to $h(x)$ for $x \in$ a grid of points.

Exercise 4.1.2 Show that if (S, h) is a dynamical system, if $x' \in S$ is the limit of some trajectory (i.e., $h^t(x) \to x'$ as $t \to \infty$ for some $x \in S$), and if h is continuous at x', then x' is a fixed point of h.[2]

Exercise 4.1.3 Prove that if h is continuous on S and $h(A) \subset A$ (i.e., h maps $A \to A$), then $h(\operatorname{cl} A) \subset \operatorname{cl} A$.

Let x^* be a fixed point of h on S. By the *stable set* $\Lambda(x^*)$ of x^* we refer to all $x \in S$ such that $\lim_{t \to \infty} h^t(x) = x^*$. Clearly, $\Lambda(x^*)$ is nonempty. (Why?) The fixed point x^* is said to be *locally stable*, or an *attractor*, whenever there exists an open set G with $x^* \in G \subset \Lambda(x^*)$. Equivalently x^* is locally stable whenever there exists an ϵ-ball around x^* such that every trajectory starting in that ball converges to x^*:

Exercise 4.1.4 Prove that x^* is locally stable if and only if there exists an $\epsilon > 0$ such that $B(\epsilon, x^*) \subset \Lambda(x^*)$.

In this book we will be interested primarily in *global* stability:

Definition 4.1.5 A dynamical system (S, h) is called *globally stable* or *asymptotically stable* if

1. h has a *unique* fixed point x^* in S, and

2. $h^t(x) \to x^*$ as $t \to \infty$ for all $x \in S$.

---

[2]Hint: Let $x_t := h^t(x)$, so $x_t \to x'$. Consider the sequence $(h(x_t))_{t \geq 1}$. Argue that $h(x_t) \to h(x')$. Now show that $h(x_t) \to x'$ also holds. (Why?!) What do you conclude?

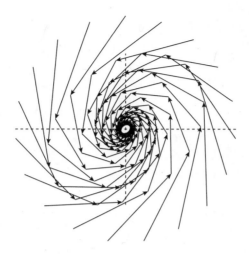

Figure 4.2 Global stability

Exercise 4.1.6 Prove that if x^* is a fixed point of (S, h) to which every trajectory converges, then x^* is the only fixed point of (S, h).

Figure 4.2 helps to visualize the concept of global stability, plotting 20 individual trajectories of a stable map h on $\mathbb{R}^2$. Figure 4.3 also illustrates global stability, in this case for the one-dimensional system (S, h), where $S := (0, \infty)$ and $h(k) := sAk^\alpha$ with $s \in (0, 1]$, $A > 0$ and $\alpha \in (0, 1)$. The system represents a simple Solow–Swan growth model, where next period's capital stock $h(k)$ is the savings rate s times current output Ak^α. The value A is a productivity parameter and α is the capital intensity. Figure 4.3 is called a 45 degree diagram. When the curve h lies above (resp., below) the 45 degree line we have $h(k) > k$ (resp., $h(k) < k$), and hence the trajectory moves to the right (resp., left). Two trajectories are shown, converging to the unique fixed point k^*.

Regarding local stability of (S, h) when S is an open subset of $\mathbb{R}$, it is well-known that

Lemma 4.1.7 *If h is a map with continuous derivative h' and x^* is a fixed point of h with $|h'(x^*)| < 1$, then x^* is locally stable.*

The most enthusiastic readers might like to attempt a proof, although the lemma is not particularly central to what follows.

Example 4.1.8 Consider a growth model with "threshold" nonconvexities of the form $k_{t+1} = sA(k_t)k_t^\alpha$, where $s \in (0, 1]$ and $k \mapsto A(k)$ is some increasing function with

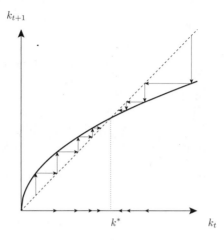

Figure 4.3 45 degree diagram

$A(k) > 0$ when $k > 0$. Suppose, for example, that A is a step function of the form

$$A(k) = \begin{cases} A_1 & \text{if } 0 < k < k_b \\ A_2 & \text{if } k_b \leq k < \infty \end{cases}$$

Here k_b is a "threshold" value of capital stock, and $0 < A_1 < A_2$. Let k_i^* be the solution to $k = sA_ik^\alpha$ for $i = 1, 2$ when it exists. A plot is given in figure 4.4 for the case where $k_1^* < k_b < k_2^*$. The two fixed points k_1^* and k_2^* are local attractors, as can be verified from lemma 4.1.7. Long-run outcomes depend on initial conditions, and for this reason the model is said to exhibit *path dependence*.

Exercise 4.1.9 A dynamical system (S, h) is called *Lagrange stable* if every trajectory is precompact in S. In other words, the set $\{h^n(x) : n \in \mathbb{N}\}$ is precompact for every $x \in S$.[3] Show that if S is a closed and bounded subset of $\mathbb{R}^n$, then (S, h) is Lagrange stable for any choice of h.

Exercise 4.1.10 Give an example of a dynamical system (S, h) where S is unbounded but (S, h) is Lagrange stable.

Exercise 4.1.11 Let $S = \mathbb{R}$, and let $h \colon \mathbb{R} \to \mathbb{R}$ be an increasing function, in the sense that if $x \leq y$, then $h(x) \leq h(y)$. Show that every trajectory of h is a monotone sequence in $\mathbb{R}$ (either increasing or decreasing).

---

[3]Equivalently every subsequence of the trajectory has a convergent subsubsequence.

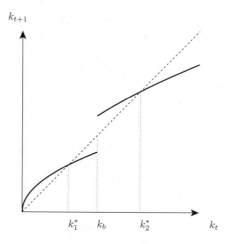

Figure 4.4 Threshold externalities

Exercise 4.1.12 Now order points in $\mathbb{R}^n$ by setting $x \leq y$ whenever $x_i \leq y_i$ for i in $\{1, \ldots, n\}$ (i.e., each component of x is dominated by the corresponding component of y). Let $S = \mathbb{R}^n$, and let $h \colon S \to S$ be monotone increasing. (The definition is the same.) Show that the same result no longer holds—h does not necessarily generate monotone trajectories.

4.1.2 Global Stability

Global stability will be a key concept for the remainder of the text. Let's start our investigation of global stability by looking at linear (more correctly, affine) systems in one dimension.

Exercise 4.1.13 Let $S = (\mathbb{R}, |\cdot|)$ and $h(x) = ax + b$. Prove that

$$h^t(x) = a^t x + b \sum_{i=0}^{t-1} a^i \qquad (x \in S, \ t \in \mathbb{N})$$

(Hint: Use induction.) From this expression, prove that (S, h) is globally stable whenever $|a| < 1$, and exhibit the fixed point.

Exercise 4.1.14 Show that the condition $|a| < 1$ is also necessary, in the sense that if $|a| \geq 1$, then (S, h) is not globally stable. Show, in particular, that $h^t(x_0)$ converges to $x^* := b/(1 - a)$ only if $x_0 = x^*$.

In exercise 4.1.13 we found a direct proof of global stability for our affine system when $|a| < 1$. For more complex systems direct methods are usually unavailable, and we must deploy more powerful machinery, such as Banach's fixed point theorem (theorem 3.2.36 on page 53).

Exercise 4.1.15 Let (S, h) be as in exercise 4.1.13. Using theorem 3.2.36, prove that (S, h) is globally stable whenever $|a| < 1$.

Exercise 4.1.16 Let $S := (0, \infty)$ with $\rho(x, y) := |\ln x - \ln y|$. Prove that ρ is a metric on S and that (S, ρ) is a complete metric space.[4] Consider the growth model $k_{t+1} = h(k_t) = sAk_t^\alpha$ in figure 4.3, where $s \in (0, 1]$, $A > 0$ and $\alpha \in (0, 1)$. Convert this into a dynamical system on (S, ρ), and prove global stability using theorem 3.2.36.

Next we consider linear systems in $\mathbb{R}^n$. In general, a function $h \colon \mathbb{R}^n \to \mathbb{R}^n$ is called linear if

$$h(\alpha x + \beta y) = \alpha h(x) + \beta h(y) \qquad \forall\, x, y \in \mathbb{R}^n \quad \forall\, \alpha, \beta \in \mathbb{R} \tag{4.1}$$

It can be shown that every such h is continuous. If E is an $n \times n$ matrix, then the map on $\mathbb{R}^n$ defined by $x \mapsto Ex$ is linear. In fact it can be shown that for *all* linear maps $h \colon \mathbb{R}^n \to \mathbb{R}^n$ there exists a matrix E_h with $h(x) = E_h x$ for all $x \in \mathbb{R}^n$. An *affine* system on $\mathbb{R}^n$ is a map $h \colon \mathbb{R}^n \to \mathbb{R}^n$ given by

$$h(x) = Ex + b \quad \text{where } E \text{ is an } n \times n \text{ matrix and } b \in \mathbb{R}^n$$

To investigate this system, let $\|\cdot\|$ be any norm on $\mathbb{R}^n$, and define

$$\lambda := \max\{\|Ex\| : x \in \mathbb{R}^n,\ \|x\| = 1\} \tag{4.2}$$

Exercise 4.1.17 If you can, prove that the maximum exists. Using the properties of norms and linearity of E, show that $\|Ex\| \le \lambda \|x\|$ for all $x \in \mathbb{R}^n$. Show in addition that if $\lambda < 1$, then $(\mathbb{R}^n, h)$ is globally stable.

Let's look at an application of these ideas. In a well-known paper, Long and Plosser (1983) studied the modeling of business cycles using multisector growth models. After solving their model, they find a system for log output given by $y_{t+1} = Ay_t + b$. Here $A = (a_{ij})$ is a matrix of input/output elasticities across sectors, and y_t is a 6×1 vector recording output in agriculture, mining, construction, manufacturing, transportation and services.[5] Using cost share data and the hypothesis of perfect

---

[4]Hint: Take a Cauchy sequence (x_n) in (S, ρ), and map it into $(\ln x_n)$, which is a sequence in $(\mathbb{R}, |\cdot|)$. Prove this sequence is Cauchy in $(\mathbb{R}, |\cdot|)$, and therefore converges to some $y \in \mathbb{R}$. Now prove that $x_n \to e^y$ in (S, ρ).

[5]In their model b is random, but let's ignore this complication for now.

competition, the authors calculate A to be given by

$$A = (a_{ij}) = \begin{pmatrix} 0.45 & 0.00 & 0.01 & 0.21 & 0.10 & 0.16 \\ 0.00 & 0.09 & 0.04 & 0.17 & 0.05 & 0.49 \\ 0.00 & 0.01 & 0.00 & 0.42 & 0.12 & 0.09 \\ 0.06 & 0.03 & 0.01 & 0.46 & 0.06 & 0.13 \\ 0.00 & 0.00 & 0.02 & 0.12 & 0.10 & 0.32 \\ 0.02 & 0.02 & 0.06 & 0.20 & 0.09 & 0.38 \end{pmatrix}$$

Exercise 4.1.18 Prove that Long and Plosser's system is stable in the following way: Let $A = (a_{ij})$ be an $n \times n$ matrix where the sum of any of the rows of A is strictly less than 1 (i.e., $\max_i \alpha_i < 1$, where $\alpha_i := \sum_j |a_{ij}|$). Using the norm $\| \cdot \|_\infty$ in (4.2), show that for A we have $\lambda < 1$. Now argue that in Long and Plosser's model, (y_t) converges to a limit y^*, which is independent of initial output y_0, and, moreover, is the unique solution to the equation $y^* = Ay^* + b$.[6]

Exercise 4.1.19 Let $B = (b_{ij})$ be an $n \times n$ matrix where the sum of any of the *columns* of B is strictly less than 1 (i.e., $\max_j \beta_j < 1$, where $\beta_j := \sum_i |b_{ij}|$). Using the norm $\| \cdot \|_1$ in (4.2), show that for B we have $\lambda < 1$. Conclude that if $h(x) = Bx + b$, then $(\mathbb{R}^n, h)$ is globally stable.

The following results will be needed later in the text:

Exercise 4.1.20 Suppose that h is uniformly contracting on complete space S, so (S, h) is globally stable. Prove that if $A \subset S$ is nonempty, closed and invariant under h (i.e., $h(A) \subset A$), then the fixed point of h lies in A.

Lemma 4.1.21 *Let (S, h) be a dynamical system. If h is nonexpansive and (S, h^N) is globally stable for some $N \in \mathbb{N}$, then (S, h) is globally stable.*

Proof. By hypothesis, h^N has a unique fixed point x^* in S, and $h^{kN}(x) \to x^*$ as $k \to \infty$ for all $x \in S$. Pick any $\epsilon > 0$, and choose $k \in \mathbb{N}$ so that $\rho(h^{kN}(h(x^*)), x^*) < \epsilon$. Then

$$\rho(h(x^*), x^*) = \rho(h(h^{kN}(x^*)), x^*) = \rho(h^{kN}(h(x^*)), x^*) < \epsilon$$

It follows that x^* is a fixed point of h. (Why?)

Stability: Fix $x \in S$ and $\epsilon > 0$. Choose $k \in \mathbb{N}$ so that $\rho(h^{kN}(x), x^*) < \epsilon$. Then nonexpansiveness implies that for each $n \geq kN$,

$$\rho(h^n(x), x^*) = \rho(h^{n-kN}(h^{kN}(x)), x^*) \leq \rho(h^{kN}(x), x^*) < \epsilon$$

In other words, (S, h) is globally stable.   □

---

[6]You are proving d_∞-convergence of trajectories, but this is equivalent to d_2-convergence by theorem 3.2.30.

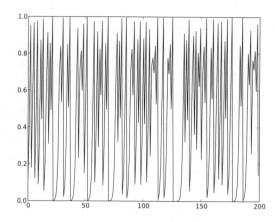

Figure 4.5 Trajectory of the quadratic map

4.1.3 Chaotic Dynamic Systems

In this section we make a very brief foray into complex dynamical systems. Chaotic dynamics is an area that has suffered from excessive hype but is nonetheless interesting for any student of dynamics. We will not be covering much theory, but we will get some practice with programming.

To begin, consider first the dynamical system (S, h) defined by

$$h(x) = 4x(1 - x) \qquad (x \in S := [0, 1]) \tag{4.3}$$

The function h is called the quadratic (or logistic) map, as is often found in biological models related to population dynamics. Readers can check that h maps S into itself.

In the previous section we defined global stability. For stable systems the trajectories converge to a single point, and long series will have an average value close to that point. Other systems can have several attractors, so the point where the trajectory settles down to depends on the initial condition. We will see that for (4.3) dynamics are still more complicated.

Figure 4.5 shows one trajectory starting at initial condition 0.11. Code used to generate the figure is given in listing 4.1. It uses Matplotlib's `pylab`, a module for producing graphs in Python.

Listing 4.2 provides an alternative method of generating trajectories using OOP (see §2.2.3 for an introduction to OOP design). Dynamical systems are naturally represented as objects because they combine data (current state) and action (iteration of

Listing 4.1 (`quadmap1.py`) Trajectories of the quadratic map

```python
from pylab import plot, show    # Requires Matplotlib

datapoints = []                 # Stores trajectory
x = 0.11                        # Initial condition
for t in range(200):
    datapoints.append(x)
    x = 4 * x * (1 - x)

plot(datapoints)
show()
```

the map). Here we define a class `DS` that is an abstraction of a dynamical system. The class definition is relatively self-explanatory. Although the use of OOP is perhaps excessive for this simple problem, one can see how abstraction facilitates code reuse: `DS` can be used to generate trajectories for any dynamical system. An example of usage is given in listing 4.3.

Notice that in figure 4.5 the trajectory continues to traverse through the space without settling down. Some experimentation shows that this happens for many initial conditions (but not all—does the map have any fixed points?). Moreover a slight variation in the initial condition typically leads to a time series that bears no clear resemblance to the previous one.

Science and mathematics are all about simplification and reduction. For example, with a globally stable system we can usually focus our attention on the steady state. (How does this state fit with the data?) From this perspective figure 4.5 is a little distressing. Unless the initial conditions are very special and can be known exactly, it seems that long run outcomes cannot be predicted.[7] However, this conclusion is too pessimistic, as the next exercise illustrates.

Exercise 4.1.22 Using your preferred plotting tool, histogram some trajectories generated by the quadratic map, starting at different initial conditions. Use relatively long trajectories (e.g., around 5,000 points), and a fine histogram (about 40 bins). What regularities can you observe?

Incidentally, the time series in figure 4.5 looks quite random, and in exercise 4.1.22 we treated the trajectory in a "statistical" way, by computing its histogram. Is there in

---

[7]Which is problematic for a scientific study—what falsifiable implications can be drawn from these models?

Listing 4.2 (ds.py) An abstract dynamical system

```
class DS:

    def __init__(self, h=None, x=None):
        """Parameters: h is a function and x is an
        element of S representing the current state."""
        self.h, self.x = h, x

    def update(self):
        "Update the state of the system by applying h."
        self.x = self.h(self.x)

    def trajectory(self, n):
        """Generate a trajectory of length n, starting
        at the current state."""
        traj = []
        for i in range(n):
            traj.append(self.x)
            self.update()
        return traj
```

Listing 4.3 (testds.py) Example application

```
from ds import DS            # Import from listing 4.2

def quadmap(x):
    return 4 * x * (1 - x)

q = DS(h=quadmap, x=0.1)    # Create an instance q of DS
T1 = q.trajectory(100)      # T1 holds trajectory from 0.1

q.x = 0.2                   # Reset current state to 0.2
T2 = q.trajectory(100)      # T2 holds trajectory from 0.2
```

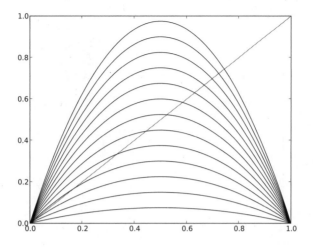

Figure 4.6 Quadratic maps, $r \in [0, 4]$

fact any formal difference between this kind of complex dynamics and the dynamics produced in systems perturbed by random variables?

One answer is that proposed by Kolmogorov, who suggested measuring the "randomness" of a string of numbers by the size of the shortest computer program that can replicate it.[8] The upper bound of this measure is the size of the string itself because, if necessary, one can simply enumerate the string. This upper bound is associated with complete randomness. On the other hand, our code used to produce the time series for the quadratic map was only a few lines, and therefore has a low Kolmogorov score. In this sense we can differentiate it from a random string.

How does the quadratic map behave when we let the multiplicative parameter take values other than 4? Consider the more general map $x \mapsto rx(1 - x)$, where $0 \leq r \leq 4$. A subset of these maps is plotted in figure 4.6, along with a 45 degree line. More curvature corresponds to greater r. It turns out that for some values of r this system is globally stable. For others, like 4, the behavior is highly complex.

The *bifurcation diagram* shown in figure 4.7 helps to give an understanding of the dynamics. On the x-axis the parameter r ranges from 2.7 to 4. The y-axis corresponds to the state space S. For each value r in a grid over $[2.7, 4]$, a trajectory of length 1000 was generated. The first 950 points were discarded, and the last 50 were plotted.

---

[8]Put differently, by how much can we *compress* such a string of numbers?

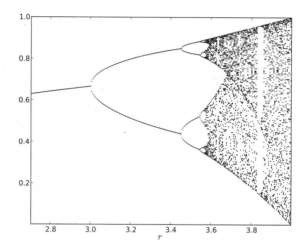

Figure 4.7 Bifurcation diagram

For $r \leq 3$, interior points converge to a unique interior steady state. For $r \in (3, 1 + \sqrt{6}]$, the state eventually oscillates between two "periodic attractors." From there the number of periodic attractors increases rapidly, and the behavior of the system becomes correspondingly more "chaotic."

Exercise 4.1.23 Reproduce figure 4.7 using Python to generate the data and your preferred plotting tool.

4.1.4 Equivalent Dynamics and Linearization

In general, nonlinear models are much more difficult to analyze than linear models, leading researchers to approximate nonlinear models with linearized versions. The latter are usually obtained by a first-order Taylor expansion. Since fixed points are the natural focus of analysis, it is standard to take expansions around fixed points.

Let us see how this is done in the one-dimensional case. Let (S, h) be a dynamical system where S is an open subset of $\mathbb{R}$, and h is continuously differentiable, with derivative h' on S. Pick any $a \in S$. The first-order Taylor expansion around a is the map h_1 defined by

$$h_1(x) = h(a) + h'(a)(x - a) \tag{4.4}$$

Notice that h_1 is an affine function on $\mathbb{R}$ with $h(a) = h_1(a)$. Clearly, h_1 approximates

h in some sense when $|x - a|$ is small. For this reason it is regarded as a "linear" approximation to h around a.

Now let x^* be a fixed point of h, so

$$h_1(x) = x^* + h'(x^*)(x - x^*) \tag{4.5}$$

You can check that x^* is also a fixed point of the approximating map h_1. Note also that x^* will be stable for h_1 whenever $|h'(x^*)| < 1$. But this is precisely the condition for x^* to be a local attractor for h (lemma 4.1.7). So it seems that we can learn something about how $h^t(x)$ will behave when $|x - x^*|$ is small by studying the simple affine map h_1 and the trajectory $h_1^t(x)$ that it generates.

The well-known Hartman–Grobman theorem formalizes this idea. To state the theorem, it is necessary to introduce the abstract but valuable notion of topological conjugacy. First, let S and $\hat{S}$ be two metric spaces. A function τ from S to $\hat{S}$ is called a *homeomorphism* if it is continuous, a bijection, and its inverse τ^{-1} is also continuous. Two dynamical systems (S, g) and $(\hat{S}, \hat{g})$ are said to be *topologically conjugate* if there exists a homeomorphism τ from S into $\hat{S}$ such that g and $\hat{g}$ commute with τ in the sense that $\hat{g} = \tau \circ g \circ \tau^{-1}$ everywhere on $\hat{S}$. In other words, shifting a point $\hat{x} \in \hat{S}$ to $\hat{g}(\hat{x})$ using the map $\hat{g}$ is equivalent to moving $\hat{x}$ into S with τ^{-1}, applying g, and then moving the result back using τ:

$$
\begin{array}{ccc}
x & \xrightarrow{\;g\;} & g(x) \\[4pt]
\uparrow{\scriptstyle\tau^{-1}} & & \downarrow{\scriptstyle\tau} \\[4pt]
\hat{x} & \xrightarrow{\;\hat{g}\;} & \hat{g}(\hat{x})
\end{array}
$$

Exercise 4.1.24 Let $S := ((0, \infty), |\cdot|)$ and $\hat{S} := (\mathbb{R}, |\cdot|)$. Let $g(x) = Ax^\alpha$, where $A > 0$ and $\alpha \in \mathbb{R}$, and let $\hat{g}(\hat{x}) = \ln A + \alpha \hat{x}$. Show that g and $\hat{g}$ are topologically conjugate under $\tau := \ln$.

Exercise 4.1.25 Show that if (S, g) and $(\hat{S}, \hat{g})$ are topologically conjugate, then $x^* \in S$ is a fixed point of g on S if and only if $\tau(x^*) \in \hat{S}$ is a fixed point of $\hat{g}$ on $\hat{S}$.

Exercise 4.1.26 Let $x^* \in S$ be a fixed point of g and let x be any point in S. Show that $\lim_{t \to \infty} g^t(x) = x^*$ if and only if $\lim_{t \to \infty} \hat{g}^t(\tau(x)) = \tau(x^*)$.

Exercise 4.1.27 Let $x^* \in S$ be a fixed point of g. Show that if x^* is a local attractor for (S, g), then $\tau(x^*)$ is a local attractor for $(\hat{S}, \hat{g})$. Show that if (S, g) is globally stable, then $(\hat{S}, \hat{g})$ is globally stable.

We can now state the theorem of Hartman and Grobman:

Theorem 4.1.28 (Hartman–Grobman) *Let S be an open subset of $\mathbb{R}$, let $h\colon S \to S$ be a continuously differentiable function, let $x^* \in S$ be a fixed point of h, and let h_1 be the Taylor approximation in (4.5). If $|h'(x^*)| \neq 1$, then there exists an open set G containing x^* such that h and h_1 are topologically conjugate on G.*[9]

Be careful when applying this theorem, which is one of the most misused mathematical results in all of economics. It provides only a *neighborhood* of S such that behavior of the approximation is *qualitatively* similar to that of the original system. As it stands, the Hartman–Grobman theorem provides no basis for *quantitative* analysis.

4.2   Finite State Markov Chains

Next we start our journey into the world of stochastic dynamics. Stochastic dynamics is a technically demanding field, mainly because any serious attempt at probability theory on general state spaces requires at least a modicum of measure theory. For now we concentrate on *finite* state spaces, where measure theory is unnecessary. However, our treatment of finite state stochastic dynamics is geared in every way toward building understanding, intuition, notation and tools that will be used in the general (uncountable state) case.

4.2.1   Definition

We begin with a state space S, which is a finite set $\{x_1, \ldots, x_N\}$. A typical element of S is usually denoted by x, rather than a symbol such as x_i or x_n. This makes our notation more consistent with the continuous state theory developed below. The set of *distributions* on S will be denoted $\mathscr{P}(S)$, and consists of all functions $\phi\colon S \to \mathbb{R}$ with $\phi(x) \geq 0$ for all $x \in S$, and $\sum_{x \in S} \phi(x) = 1$. In general, $\phi(x)$ will correspond to the probability attached to the point x in the state space under some given scenario.[10]

A quick digression: Although ϕ has been introduced as a function from S to $\mathbb{R}$, one can also think of it as a *vector* under the one-to-one correspondence

$$\mathscr{P}(S) \ni \phi \leftrightarrow (\phi(x))_{x \in S} := (\phi(x_1), \ldots, \phi(x_N)) \in \mathbb{R}^N \tag{4.6}$$

Under the correspondence (4.6), the collection of functions $\mathscr{P}(S)$ becomes a subset of the vector space $\mathbb{R}^N$—in particular, the elements of $\mathbb{R}^N$ that are nonnegative and sum to one. This set is called the unit simplex, and is illustrated for the case of $N = 3$ in figure 4.8.

The basic primitive for a discrete time Markov process on S is a *stochastic kernel*, the definition of which is as follows.

---

[9] To see why $|h'(x^*)| \neq 1$ is important, consider the case of $h(x) = \arctan(x)$.

[10] What we call a distribution here is also referred to as a probability mass function.

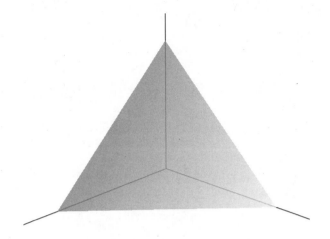

Figure 4.8 The unit simplex with $N = 3$

Definition 4.2.1 A *stochastic kernel* p is a function from $S \times S$ into $[0, 1]$ such that

1. $p(x, y) \geq 0$ for each (x, y) in $S \times S$, and

2. $\sum_{y \in S} p(x, y) = 1$ for each $x \in S$.

In other words, the function $S \ni y \mapsto p(x, y) \in \mathbb{R}$ is an element of $\mathscr{P}(S)$ for all $x \in S$. This distribution is represented by the symbol $p(x, dy)$ in what follows.

As well as being a function, the distribution $p(x, dy)$ can be viewed as a row[11] vector $(p(x, x_1), \dots, p(x, x_N))$ in $\mathbb{R}^N$, located in the unit simplex, and these rows can be stacked horizontally to produce an $N \times N$ matrix with the property that each row is nonnegative and sums to one:

$$p = \begin{pmatrix} p(x_1, dy) \\ \vdots \\ p(x_N, dy) \end{pmatrix} = \begin{pmatrix} p(x_1, x_1) & \cdots & p(x_1, x_N) \\ \vdots & & \vdots \\ p(x_N, x_1) & \cdots & p(x_N, x_N) \end{pmatrix} \qquad (4.7)$$

Conversely, any square $N \times N$ matrix that is nonnegative and has all rows summing to one defines a stochastic kernel. However, when we move on to infinite state spaces there is no concept of matrices, and hence most of the theory is stated in terms of kernels.

---

[11]When treating distributions as vectors it is traditional in the Markov chain literature to regard them as row vectors.

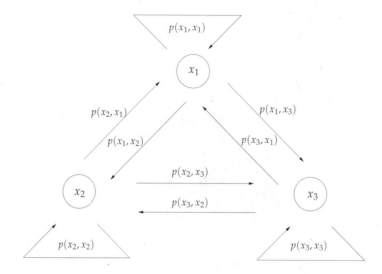

Figure 4.9 Finite Markov chain

In this chapter we are going to study a sequence of random variables $(X_t)_{t \geq 0}$, where each X_t takes values in S. The sequence updates according to the following rule: If $X_t = x$, then, in the following period X_{t+1} takes the value y with probability $p(x, y)$. In other words, once the current state X_t is realized, the probabilities for X_{t+1} are given by $p(X_t, dy)$. Figure 4.9 depicts an example of a simple Markov system, where $S = \{x_1, x_2, x_3\}$, and $p(x_i, x_j)$ is the probability that X_t moves from state x_i at time t to x_j at time $t + 1$.

The transition probabilities at each time depend on nothing other than the *current* location of the state. This is the "Markov" assumption. Moreover the transition probabilities do not depend on time. This is called time homogeneity. These are assumptions, which fit the motion of some models well and others poorly. Although they may seem restrictive, it turns out that, with some manipulation, a large class of systems can be embedded in the basic Markov framework.

An example of a stochastic kernel is the one used in a recent study by Hamilton (2005), who investigates a nonlinear statistical model of the business cycle based on US unemployment data. As part of his calculation he estimates the kernel

$$p_H := \begin{pmatrix} 0.971 & 0.029 & 0 \\ 0.145 & 0.778 & 0.077 \\ 0 & 0.508 & 0.492 \end{pmatrix} \tag{4.8}$$

Here $S = \{x_1, x_2, x_3\} = \{NG, MR, SR\}$, where NG corresponds to normal growth,

MR to mild recession, and *SR* to severe recession. For example, the probability of transitioning from severe recession to mild recession in one period is 0.508. The length of the period is one month.

For another example of a Markov model, consider the growth dynamics study of Quah (1993), who analyzes the evolution of real GDP per capita relative to the world average for a "typical" country (e.g., $X_t = 2$ implies that income per capita for the country in question is twice the world average at time t). A natural state space is $\mathbb{R}_+$, but to simplify matters Quah discretizes this space into five bins that correspond to values for relative GDP of 0 to 0.25, 0.25 to 0.5, 0.5 to 1, 1 to 2 and 2 to ∞ respectively. He then calculates the stochastic kernel by setting $p(x, y)$ equal to the fraction of times that a country, finding itself in state x, subsequently makes the transition to state y.[12] The result of this calculation is

$$p_Q := \begin{pmatrix} 0.97 & 0.03 & 0.00 & 0.00 & 0.00 \\ 0.05 & 0.92 & 0.03 & 0.00 & 0.00 \\ 0.00 & 0.04 & 0.92 & 0.04 & 0.00 \\ 0.00 & 0.00 & 0.04 & 0.94 & 0.02 \\ 0.00 & 0.00 & 0.00 & 0.01 & 0.99 \end{pmatrix} \qquad (4.9)$$

For example, the probability of our typical country transitioning from the lowest bin to the second lowest bin in one year is 0.03.

Algorithm 4.1: Simulation of a Markov chain

---

draw $X_0 \sim \psi$ and set $t = 0$
while True **do**   // "while True" means repeat forever
 | draw $X_{t+1} \sim p(X_t, dy)$
 | set $t = t + 1$
end

Let us now try to pin down the definition of a Markov chain $(X_t)_{t \geq 0}$ corresponding to a given stochastic kernel p. It is helpful to imagine that we wish to simulate $(X_t)_{t \geq 0}$ on a computer. First we draw X_0 from some predetermined *initial condition* $\psi \in \mathscr{P}(S)$. As $p(x, dy)$ gives the transition probabilities for X_{t+1} conditional on $X_t = x$, we now draw X_1 from $p(X_0, dy)$. Taking the result X_1, we then draw X_2 from $p(X_1, dy)$, and so on. This is the content of algorithm 4.1, as well as the next definition.

Definition 4.2.2 Let $\psi \in \mathscr{P}(S)$. A sequence of S-valued random variables $(X_t)_{t \geq 0}$ is called *Markov-(p, ψ)* if

---

[12]His data span the period 1962 to 1984, and have a sample of 118 countries. The transitions are over a one year period. The model is assumed to be stationary (transition probabilities do not vary with time), so all of the transitions (1962 to 1963, 1963 to 1964, etc.) can be pooled when calculating transition probabilities.

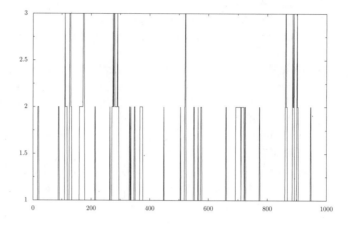

Figure 4.10 Simulation of the chain

1. at time zero, X_0 is drawn from ψ, and

2. at time $t + 1$, X_{t+1} is drawn from $p(X_t, dy)$.

If $\psi(x) = 1$ for some $x \in S$, then $(X_t)_{t \geq 0}$ is called Markov-(p, x).

Let's try simulating a Markov-(p, ψ) chain. One method is given in listing 4.4. The function `sample()` defined at the top is used to generate draws from a given finite distribution using the inverse transform algorithm.[13] Next we define a class `MC`, which is similar to the class `DS` in listing 4.2 (page 64). An instance is initialized with a stochastic kernel p and an initial state X. The kernel p should be such that p[x] is a sequence representing $p(x, dy)$—see listing 4.5 for an example. The method `update()` updates the current state using `sample()`, drawing X_{t+1} from $p(X_t, dy)$.[14]

In listing 4.5 an instance h of `MC` is created for Hamilton's stochastic kernel p_H, and the `sample_path()` method is used to produce some time series. In the code, `genfinitemc` is the name of the file in listing 4.4. Figure 4.10 shows a simulated series.

4.2.2 Marginal Distributions

Let $(X_t)_{t \geq 0}$ be Markov-(p, ψ). For every $t \in \mathbb{N}$, let $\psi_t \in \mathscr{P}(S)$ denote the distribution of X_t. That is, $\psi_t(y)$ is the probability that $X_t = y$, given that X_0 is drawn from

---

[13] There are existing libraries that can do this job, but the technique used in `sample()` is something we will revisit later on (see algorithm 5.3 on page 109), and we present it here for future reference. Don't be concerned if the logic is not yet clear.

[14] All of this code is written for clarity rather than speed, and is suitable only for small state spaces. See the text home page for more efficient methods.

Listing 4.4 (genfinitemc.py) Finite Markov chain

```python
from random import uniform

def sample(phi):
    """Returns i with probability phi[i], where phi is an
    array (e.g., list or tuple)."""
    a = 0.0
    U = uniform(0,1)
    for i in range(len(phi)):
        if a < U <= a + phi[i]:
            return i
        a = a + phi[i]


class MC:
    """For generating sample paths of finite Markov chains
    on state space S = {0,...,N-1}."""

    def __init__(self, p=None, X=None):
        """Create an instance with stochastic kernel p and
        current state X. Here p[x] is an array of length N
        for each x, and represents p(x,dy).
        The parameter X is an integer in S."""
        self.p, self.X = p, X

    def update(self):
        "Update the state by drawing from p(X,dy)."
        self.X = sample(self.p[self.X])

    def sample_path(self, n):
        """Generate a sample path of length n, starting from
        the current state."""
        path = []
        for i in range(n):
            path.append(self.X)
            self.update()
        return path
```

Listing 4.5 (`testgenfinitemc.py`) Example application

```
from genfinitemc import sample, MC   # Import from listing 4.4

pH = ((0.971, 0.029, 0.000),
      (0.145, 0.778, 0.077),
      (0.000, 0.508, 0.492))

psi = (0.3, 0.4, 0.3)               # Initial condition
h = MC(p=pH, X=sample(psi))         # Create an instance of class MC
T1 = h.sample_path(1000)            # Series is Markov-(p, psi)

psi2 = (0.8, 0.1, 0.1)              # Alternative initial condition
h.X = sample(psi2)                  # Reset the current state
T2 = h.sample_path(1000)            # Series is Markov-(p, psi2)
```

initial distribution ψ, and that the chain subsequently follows $X_{t+1} \sim p(X_t, dy)$. This distribution is sometimes called the *marginal* or *unconditional* distribution of X_t. We can understand it as follows: Generate n independent realizations of X_t, and calculate the fraction that takes the value y. Call this number $\psi_t^n(y)$. The probability $\psi_t(y)$ can be thought of as the limit of $\psi_t^n(y)$ as $n \to \infty$.

A method for computing the fraction $\psi_t^n(y)$ is given in algorithm 4.2. In the algorithm, the instruction draw $X \sim p(X, dy)$ should be interpreted as: Draw a random variable Y according to the distribution $p(X, dy)$ and then set $X = Y$. Also, $\mathbb{1}\{X_t^i = y\}$ is an indicator function, equal to one when $X_t^i = y$ and zero otherwise.

Algorithm 4.2: Approximate marginal distribution

---

for i *in 1 to n* **do**
    draw $X \sim \psi$
    for j *in 1 to t* **do**
        draw $X \sim p(X, dy)$
    end
    set $X_t^i = X$
end
return $(1/n) \sum_{i=1}^{n} \mathbb{1}\{X_t^i = y\}$

Exercise 4.2.3 Implement algorithm 4.2 for Hamilton's Markov chain.[15] Set $\psi = (0, 0, 1)$, so the economy starts in severe recession with probability one. Compute an approximation to $\psi_t(y)$, where $t = 10$ and $y = NG$. For sufficiently large n you should get an answer close to 0.6.

Exercise 4.2.4 Rewrite algorithm 4.2 using a counter that increments by one whenever the output of the inner loop produces a value equal to y instead of recording the value of each X_t^i.

Now consider again a Markov-(p, ψ) chain $(X_t)_{t \geq 0}$ for arbitrary stochastic kernel p and initial condition ψ. As above, let $\psi_t \in \mathscr{P}(S)$ be the marginal distribution of X_t. From ψ_t and our complete description of the dynamics in p, it seems possible that we will be able to calculate the distribution of X_{t+1}. That is to say, we might be able to link ψ_t and ψ_{t+1} using p. That we can in fact construct such a recursion is one of the most fundamental and important properties of Markov chains.

To begin, pick any $y \in S$. Using the law of total probability (see §A.1.3), we can decompose the probability that $X_{t+1} = y$ into conditional parts as follows:

$$\mathbb{P}\{X_{t+1} = y\} = \sum_{x \in S} \mathbb{P}\{X_{t+1} = y \mid X_t = x\} \cdot \mathbb{P}\{X_t = x\}$$

Rewriting this statement in terms of our marginal and conditional probabilities gives

$$\psi_{t+1}(y) = \sum_{x \in S} p(x, y)\psi_t(x) \qquad (y \in S) \tag{4.10}$$

This is precisely the kind of recursion we are looking for. Let's introduce some additional notation to help manipulate this expression.

Definition 4.2.5 Given stochastic kernel p, the *Markov operator* corresponding to p is the map $\mathbf{M}$ sending $\mathscr{P}(S) \ni \psi \mapsto \psi\mathbf{M} \in \mathscr{P}(S)$, where $\psi\mathbf{M}$ is defined by

$$\psi\mathbf{M}(y) = \sum_{x \in S} p(x, y)\psi(x) \qquad (y \in S) \tag{4.11}$$

The notation appears unusual, in the sense that we normally write $\mathbf{M}(\psi)$ instead of $\psi\mathbf{M}$ for the image of ψ under a mapping $\mathbf{M}$. However, such notation is traditional in the Markov literature. It reminds us that applying the Markov operator to a distribution $\psi \in \mathscr{P}(S)$ is just postmultiplication of the row vector $(\psi(x))_{x \in S}$ by the stochastic kernel (viewed as a matrix).

---

[15]Ideally you should add the extra functionality to the class MC rather than writing it specifically for Hamilton's chain. Note that in Python, if Y is a sequence containing observations of X_t, then Y.count(y) gives the number of elements equal to y.

Combining (4.10) and (4.11), we obtain the fundamental recursion

$$\psi_{t+1} = \psi_t \mathbf{M} \qquad (4.12)$$

Check this carefully until you feel comfortable with the notation.

This representation (4.12) is easy to manipulate. For example, suppose that we want to calculate ψ_{j+k} from ψ_j. Clearly,

$$\psi_{j+k} = \psi_{j+k-1}\mathbf{M} = (\psi_{j+k-2}\mathbf{M})\mathbf{M} = \psi_{j+k-2}\mathbf{M}^2 = \cdots = \psi_j \mathbf{M}^k$$

where $\mathbf{M}^m$ is the m-th composition of the map $\mathbf{M}$ with itself. In particular, setting $j = 0$ and $k = t$, we have $X_t \sim \psi\mathbf{M}^t$ when $X_0 \sim \psi$. Let's state these results as a theorem:

Theorem 4.2.6 *Let $(X_t)_{t \geq 0}$ be Markov-(p, ψ), and let $\mathbf{M}$ be the Markov operator corresponding to p. If ψ_t is the marginal distribution of X_t for each t, then $\psi_{t+1} = \psi_t \mathbf{M}$ and $\psi_t = \psi\mathbf{M}^t$.*

To illustrate these ideas, consider again the kernel p_Q calculated by Danny Quah, and let $\mathbf{M}_Q$ be the Markov operator. We can evaluate probabilities of different outcomes for a given country over time by iteratively applying $\mathbf{M}_Q$ to an initial condition ψ, generating the sequence $(\psi\mathbf{M}_Q^t)$. Figure 4.11 shows the elements $\psi\mathbf{M}_Q^{10}$, $\psi\mathbf{M}_Q^{60}$, and $\psi\mathbf{M}_Q^{160}$ of this sequence. In the top graph, the country in question is initially in the poorest group, so $\psi = (1, 0, 0, 0, 0)$. The bottom graph shows the corresponding elements when the initial condition is reset to $\psi = (0, 0, 0, 1, 0)$.

4.2.3   Other Identities

Let's think a bit more about the iterates of the Markov operator $\mathbf{M}$. To begin, fix a kernel p with Markov operator $\mathbf{M}$ and define the t-th order kernel p^t by

$$p^1 := p, \quad p^t(x, y) := \sum_{z \in S} p^{t-1}(x, z)p(z, y) \qquad ((x, y) \in S \times S, \, t \in \mathbb{N})$$

Exercise 4.2.7 Show that p^t is in fact a stochastic kernel on S for each $t \in \mathbb{N}$. (Hint: Use induction.)

To interpret p^t, we can use the following lemma:

Lemma 4.2.8 *If $\mathbf{M}$ is the Markov operator defined by stochastic kernel p on S, then its t-th iterate $\mathbf{M}^t$ is the Markov operator defined by p^t, the t-th order kernel of p. In other words, for any $\psi \in \mathscr{P}(S)$ we have*

$$\psi\mathbf{M}^t(y) = \sum_{x \in S} p^t(x, y)\psi(x) \qquad (y \in S)$$

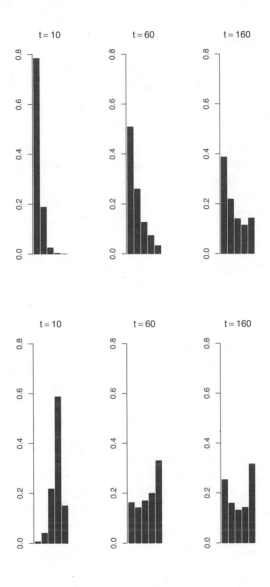

Figure 4.11 Top: $X_0 = 1$. Bottom: $X_0 = 4$

We prove only the case $t = 2$ here, and leave the full proof for the reader. (Hint: Use induction.) Pick any $\psi \in \mathscr{P}(S)$ and any y in S. Then

$$\psi \mathbf{M}^2(y) = ((\psi \mathbf{M})\mathbf{M})(y) = \sum_{z \in S} p(z, y)\psi \mathbf{M}(z)$$

$$= \sum_{z \in S} p(z, y) \sum_{x \in S} p(x, z)\psi(x)$$

$$= \sum_{x \in S} \sum_{z \in S} p(x, z)p(z, y)\psi(x) = \sum_{x \in S} p^2(x, y)\psi(x)$$

Now let $\delta_x \in \mathscr{P}(S)$ be the distribution that puts all mass on $x \in S$ (i.e., $\delta_x(y) = 1$ if $y = x$ and zero otherwise). Applying lemma 4.2.8 with $\psi = \delta_x$, we obtain $\delta_x \mathbf{M}^t(y) = p^t(x, y)$ for all $y \in S$. In other words, the distribution $p^t(x, dy)$ is precisely $\delta_x \mathbf{M}^t$, which we know is the distribution of X_t when $X_0 = x$. More generally, $p^k(x, y)$ is the probability that the state moves from x now to y in k steps:

$$p^k(x, y) = \mathbb{P}\{X_{t+k} = y \mid X_t = x\} \qquad (x, y \in S, \ k \in \mathbb{N})$$

and $p^k(x, dy)$ is the conditional distribution of X_{t+k} given $X_t = x$.

Exercise 4.2.9 Let $t \in \mathbb{N}$. Show that if p is interpreted as the matrix in (4.7), then $p^t(x, y)$ is the (x, y)-th element of its t-th power.

Now let's introduce another operation for the Markov operator $\mathbf{M}$. So far we have $\mathbf{M}$ acting on distributions to the left, as in $\psi \mathbf{M}(y) = \sum_x p(x, y)\psi(x)$. We are also going to let $\mathbf{M}$ act on functions to the right, as in

$$\mathbf{M}h(x) = \sum_{y \in S} p(x, y)h(y) \qquad (x \in S) \tag{4.13}$$

where $h \colon S \to \mathbb{R}$ is any function. Thus $\mathbf{M}$ takes a given function h on S and sends it into a new function $\mathbf{M}h$ on S. In terms of matrix algebra, this is pre-multiplication of the column vector $(h(y))_{y \in S}$ by the matrix (4.7).

To understand (4.13), recall that if Y is a random variable on S with distribution $\phi \in \mathscr{P}(S)$ (i.e., $\mathbb{P}\{Y = y\} = \phi(y)$ for all $y \in S$) and h is a real-valued function on S, then the expectation $\mathbb{E}h(Y)$ of $h(Y)$ is the sum of all values $h(y)$ weighted by the probabilities $\mathbb{P}\{Y = y\}$:

$$\mathbb{E}h(Y) := \sum_{y \in S} h(y)\mathbb{P}\{Y = y\} = \sum_{y \in S} \phi(y)h(y)$$

In terms of vectors we are just computing inner products.

It is now clear that $\mathbf{M}h(x) = \sum_{y \in S} p(x, y)h(y)$ should be interpreted as the expectation of $h(X_{t+1})$ given $X_t = x$. Analogous to the result in lemma 4.2.8, we have

$$\mathbf{M}^t h(x) = \sum_{y \in S} p^t(x, y)h(y) \qquad (x \in S, \ t \in \mathbb{N}) \tag{4.14}$$

Since $p^t(x, dy)$ is the distribution of X_t given $X_0 = x$, it follows that $\mathbf{M}^t h(x)$ is the expectation of $h(X_t)$ given $X_0 = x$.

Exercise 4.2.10 Using induction, confirm the claim in (4.14).

Now the finishing touches. Fix an initial condition $\psi \in \mathscr{P}(S)$, a function h as above and a $k \in \mathbb{N}$. Define

$$\psi \mathbf{M}^k h := \sum_{y \in S} \sum_{x \in S} p^k(x, y) \psi(x) h(y) \tag{4.15}$$

In terms of linear algebra, this expression can be thought of as the inner product of $\psi \mathbf{M}^k$ and h. Since $\psi \mathbf{M}^k$ is the distribution of X_{t+k} when $X_t \sim \psi$, (4.15) gives us the expectation of $h(X_{t+k})$ given $X_t \sim \psi$. In symbols,

$$\psi \mathbf{M}^k h = \mathbb{E}[h(X_{t+k}) \mid X_t \sim \psi] \tag{4.16}$$

Exercise 4.2.11 Confirm the *Chapman–Kolmogorov equation*, which states that for any $k, j \in \mathbb{N}$,
$$p^{k+j}(x, y) = \sum_{z \in S} p^j(z, y) p^k(x, z) \qquad ((x, y) \in S \times S)$$

The listing below provides suggested code for implementing the operator $\psi \mapsto \psi \mathbf{M}$ given a kernel p. (The map $h \mapsto \mathbf{M}h$ can be implemented in similar fashion.) From NumPy we import dot(), which performs matrix multiplication (NumPy must be installed for this to work). The second line imports the class DS from listing 4.2. Next we set up Hamilton's kernel and use it to define the corresponding Markov operator as a function acting on distributions. Using the class DS, we compute 100 elements of the sequence $(\psi \mathbf{M}^t)_{t \geq 0}$.

```
from numpy import dot              # Matrix multiplication
from ds import DS                  # Dynamical system class
pH = ((0.971, 0.029, 0.000),       # Hamilton's kernel
      (0.145, 0.778, 0.077),
      (0.000, 0.508, 0.492))
psi = (0.3, 0.4, 0.3)              # Initial condition
M = lambda phi: dot(phi, pH)       # Define Markov operator
markovds = DS(h=M, x=psi)          # Instance of class DS
T = markovds.trajectory(100)       # Compute trajectory
```

Exercise 4.2.12 Suppose that the business cycle evolves according to Hamilton's kernel p_H on $S = \{NG, MR, SR\}$, and that a firm makes profits $\{1000, 0, -1000\}$ in these three states. Compute expected profits at $t = 5$, given that the economy starts in NG. How much do profits change when the economy starts in SR?

Exercise 4.2.13 Compute expected profits at $t = 1000$ for each of the three possible initial states. What do you notice?

Exercise 4.2.14 Suppose now that the initial state will be drawn according to $\psi = (0.2, 0.2, 0.6)$. Compute expected profits at $t = 5$ using (4.16).

4.2.4   Constructing Joint Distributions

Let's now consider the joint distributions of a Markov-(p, ψ) process $(X_t)_{t \geq 0}$. We would like to understand more about probabilities not just for individual elements of the sequence such as X_t, but rather for a collection of elements. For example, how do we compute the probability that $(X_t, X_{t+1}) = (x, y)$, or that $X_j \leq x$ for $j \leq t$?

Consider first the pair (X_0, X_1), which can be thought of as a single bivariate random variable taking values in $S^2 := S \times S$. Thus the joint distribution is an element of $\mathscr{P}(S^2)$. A typical element of S^2 is a pair (x^0, x^1), where $x^i \in S$.[16] We wish to find the probability $\mathbb{P}\{X_0 = x^0, X_1 = x^1\}$.

To begin, pick any $(x^0, x^1) \in S^2$, and let

$$q_2(x^0, x^1) := \mathbb{P}\{X_0 = x^0, X_1 = x^1\} = \mathbb{P}\{X_0 = x^0\} \cap \{X_1 = x^1\}$$

From (A.2) on page 325, for any events A and B we have $\mathbb{P}(A \cap B) = \mathbb{P}(A)\mathbb{P}(B \mid A)$. It follows that

$$q_2(x^0, x^1) = \mathbb{P}\{X_0 = x^0\}\mathbb{P}\{X_1 = x^1 \mid X_0 = x^0\} = \psi(x^0)p(x^0, x^1)$$

Similarly, the distribution $q_3 \in \mathscr{P}(S^3)$ of (X_0, X_1, X_2) is

$$\begin{aligned}
q_3(x^0, x^1, x^2) &= \mathbb{P}\{X_0 = x^0, X_1 = x^1, X_2 = x^2\} \\
&= \mathbb{P}\{X_0 = x^0, X_1 = x^1\}\mathbb{P}\{X_2 = x^2 \mid X_0 = x^0, X_1 = x^1\} \\
&= \psi(x^0)p(x^0, x^1)p(x^1, x^2)
\end{aligned}$$

Notice that we are using $\mathbb{P}\{X_2 = x^2 \mid X_0 = x^0, X_1 = x^1\} = p(x^1, x^2)$. This is reasonable because, if $X_1 = x^1$, then $X_2 \sim p(x^1, dy)$.

Continuing along the same lines yields the general expression

$$q_{T+1}(x^0, \ldots, x^T) = \psi(x^0) \prod_{t=0}^{T-1} p(x^t, x^{t+1}) \tag{4.17}$$

To evaluate the probability of a given path $x^0, \ldots, x^T$ for some stochastic kernel p and initial condition ψ, we can use a function such as

---

[16] A word on notation: Superscripts represent time, so $x^0 \in S$ is a typical realization of X_0, $x^1 \in S$ is a typical realization of X_1, and so on.

```
def path_prob(p, psi, X):    # X a sequence giving the path
    prob = psi[X[0]]
    for t in range(len(X) - 1):
        prob = prob * p[X[t], X[t+1]]
    return prob
```

Here p is such that p[x, y] corresponds to $p(x, y)$,[17] while psi[x] represents the probability of initial state x according to ψ, and X is a sequence representing the path we wish to evaluate.

Exercise 4.2.15 Show that for Hamilton's kernel p_H and $\psi = (0.2, 0.2, 0.6)$, the probability of path (NG, MR, NG) is 0.000841.

Exercise 4.2.16 Note that $q_{T+1}(x^0, \ldots, x^T) = q_T(x^0, \ldots, x^{T-1}) p(x^{T-1}, x^T)$ also holds for each T and each path. Readers familiar with recursive function calls can try rewriting path_prob() using recursion.

From our expression for q_{T+1} in (4.17) we can also compute the probabilities of more complex events. By an event is meant any subset B of S^{T+1}. For example,

$$B := \{(x^0, \ldots, x^T) \in S^{T+1} : x^t \le x^{t+1} \text{ for } t = 0, \ldots, T-1\}$$

is an event. It consists of all paths $(x^0, \ldots, x^T)$ in S^{T+1} that are increasing (i.e., nondecreasing). To obtain the probability of any such event B we just sum $q_{T+1}(x^0, \ldots, x^T)$ over all distinct paths in B.

One important special case is events of the form

$$D^0 \times \cdots \times D^T = \{(x^0, \ldots, x^T) \in S^{T+1} : x^t \in D^t \text{ for } t = 0, \ldots, T\}$$

where $D^t \subset S$ for each t. Then $\mathbb{P}\{(X_0, \ldots, X_T) \in D^0 \times \cdots \times D^T\} = \mathbb{P} \cap_{t \le T} \{X_t \in D^t\}$, and for this kind of event the following lemma applies:

Lemma 4.2.17 *If $D^0, \ldots, D^T$ is any collection of subsets of S, then*

$$\mathbb{P} \cap_{t \le T} \{X_t \in D^t\} = \sum_{x^0 \in D^0} \psi(x^0) \sum_{x^1 \in D^1} p(x^0, x^1) \cdots \sum_{x^T \in D^T} p(x^{T-1}, x^T)$$

Proof. For any such sets D^t, the probability $\mathbb{P} \cap_{t \le T} \{X_t \in D^t\}$ can be computed by summing over distinct paths:

$$\mathbb{P} \cap_{t \le T} \{X_t \in D^t\} = \sum_{(x^0, \ldots, x^T) \in D^0 \times \cdots \times D^T} q_{T+1}(x^0, \ldots, x^T)$$

$$= \sum_{x^0 \in D^0} \cdots \sum_{x^T \in D^T} q_{T+1}(x^0, \ldots, x^T)$$

The last step now follows from the expression for q_{T+1} in (4.17).   □

---

[17]For example, p might be implemented as a 2D NumPy array. Alternatively, you can use a sequence of sequences (e.g., list of lists), in which case change p[x, y] to p[x][y].

Exercise 4.2.18 Returning to Hamilton's kernel p_H, and using the same initial condition $\psi = (0.2, 0.2, 0.6)$ as in exercise 4.2.15, compute the probability that the economy starts and remains in recession through periods 0, 1, 2 (i.e., that $x^t \neq NG$ for $t = 0, 1, 2$). [Answer: 0.704242]

Another way to compute this probability is via Monte Carlo:

Exercise 4.2.19 Generate 10,000 observations of (X_0, X_1, X_2), starting at the same initial condition $\psi = (0.2, 0.2, 0.6)$. Count the number of paths that do not enter state NG and divide by 10,000 to get the fraction of paths that remain in recession. This fraction converges to the probability of the event, so you should get approximately the same number as you found in exercise 4.2.18.

Now let's think a little bit about computing expectations. Recall the firm in exercise 4.2.12. If the firm operates up until period T, and if the interest rate is equal to r, then the net present value (NPV) of the firm is the expected sum of discounted profits

$$\mathbb{E}\,\Pi(X_0, \ldots, X_T) \quad \text{where} \quad \Pi(X_0, \ldots, X_T) := \sum_{t=0}^{T} \rho^t h(X_t)$$

and $\rho := 1/(1+r)$. Expectations for finite state spaces are found by summing values weighted by probabilities. In this case,

$$\mathbb{E}\,\Pi(X_0, \ldots, X_T) = \sum \Pi(x^0, \ldots, x^T) q_{T+1}(x^0, \ldots, x^T) =: \sum \Pi(\mathbf{x}) q_{T+1}(\mathbf{x})$$

where the sum is over all $\mathbf{x} \in S^{T+1}$.

Exercise 4.2.20 Compute the NPV when $T = 2$ and $r = 0.05$. Take the same initial condition as in exercise 4.2.15. [Answer: -396.5137]

For larger T and S this kind of computation is problematic. For example, if S has ten elements and $T = 100$, then we must sum $\Pi(\mathbf{x}) q_{T+1}(\mathbf{x})$ over 10^{100} paths.

Exercise 4.2.21 If the computer can evaluate one billion (10^9) paths per second, how may years will it take to evaluate all of the paths? Compare this with current estimates of the age of the universe.

In high dimensions, expectations are often best evaluated via Monte Carlo.

Exercise 4.2.22 Redo exercise 4.2.20 using Monte Carlo. Generate 10,000 observations of the path (X_0, X_1, X_2) and evaluate average profits.

Actually the computational problem we have been discussing can be greatly simplified in this particular case by linearity of expectations, which gives

$$\mathbb{E}\Pi = \mathbb{E}\left[\sum_{t=0}^{T} \rho^t h(X_t)\right] = \sum_{t=0}^{T} \rho^t \mathbb{E}h(X_t) = \sum_{t=0}^{T} \rho^t \psi \mathbf{M}^t h$$

The second equality (linearity of $\mathbb{E}$) can be proved from the definition of the joint distribution, but we treat it in much greater generality below. The third equality follows from (4.16) on page 79.

Exercise 4.2.23 Redo exercise 4.2.20 using this last expression, taking parameters T, r, and ψ as given in that exercise. Next plot expected profits against T. After how many periods (for what value of T) will the firm's expected profits be positive?

4.3   Stability of Finite State MCs

In chapter 1 we investigated a Markovian model where the distribution for log income converges to a unique distribution $N(\mu^*, v^*)$, independent of initial conditions. This behavior means that knowledge of the limiting distribution gives us a great deal of predictive power in terms of likelihoods for long-run outcomes. In fact stability also gives us a number of statistical properties that are central to time series econometrics. As a result, we are motivated to study when one does observe stability, beginning with the case of finite state Markov chains.

To start the ball rolling, consider again the sequences of distributions in figure 4.11 (page 77). What happens if we extend the time horizon? In other words, what sort of limiting properties, if any, do these sequences possess? Figure 4.12 repeats the same distribution projections, but this time for dates $t = 160$, $t = 500$, and $t = 1,000$. Looking at the top graph for starters, note that after about $t = 500$ there seems to be very little change in ψ_t. In other words, it appears that the sequence (ψ_t) is converging. Interestingly, the sequence in the bottom graph seems to be converging to the same limit.

Perhaps we are again observing a form of global stability? It turns out that we are, but to show this we must first define stability for Markov chains and derive theorems that allow us to establish this property.

4.3.1   Stationary Distributions

Recall that a dynamical system (U, h) consists of a metric space U and a map $h\colon U \to U$. Recall also the definition of the Markov operator $\mathbf{M}$ corresponding to a given stochastic kernel p: Given $\psi \in \mathscr{P}(S)$, the operator $\mathbf{M}$ is a map sending ψ into $\psi\mathbf{M}$, where $\psi\mathbf{M}(y) = \sum_{x \in S} p(x, y)\psi(x)$ for each $y \in S$. What we are going to do now is view $(\mathscr{P}(S), \mathbf{M})$ as a dynamical system in its own right (recalling that trajectories of the form $(\psi\mathbf{M}^t)_{t \geq 0}$ correspond to the sequence of marginal distributions for a Markov-(p, ψ) process $(X_t)_{t \geq 0}$; see page 76). To do this, we need to introduce a metric on $\mathscr{P}(S)$, and also establish that $\mathbf{M}$ does indeed send $\mathscr{P}(S)$ into itself.

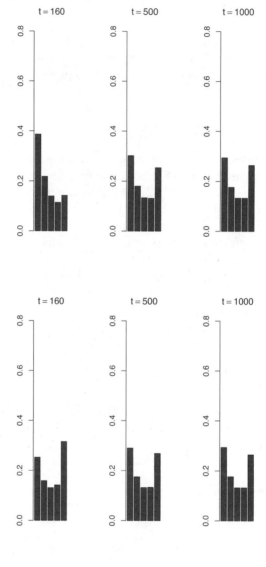

Figure 4.12 Top: $X_0 = 1$. Bottom: $X_0 = 4$

Exercise 4.3.1 Confirm that $\psi \mathbf{M} \in \mathscr{P}(S)$ whenever $\psi \in \mathscr{P}(S)$.

To set $\mathscr{P}(S)$ up as a metric space, we define

$$\|\psi\|_1 := \sum_{x \in S} |\psi(x)| \text{ for each } \psi \in \mathscr{P}(S), \quad \text{and} \quad d_1(\psi, \psi') := \|\psi - \psi'\|_1$$

If one views $\mathscr{P}(S)$ as the unit simplex in $\mathbb{R}^N$ rather than as a space of functions (see the correspondence (4.6) on page 68), then our norm and distance are just the regular $\|\cdot\|_1$ norm (see page 37) and d_1 distance on $\mathbb{R}^N$. Viewed in this way, $\mathscr{P}(S)$ is a closed and bounded subset of $(\mathbb{R}^N, d_1)$, and therefore both compact and complete.[18]

Exercise 4.3.2 Let $\psi_1, \psi_2 \in \mathscr{P}(S)$, and for each $A \subset S$ let $\Psi_i(A) := \sum_{x \in A} \psi_i(x) =$ the probability of $A \subset S$ according to distribution ψ_i. Show that $\|\psi_1 - \psi_2\|_1 = 2 \sup_{A \subset S} |\Psi_1(A) - \Psi_2(A)|$.[19]

To illustrate the dynamical system $(\mathscr{P}(S), \mathbf{M})$ and its trajectories, consider Hamilton's kernel p_H and the corresponding operator $\mathbf{M}_H$. Here $\mathscr{P}(S)$ can be visualized as the unit simplex in $\mathbb{R}^3$. Figure 4.13 shows four trajectories $(\psi \mathbf{M}_H^t)$ generated by iterating $\mathbf{M}_H$ on four different initial conditions ψ. All trajectories converge toward the bottom right-hand corner. Indeed, we will prove below that $(\mathscr{P}(S), \mathbf{M}_H)$ is globally stable.

Exercise 4.3.3 Let $\mathbf{M}$ be the Markov operator determined by an arbitrary stochastic kernel p. Show that $\mathbf{M}$ is d_1-nonexpansive on $\mathscr{P}(S)$, in the sense that for any $\psi, \psi' \in \mathscr{P}(S)$ we have $d_1(\psi \mathbf{M}, \psi' \mathbf{M}) \leq d_1(\psi, \psi')$.

Let us now turn to the existence of fixed points for the system $(\mathscr{P}(S), \mathbf{M})$. For Markov chains, fixed points are referred to as stationary distributions:

Definition 4.3.4 A distribution $\psi^* \in \mathscr{P}(S)$ is called *stationary* or *invariant* for $\mathbf{M}$ if $\psi^* \mathbf{M} = \psi^*$. In other words, ψ^* is a stationary distribution for $\mathbf{M}$ if it is a fixed point of the dynamical system $(\mathscr{P}(S), \mathbf{M})$.

If ψ^* is stationary for $\mathbf{M}$, if $\mathbf{M}$ corresponds to kernel p, if $(X_t)_{t \geq 0}$ is Markov-(p, ψ), and if X_t has distribution ψ^* for some t, then X_{t+1} has distribution $\psi_{t+1} = \psi_t \mathbf{M} =$

---

[18]Interested readers are invited to supply the details of the argument. The connection between the function space $(\mathscr{P}(S), d_1)$ and the unit simplex in $(\mathbb{R}^N, d_1)$ can be made precise using the concept of isomorphisms. Metric spaces (S, ρ) and (S', ρ') are said to be *isometrically isomorphic* if there exists a bijection $\tau \colon S \to S'$ such that $\rho(x, y) = \rho'(\tau(x), \tau(y))$ for all $x, y \in S$. In our case, the bijection in question is (4.6) on page 68. If (S, ρ) and (S', ρ') are isometrically isomorphic, then (S, ρ) is complete if and only if (S', ρ') is complete, and compact if and only if (S', ρ') is compact.

[19]Hint: The set that attains the supremum in this expression is $A := \{x \in S : \psi_1(x) \geq \psi_2(x)\}$. Make use of the fact that $\sum_{x \in A} (\psi_1(x) - \psi_2(x)) = \sum_{x \in A^c} (\psi_2(x) - \psi_1(x))$. (Why?) Now decompose the sum in $\|\psi_1 - \psi_2\|_1$ into a sum over A and A^c.

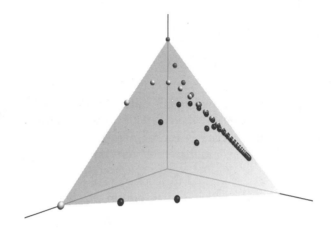

Figure 4.13 Trajectories of $(\mathscr{P}(S), \mathbf{M}_H)$

$\psi^*\mathbf{M} = \psi^*$. In fact iteration shows that X_{t+k} has distribution ψ^* for every $k \in \mathbb{N}$, so probabilities are stationary over time. Moreover if $(X_t)_{t\geq 0}$ is Markov-(p, ψ^*), then $X_t \sim \psi^*$ for all t, and the random variables $(X_t)_{t\geq 0}$ are identically distributed (but not IID—why?).

On the other hand, stationary distributions are just fixed points of a dynamical system $(\mathscr{P}(S), \mathbf{M})$. This is convenient for analysis because we already know various techniques for studying fixed points and stability properties of deterministic dynamical systems. For example, suppose that we view $\mathscr{P}(S)$ as the unit simplex in $\mathbb{R}^N$, and $\psi \mapsto \psi\mathbf{M}$ as postmultiplication of vector $\psi \in \mathbb{R}^N$ by the matrix corresponding to p. This mapping is d_1-nonexpansive (recall exercise 4.3.3), and hence d_1-continuous (exercise 3.2.33, page 52). The unit simplex is a compact, convex subset of $(\mathbb{R}^N, d_1)$. (Proof?) Applying Brouwer's theorem (theorem 3.2.32, page 52) we obtain our first major result for Markov chains:

Theorem 4.3.5 *Every Markov chain on a finite state space has at least one stationary distribution.*

There may, of course, be many stationary distributions, just as other dynamical systems can have many fixed points.

Exercise 4.3.6 For which kernel p is every $\psi \in \mathscr{P}(S)$ stationary?

Let's consider a technique for computing fixed points using matrix inversion. In terms of linear algebra, row vector $\psi \in \mathscr{P}(S)$ is stationary if and only if $\psi(\mathbf{I}_N - p) = 0$,

Listing 4.6 (fphamilton.py) Computing stationary distributions

```python
from numpy import ones, identity, transpose
from numpy.linalg import solve

pH = ((0.971, 0.029, 0.000),        # Hamilton's kernel
      (0.145, 0.778, 0.077),
      (0.000, 0.508, 0.492))

I = identity(3)                     # 3 by 3 identity matrix
Q, b = ones((3, 3)), ones((3, 1))   # Matrix and vector of ones
A = transpose(I - pH + Q)
print(solve(A, b))
```

where $\mathbf{I}_N$ is the $N \times N$ identity matrix, and p is the matrix in (4.7). One idea would be to try to invert $(\mathbf{I}_N - p)$. However, this does not impose the restriction that the solution ψ is an element of $\mathscr{P}(S)$. That restriction can be imposed in the following way.

Exercise 4.3.7 Let $\mathbb{1}_N$ be the $1 \times N$ row vector $(1, \ldots, 1)$. Let $\mathbb{1}_{N \times N}$ be the $N \times N$ matrix of ones. Show that if ψ is stationary, then

$$\mathbb{1}_N = \psi(\mathbf{I}_N - p + \mathbb{1}_{N \times N}) \qquad (4.18)$$

Explain how this imposes the restriction that the elements of ψ sum to 1.

Taking the transpose of (4.18) we get $(\mathbf{I}_N - p + \mathbb{1}_{N \times N})^\top \psi^\top = \mathbb{1}_N^\top$. This is a linear system of the form $Ax = b$, which can be solved for $x = A^{-1}b$. (The solution is not necessarily unique. We return to the issue of uniqueness below.) Listing 4.6 shows how to do this in Python using NumPy.

Exercise 4.3.8 Use this technique to solve for the stationary distribution of Quah's kernel p_Q.[20] Plot it as a bar plot, and compare with the $t = 1000$ distributions in figure 4.12.

Exercise 4.3.9 Recall the firm introduced on page 79. Compute expected profits at the stationary distribution. Compare it with profits at $t = 1000$, as computed in exercise 4.2.13.

---

[20]We prove below that the fixed point is unique.

Exercise 4.3.10 According to the definition of the stationary distribution, if ψ^* is stationary for p_H and if $X_0 \sim \psi^*$, then $X_t \sim \psi^*$ for all t. Using algorithm 4.2 (page 74), check this for p_H by drawing X_0 from the stationary distribution calculated in listing 4.6 and then computing an approximation to the distribution ψ_T of X_T when $T = 20$.

4.3.2   The Dobrushin Coefficient

Now let's consider convergence to the stationary distribution. We continue to impose on $\mathscr{P}(S)$ the distance d_1 and study the dynamical system $(\mathscr{P}(S), \mathbf{M})$. By definition 4.1.5, the system $(\mathscr{P}(S), \mathbf{M})$ is globally stable if

1. it has a unique fixed point (stationary distribution) $\psi^* \in \mathscr{P}(S)$, and

2. $d_1(\psi \mathbf{M}^t, \psi^*) := \|\psi \mathbf{M}^t - \psi^*\|_1 \to 0$ as $t \to \infty$ for all $\psi \in \mathscr{P}(S)$.

The second condition implies that if $(X_t)_{t \geq 0}$ is Markov-(p, ψ) for some $\psi \in \mathscr{P}(S)$, then the distribution of X_t converges to ψ^*.[21]

Exercise 4.3.11 Exercise 4.3.6 asked you to provide an example of a kernel where global stability fails. Another is the "periodic" Markov chain

$$p = \begin{pmatrix} 0 & 1 \\ 1 & 0 \end{pmatrix}$$

Show that $\psi^* := (1/2, 1/2)$ is the unique stationary distribution. Give a counterexample to the claim $\|\psi \mathbf{M}^t - \psi^*\|_1 \to 0$ as $t \to \infty$, $\forall \psi \in \mathscr{P}(S)$.

How might one check for stability of a given kernel p and associated dynamical system $(\mathscr{P}(S), \mathbf{M})$? Exercise 4.3.3 suggests the way forward: $\mathbf{M}$ is nonexpansive on $\mathscr{P}(S)$, and if we can upgrade this to a uniform contraction then Banach's fixed point theorem (page 53) implies that $(\mathscr{P}(S), \mathbf{M})$ is globally stable, and that convergence to equilibrium takes place at a geometric rate.

Which kernels will we be able to upgrade? Intuitively, stable kernels are those where current states have little influence on future states. An extreme example is where the distributions $p(x, dy)$ are all equal: $p(x, dy) = q \in \mathscr{P}(S)$ for all $x \in S$. In this case the current state has no influence on tomorrow's state—indeed, the resulting process is IID with $X_t \sim q$ for all t. The Markov operator satisfies $\psi \mathbf{M} = q$ for all $\psi \in \mathscr{P}(S)$ (check it), and $(\mathscr{P}(S), \mathbf{M})$ is globally stable.

A less extreme case is when the distributions $p(x, dy)$ are "similar" across $x \in S$. One similarity measure for two distributions $p(x, dy)$ and $p(x', dy)$ is $\sum_y p(x, y) \wedge$

---

[21]In this context, global stability is sometimes referred to as *ergodicity*.

$p(x',y)$, where $a \wedge b := \min\{a,b\}$. If $p(x,dy) = p(x',dy)$ then the value is one. If the supports[22] of $p(x,dy)$ and $p(x',dy)$ are disjoint, then the value is zero. This leads us to the Dobrushin coefficient, which measures the stability properties of a given kernel p.

Definition 4.3.12 Given stochastic kernel p, the *Dobrushin coefficient* $\alpha(p)$ is defined by

$$\alpha(p) := \min \left\{ \sum_{y \in S} p(x,y) \wedge p(x',y) : (x,x') \in S \times S \right\} \tag{4.19}$$

Exercise 4.3.13 Prove that $0 \leq \alpha(p) \leq 1$ always holds.

Exercise 4.3.14 Show that $\alpha(p) = 1$ if and only if $p(x,dy)$ is equal to a constant distribution $q \in \mathscr{P}(S)$ for every $x \in S$.

Exercise 4.3.15 Show that $\alpha(p) = 0$ for the periodic kernel in exercise 4.3.11, and for p corresponding to the identity matrix.

Exercise 4.3.16 Distributions ϕ and ψ are said to *overlap* if there exists a y such that $\phi(y) > 0$ and $\psi(y) > 0$. Show that $\alpha(p) > 0$ if and only if for each pair $(x,x') \in S \times S$ the distributions $p(x,dy)$ and $p(x',dy)$ overlap.

The following result links the Dobrushin coefficient to stability via Banach's fixed point theorem (page 53).

Theorem 4.3.17 *If p is a stochastic kernel on S with Markov operator $\mathbf{M}$, then*

$$\|\phi\mathbf{M} - \psi\mathbf{M}\|_1 \leq (1 - \alpha(p))\|\phi - \psi\|_1 \qquad \forall \, \phi, \psi \in \mathscr{P}(S)$$

Moreover this bound is the best available, in the sense that if $\lambda < 1 - \alpha(p)$, then there exists a pair ϕ, ψ in $\mathscr{P}(S)$ such that $\|\phi\mathbf{M} - \psi\mathbf{M}\|_1 > \lambda\|\phi - \psi\|_1$.

The first half of the theorem says that if $\alpha(p) > 0$, then $\mathbf{M}$ is uniformly contracting (for the definition see page 52) with modulus $1 - \alpha(p)$. Since $(\mathscr{P}(S), d_1)$ is complete, Banach's fixed point theorem then implies global stability of $(\mathscr{P}(S), \mathbf{M})$. The second part of the theorem says that this rate $1 - \alpha(p)$ *is the best available*, which in turn suggests that the Dobrushin coefficient is a good measure of the stability properties of $\mathbf{M}$. For example, if $\alpha(p) = 0$, then we can be certain $\mathbf{M}$ is not a uniform contraction.

Some intuition for theorem 4.3.17 and it's stability implications was discussed above. The coefficient is large (close to one) when all distributions $p(x,dy)$ are similar across x, and the current state has little influence on future states. This is the stable case. The coefficient is zero when there exists states x and x' such that $p(x,dy)$ and

---

[22]The support of $\phi \in \mathscr{P}(S)$ is $\{y \in S : \phi(y) > 0\}$.

$p(x', dy)$ have disjoint support, as with the identity kernel and the periodic kernel. More intuition on the link between positivity of $\alpha(p)$ and stability is given in the next section.

The proof of theorem 4.3.17 is given in the appendix to this chapter. The fact that $1 - \alpha(p)$ is the best rate possible may suggest to you that the proof is not entirely trivial. Indeed this is the case. We have to do better than crude inequalities. All but the most enthusiastic readers are encouraged to skip the proof and move to the next section.

4.3.3   Stability

Let p be a stochastic kernel on S. If $\alpha(p) > 0$, then $(\mathscr{P}(S), \mathbf{M})$ is globally stable by Banach's fixed point theorem. In fact we can say a bit more. We now present our main stability result for finite chains, which clarifies the relationship between the Dobrushin coefficient and stability.

Theorem 4.3.18 *Let p be a stochastic kernel on S with Markov operator $\mathbf{M}$. The following statements are equivalent:*

1. *The dynamical system $(\mathscr{P}(S), \mathbf{M})$ is globally stable.*

2. *There exists a $t \in \mathbb{N}$ such that $\alpha(p^t) > 0$.*

Another way to phrase the theorem is that $(\mathscr{P}(S), \mathbf{M})$ is globally stable if and only if there is a $t \in \mathbb{N}$ such that, given any pair of states x, x', one can find at least one state y such that $p^t(x, y)$ and $p^t(x', y)$ are both positive. Thus, if we run two Markov chains from any two starting points x and x', there is a positive probability that the chains will meet. This is connected with global stability because it rules out the kind of behavior seen in example 4.1.8 (page 57), where initial conditions determine long-run outcomes.

Exercise 4.3.19 Consider the periodic kernel in exercise 4.3.11. Show that $\alpha(p^t) = 0$ for every $t \in \mathbb{N}$.

Exercise 4.3.20 Prove that if $\min_{x \in S} p^t(x, \bar{y}) =: \epsilon > 0$ for some $\bar{y} \in S$, then $\alpha(p^t) \geq \epsilon$, and hence $(\mathscr{P}(S), \mathbf{M})$ is globally stable.

Exercise 4.3.21 Stokey and Lucas (1989, thm. 11.4) prove that $(\mathscr{P}(S), \mathbf{M})$ is globally stable if there exists a $t \in \mathbb{N}$ such that $\sum_{y \in S} \min_{x \in S} p^t(x, y) > 0$. Show how this result is implied by theorem 4.3.18.

Exercise 4.3.22 Prove theorem 4.3.18. To show that (2) implies (1), use Lemma 4.2.8 (page 76) and Theorem 4.3.17. To show that (1) implies (2), let ψ^* be the stationary distribution. Note that $\exists \bar{y} \in S$ with $\psi^*(\bar{y}) > 0$. Argue that $p^t(x, \bar{y}) \to \psi^*(\bar{y})$ for any x. Using finiteness of S, show that there is a $t \in \mathbb{N}$ with $\min_{x \in S} p^t(x, \bar{y}) > 0$. Conclude that $\alpha(p^t) > 0$.

Let's consider how to apply theorem 4.3.18. In view of exercise 4.3.20, if there exists a y with $p(x, y) > 0$ for all $x \in S$, then $\alpha(p) > 0$ and global stability holds. A case in point is Hamilton's kernel (4.8) on page 70, which is globally stable as a result of the strict positivity of column two.

Next consider Quah's kernel p_Q (page 71). We know from theorem 4.3.5 that at least one stationary distribution exists, and we calculated a stationary distribution in exercise 4.3.8. We should now check that there are not many stationary distributions— otherwise exhibiting one of them is not very interesting. Also, the stationary distribution becomes a better predictor of outcomes if we know that all trajectories converge to it.

Exercise 4.3.23 Show that the Dobrushin coefficient $\alpha(p_Q)$ is zero.

Since $\alpha(p_Q) = 0$, let's look at the higher order iterates. In his study Quah calculates the 23rd-order kernel

$$p_Q^{23} = \begin{pmatrix} 0.61 & 0.27 & 0.09 & 0.03 & 0.00 \\ 0.37 & 0.32 & 0.20 & 0.09 & 0.02 \\ 0.14 & 0.23 & 0.31 & 0.25 & 0.07 \\ 0.04 & 0.11 & 0.25 & 0.39 & 0.22 \\ 0.00 & 0.01 & 0.04 & 0.12 & 0.82 \end{pmatrix} \qquad (4.20)$$

Exercise 4.3.24 Show that $\alpha(p_Q^{23}) > 0$.

Exercise 4.3.25 As $(\mathscr{P}(S), \mathbf{M}_Q)$ is globally stable, we can iterate $\mathbf{M}_Q$ on any initial condition ψ to calculate an approximate fixed point ψ^*. Take $\psi = (1, 0, 0, 0, 0)$ as your initial condition and iterate until $d_1(\psi \mathbf{M}_Q^t, \psi \mathbf{M}_Q^{t+1}) < 0.0001$. Compare your result with that of exercise 4.3.8.

Exercise 4.3.26 Code a function that takes a kernel p as an argument and returns $\alpha(p)$. Write another function that repeatedly calls the first function to compute the smallest $t \geq 1$ such that $\alpha(p^t) > 0$, and prints that t along with the value $\alpha(p^t)$. Include a maximum value T such that if t reaches T the function terminates with a message that $\alpha(p^t) = 0$ for all $t \leq T$. Now show that the first t such that $\alpha(p_Q^t) > 0$ is 2.

One interesting fact regarding stationary distributions is as follows: Let p be a kernel such that $(\mathscr{P}(S), \mathbf{M})$ is globally stable, and let ψ^* be the unique stationary distribution. Let $(X_t)_{t \geq 0}$ be Markov-(p, x), where $\psi^*(x) > 0$. The *return time to x* is defined as the random variable

$$\tau(x) := \inf\{t \geq 1 : X_t = x\}$$

It turns out that for $\tau(x)$ so defined we have $\mathbb{E}\tau(x) = 1/\psi^*(x)$. We will skip the proof (see Norris, 1997, thm. 1.7.7), but let's try running a simulation. The pseudocode in algorithm 4.3 indicates how one might go about estimating $\mathbb{E}\tau(x)$.[23]

Algorithm 4.3: Computing the mean return time

---

for i *in 1 to n* **do**   // n is the number of replications
    set $t = 0$
    set $X = x$
    repeat
        draw $X \sim p(X, dy)$
        set $t = t + 1$
    until $X = x$
    set $\tau_i = t$
end
return $n^{-1} \sum_{i=1}^{n} \tau_i$

Exercise 4.3.27 Implement algorithm 4.3 for Hamilton's Markov chain. Examine whether for fixed $x \in S$ the output converges to $1/\psi^*(x)$ as $n \to \infty$.

Finally, let's consider a slightly more elaborate application, which concerns so-called (s, S) inventory dynamics. Inventory management is a major topic in operations research that also plays a role in macroeconomics due to the impact of inventories on aggregate demand. The discrete choice flavor of (s, S) models accord well with the data on capital investment dynamics.

Let $q, Q \in \{0\} \cup \mathbb{N}$ with $q \leq Q$, and consider a firm that, at the start of time t, has inventory $X_t \in \{0, \ldots, Q\}$. Here Q is the maximum level of inventory that the firm is capable of storing. (We are studying (q, Q) inventory dynamics because the symbol S is taken.) If $X_t \leq q$, then the firm orders inventory $Q - X_t$, bringing the current stock to Q. If $X_t > q$ then the firm orders nothing. At the end of the period t demand D_{t+1} is observed, and the firm meets this demand up to its current stock level. Any remaining inventory is carried over to the next period. Thus

$$X_{t+1} = \begin{cases} \max\{Q - D_{t+1}, 0\} & \text{if } X_t \leq q \\ \max\{X_t - D_{t+1}, 0\} & \text{if } X_t > q \end{cases}$$

If we adopt the notation $x^+ := \max\{x, 0\}$ and let $\mathbb{1}\{x \leq q\}$ be one when $x \leq q$ and zero otherwise, then this can be rewritten more simply as

$$X_{t+1} = (X_t + (Q - X_t)\mathbb{1}\{X_t \leq q\} - D_{t+1})^+$$

---

[23]If $\psi^*(x) > 0$, then $(X_t)_{t \geq 0}$ returns to x (infinitely often) with probability one, so the algorithm terminates in finite time with probability one.

or, if $h_q(x) := x + (Q - x)\mathbb{1}\{x \le q\}$ is the stock on hand after orders for inventory are completed, as

$$X_{t+1} = (h_q(X_t) - D_{t+1})^+$$

We assume throughout that $(D_t)_{t \ge 1}$ is an IID sequence taking values in $\{0\} \cup \mathbb{N}$ according to distribution $b(d) := \mathbb{P}\{D_t = d\} = (1/2)^{d+1}$.

Exercise 4.3.28 Let $S = \{0, 1, \ldots, Q\}$. Given an expression for the stochastic kernel $p(x, y)$ corresponding to the restocking policy q.[24]

Exercise 4.3.29 Let $\mathbf{M}_q$ be the corresponding Markov operator. Show that $(\mathscr{P}(S), \mathbf{M}_q)$ is always globally stable *independent* of the precise values of q and Q. Let ψ_q^* denote the stationary distribution corresponding to threshold q. Show numerically that if $Q = 5$, then

$$\psi_2^* = (0.0625, 0.0625, 0.125, 0.25, 0.25, 0.25)$$

Now consider profits of the firm. To minimize the number of parameters, suppose that the firm buys units of the product for zero dollars and marks them up by one dollar. Revenue in period t is $\min\{h_q(X_t), D_{t+1}\}$. Placing an order for inventory incurs fixed cost C. As a result profits for the firm at time t are given by

$$\pi_q(X_t, D_{t+1}) = \min\{h_q(X_t), D_{t+1}\} - C\mathbb{1}\{X_t \le q\}$$

If we now sum across outcomes for D_{t+1} taking $X_t = x$ as given, then we get

$$g_q(x) := \mathbb{E}[\pi_q(x, D_{t+1})] = \sum_{d=0}^{\infty} \pi_q(x, d)b(d) = \sum_{d=0}^{\infty} \frac{\pi_q(x, d)}{2^{d+1}}$$

which is interpreted as expected profits in the current period when the inventory state X_t is equal to x.

Exercise 4.3.30 One common performance measure for an inventory strategy (in this case, a choice of q) is long-run average profits, which is defined here as $\mathbb{E}g_q(X)$ when $X \sim \psi_q^*$ (i.e., $\sum_{x \in S} g_q(x)\psi_q^*(x)$). Show numerically that according to this performance measure, when $Q = 20$ and $C = 0.1$, the optimal policy is $q = 7$.

4.3.4 The Law of Large Numbers

Let's end our discussion of stability by investigating some probabilistic properties of sample paths. In particular, we investigate the law of large numbers (LLN) in the context of Markov chains.

---

[24]Hint: If a is any integer and D has the distribution b defined above, then what is the probability that $(a - D)^+ = y$? You might find it convenient to first consider the case $y = 0$ and then the case $y > 0$.

In algorithm 4.2 (page 74) we computed an approximation to the marginal distribution ψ_t via Monte Carlo. The basis of Monte Carlo is that if we sample independently from a fixed probability distribution and count the fraction of times that an event happens, that fraction converges to the probability of the event (as determined by this probability distribution). This is more or less the frequentist definition of probabilities, but it can also be proved from the axioms of probability theory. The theorem in question is the law of large numbers (LLN), a variation of which is as follows:

Theorem 4.3.31 *If F is a cumulative distribution function on $\mathbb{R}$, $(X_t)_{t\geq 1} \overset{\text{IID}}{\sim} F$, and h: $\mathbb{R} \to$ $\mathbb{R}$ is a measurable function with $\int |h(x)|F(dx) < \infty$, then*

$$\frac{1}{n}\sum_{i=1}^{n} h(X_i) \to \mathbb{E}h(X_1) =: \int h(x)F(dx) \quad \text{as } n \to \infty \text{ with probability one} \quad (4.21)$$

This result is fundamental to statistics. It states that for IID sequences, sample means converge to means as the sample size gets large. Later we will give a formal definition of independence and prove a version of the theorem. At that time the term "measurable function" and the nature of probability one convergence will be discussed. Suffice to know that measurability of h is never a binding restriction for the problems we consider.

Example 4.3.32 If $(X_i)_{i=1}^{n}$ are independent standard normal random variates, then according to theorem 4.3.31 we should find that $n^{-1}\sum_{i=1}^{n} X_i^2 \to 1$. (Why?) You can try this out in Python using some variation of

```
>>> from random import normalvariate
>>> Y = [normalvariate(0, 1)**2 for i in range(10000)]
>>> sum(Y) / 10000
```

Another use of the LLN: Suppose that we wish to compute $\mathbb{E}h(X)$, where h is some real function. One approach would be to use pen and paper plus our knowledge of calculus to solve the integral $\int_{-\infty}^{\infty} h(x)F(dx)$. In some situations, however, this is not so easy. If instead we have access to a random number generator that can generate independent draws $X_1, X_2, \ldots$ from F, then we can produce a large number of draws, take the mean of the $h(X_i)$ terms, and appeal to (4.21).

In (4.21) the sequence of random variables is IID. In some situations the LLN extends to sequences that are neither independent nor identically distributed. For example, we have the following result concerning stable Markov chains:

Theorem 4.3.33 *Let S be finite, let $\psi \in \mathscr{P}(S)$, let p be a stochastic kernel on S with $\alpha(p^t) >$ 0 for some $t \in \mathbb{N}$, and let h: $S \to \mathbb{R}$. If $(X_t)_{t\geq 0}$ is Markov-(p, ψ), then*

$$\frac{1}{n}\sum_{t=1}^{n} h(X_t) \to \sum_{x \in S} h(x)\psi^*(x) \quad \text{as } n \to \infty \text{ with probability one} \quad (4.22)$$

where ψ^* is the unique stationary distribution of p.

The left-hand side is the average value of $h(X_t)$, and the right-hand side is the expectation of $h(X)$ when $X \sim \psi^*$. Note that the result holds for *every* initial condition $\psi \in \mathscr{P}(S)$.

The proof of theorem 4.3.33 requires more tools than we have in hand.[25] The intuition is that when the chain is globally stable, X_t is approximately distributed according to ψ^* for large t. In addition the stability property implies that initial conditions are unimportant, and for the same reason X_t has little influence on X_{t+k} for large k. Hence there is a kind of asymptotic independence in the chain. Together, these two facts mean that our chain approximates the IID property that drives the LLN.

If $h(x) = 1$ if $x = y$ and zero otherwise (i.e., $h(x) = \mathbb{1}\{x = y\}$), then (4.22) becomes

$$\frac{1}{n} \sum_{t=1}^{n} h(X_t) = \frac{1}{n} \sum_{t=1}^{n} \mathbb{1}\{X_t = y\} \to \psi^*(y) \quad \text{as } n \to \infty \tag{4.23}$$

This provides a new technique for computing the stationary distribution, via Monte Carlo. Exercise 4.3.34 illustrates.

Exercise 4.3.34 Let p_H be Hamilton's kernel, and let $h(x) = 1$ if $x = NG$ and zero otherwise. Take any initial condition, and draw a series of length 50,000. Compute the left-hand side of (4.22). Compare it with the right-hand side, which was calculated in listing 4.6 on page 87.

When the state space is small, this is a less efficient technique for computing the stationary distribution than the algebraic method used in listing 4.6. However, the computational burden of the algebraic method increases rapidly with the size of the state space. For large or infinite state spaces, a variation of the LLN technique used in exercise 4.3.34 moves to center stage. See §6.1.3 for details.[26]

The importance of theorem 4.3.33 extends beyond this new technique for computing stationary distributions. It provides a new *interpretation* for the stationary distribution: If we turn (4.23) around, we get

$$\psi^*(y) \cong \text{fraction of time that } (X_t) \text{ spends in state } y$$

This is indeed a new interpretation of ψ^*, although it is not generally valid unless the chain in question is stable (in which case the LLN applies).

Exercise 4.3.35 Give an example of a kernel p and initial condition ψ where this interpretation fails.

---

[25] A weak version of theorem 4.3.33 is proved in §11.1.1.

[26] The look-ahead method introduced in §6.1.3 concerns infinite state spaces, but it can be applied to finite state spaces with the obvious modifications.

In the preceding discussion, h was an indicator function, which reduced the discussion of expectations to one of probabilities. Now let's consider more general expectations.

Exercise 4.3.36 Recall the firm introduced on page 79. Extending exercise 4.3.34, approximate expected profits at the stationary distribution using theorem 4.3.33. Compare your results to those of exercise 4.3.9.

Thus the LLN provides a new way to compute expectations with respect to stationary distributions. However, as was the case with probabilities above, it also provides a new interpretation of these expectations when the Markov chain is stationary. For example, if h denotes profits as above, then we have

$$\sum_{x \in S} h(x) \psi^*(x) \cong \text{long-run average profits}$$

Again, this interpretation is valid when the chain in question is stationary, but may not be valid otherwise.

4.4 Commentary

Regarding deterministic, discrete-time dynamical systems, a good mathematical introduction is provided by Holmgren (1996), who treats elementary theory, topological conjugacy, and chaotic dynamics. For dynamics from an economic perspective, see, for example, Stokey and Lucas (1989), Azariadis (1993), de la Fuente (2000), Turnovsky (2000), Shone (2003), Ljungqvist and Sargent (2004), Caputo (2005), or Gandolfo (2005).

The threshold externality model in example 4.1.8 is a simplified version of Azariadis and Drazen (1990). See Durlauf (1993) for a stochastic model with multiple equilibria.

Our discussion of chaotic dynamics lacked any economic applications, but plenty exist. The Solow–Swan model produces chaotic dynamics with some minor modifications (e.g., Böhm and Kaas 2000). Moreover rational behavior in infinite-horizon, optimizing models can lead to chaos, cycles, and complex dynamics. See, for example, Benhabib and Nishimura (1985), Boldrin and Montrucchio (1986), Sorger (1992), Nishimura et al. (1994), Venditti (1998), or Mitra and Sorger (1999). For more discussion of complex economic dynamics, see Grandmont (1985), Chiarella (1988), Galor (1994), Medio (1995), Brock and Hommes (1998), Barnett and Serletis (2000), or Kikuchi (2008).

Two good references on finite state Markov chains are Norris (1997) and Häggström (2002). These texts provide a more traditional approach to stability of Markov chains based on irreducibility and aperiodicity. It can be shown that every irreducible and aperiodic Markov chain is globally stable, and as a result satisfies the conditions of theorem 4.3.18 (in particular, $\alpha(p^t) > 0$ for some $t \in \mathbb{N}$). The converse is not true, so theorem 4.3.18 is more general.

The Dobrushin coefficient was introduced by Dobrushin (1956). For an alternative discussion of the Dobrushin coefficient in the context of finite state Markov chains, see Bremaud (1999).

The treatment of (s, S) dynamics in §4.3.3 is loosely based on Norris (1997). For another discussion of inventory dynamics see Stokey and Lucas (1989, sec. 5.14). An interesting analysis of aggregate implications is Nirei (2008). A modern treatment of discrete adjustment models can be found in Stokey (2008).

Chapter 5

Further Topics for Finite MCs

We have now covered the fundamental theory of finite state Markov chains. Next let us turn to more applied topics. In §5.1 below we consider the problem of dynamic programming, that is, of controlling Markov chains through our actions in order to achieve a given objective. In §5.2 we investigate the connection between Markov chains and stochastic recursive sequences.

5.1 Optimization

In this section we take our first look at stochastic dynamic programming. The term "dynamic programming" was coined by Richard Bellman in the early 1950s, and pertains to a class of multistage planning problems. Because stochastic dynamic programming problems typically involve Markov chains, they are also called Markov decision problems, Markov control problems, or Markov control processes. We will focus on solving a simple example problem, and defer the more difficult proofs until chapter 10.

5.1.1 Outline of the Problem

Our objective is to model the behavior of Colonel Kurtz, who can be found on a small island in the Nung river living on catfish. The local catfish bite only at dawn, and the Colonel bags a random quantity $W \in \{0, \ldots, B\}$, $W \sim \phi$. Catfish spoil quickly in the tropics, lasting only if refrigerated. The Colonel has limited refrigeration facilities: A maximum of M fish can fit into his freezer. We let X_t be the stock of fish at noon on day t, from which a quantity C_t is consumed. The remainder $R_t = X_t - C_t$ is frozen. Following the next morning's fishing trip, the state is $X_{t+1} = R_t + W_{t+1}$.

Being economists, we do not hesitate for a second to model Colonel Kurtz as a rational intertemporal maximizer. We estimate his discount factor to be ρ, and his period utility function to be U. We assume that Kurtz follows a fixed *policy function* σ: On observing state X_t, the Colonel saves quantity $R_t = \sigma(X_t)$. For the state space we take $S := \{0, \dots, M + B\}$, as $X_t \le M + B$ always holds. (Why?) The function σ is a map from S into $\{0, \dots, M\}$ that satisfies the feasibility constraint $0 \le \sigma(x) \le x$ for all $x \in S$. We denote the set of all feasible policies by Σ. For given $\sigma \in \Sigma$ the state evolves according to

$$X_{t+1} = \sigma(X_t) + W_{t+1}, \quad (W_t)_{t \ge 1} \overset{\text{IID}}{\sim} \phi, \quad X_0 = x \in S \tag{5.1}$$

Exercise 5.1.1 For each $a \in \{0, \dots, M\}$, let $\gamma(a, dy) \in \mathscr{P}(S)$ be the distribution of $a + W$. That is, $\mathbb{P}\{a + W = y\} = \gamma(a, y)$ for $y \in S$. Check that $S \times S \ni (x, y) \mapsto \gamma(\sigma(x), y) \in \mathbb{R}$ is a stochastic kernel on S.

Using this definition of γ, we can now see that the distribution of X_{t+1} given X_t is $\gamma(\sigma(X_t), dy)$, and that $(X_t)_{t \ge 0}$ in (5.1) is Markov-(p_σ, x), where $p_\sigma(x, y) := \gamma(\sigma(x), y)$. Let $\mathbf{M}_\sigma$ be the Markov operator corresponding to p_σ. For any given $h: S \to \mathbb{R}$ we have (see (4.14) on page 78)

$$\mathbf{M}_\sigma^t h(x) = \sum_{y \in S} p_\sigma^t(x, y) h(y) \qquad (t \ge 0) \tag{5.2}$$

What's important to remember from all this is that when Colonel Kurtz chooses a policy he chooses a Markov chain on S as well.

We model the optimization problem of Colonel Kurtz as

$$\max_{\sigma \in \Sigma} \mathbb{E}\left[\sum_{t=0}^{\infty} \rho^t U(X_t - \sigma(X_t))\right] \quad \text{for } (X_t)_{t \ge 0} \text{ given by (5.1)} \tag{5.3}$$

Let's try to understand this objective function as clearly as possible. For each fixed σ, the Markov chain is determined by (5.1). As discussed further in chapter 9, random variables are best thought of as *functions* on some underlying space Ω that contains the possible outcomes of a random draw. At the start of time, nature selects an element $\omega \in \Omega$ according to a given "probability" $\mathbb{P}$. The shocks $(W_t)_{t \ge 1}$ are functions of this outcome, so the draw determines the path for the shocks as $(W_t(\omega))_{t \ge 1}$.[1] From the rule $X_{t+1}(\omega) = \sigma(X_t(\omega)) + W_{t+1}(\omega)$ and $X_0(\omega) = x$ we obtain the time path $(X_t(\omega))_{t \ge 0}$ for the state. In turn each path gives us a real number $Y_\sigma(\omega)$ that corresponds to the value of the path:

$$Y_\sigma(\omega) = \sum_{t=0}^{\infty} \rho^t U(X_t(\omega) - \sigma(X_t(\omega))) \qquad (\omega \in \Omega) \tag{5.4}$$

---

[1]Note that although all random outcomes are determined at the start of time by the realization of ω, the time t value $W_t(\omega)$ is regarded as "unobservable" prior to t.

The value Y_σ is itself a random variable, being a function of ω. The objective function is the expectation of Y_σ.

For probabilities on a *finite* set Ω, the expectation of a random variable Y is given by the sum $\sum_{\omega \in \Omega} Y(\omega) \mathbb{P}\{\omega\}$. However, in the present case it turns out that Ω must be uncountable (see definition A.1.4 on page 322), and this sum is not defined. We will have to wait until we have discussed measure theory before a general definition of expectation on uncountable spaces can be constructed. We can, however, approximate (5.3) by truncating at some (large but finite) $T \in \mathbb{N}$. This takes us back to a finite scenario treated above: For given $\sigma \in \Sigma$, the chain $(X_t)_{t=0}^T$ is Markov-(p_σ, x), and we can construct its joint probabilities via (4.17) on page 80. This joint distribution is defined over the finite set S^{T+1}, and hence expectations with respect to it can be computed with sums. In particular, if

$$F \colon S^{T+1} \ni \mathbf{x} := (x^0, \ldots, x^T) \mapsto \sum_{t=0}^T \rho^t U(x^t - \sigma(x^t)) \in \mathbb{R}$$

and q_{T+1} is the joint distribution of $(X_t)_{t=0}^T$, then

$$\mathbb{E}\left[\sum_{t=0}^T \rho^t U(X_t - \sigma(X_t))\right] = \sum_{\mathbf{x} \in S^{T+1}} F(\mathbf{x})\, q_{T+1}(\mathbf{x}) \qquad (5.5)$$

In fact we can simplify further using linearity of expectations:

$$\mathbb{E}\left[\sum_{t=0}^T \rho^t U(X_t - \sigma(X_t))\right] = \sum_{t=0}^T \rho^t \mathbb{E} U(X_t - \sigma(X_t))$$

Letting $r_\sigma(x) := U(x - \sigma(x))$ and using (5.2), we can write

$$\mathbb{E} U(X_t - \sigma(X_t)) = \mathbb{E} r_\sigma(X_t) = \sum_{y \in S} p_\sigma^t(x, y) r_\sigma(y) = \mathbf{M}_\sigma^t r_\sigma(x)$$

$$\therefore \quad \mathbb{E}\left[\sum_{t=0}^T \rho^t U(X_t - \sigma(X_t))\right] = \sum_{t=0}^T \rho^t \mathbf{M}_\sigma^t r_\sigma(x) \qquad (5.6)$$

With some measure theory it can be shown (see chapter 10) that the limit of the right-hand side of (5.6) is equal to the objective function in the infinite horizon problem (5.3). In particular, if $v_\sigma(x)$ denotes total reward under policy σ when starting at initial condition $x \in S$, then

$$v_\sigma(x) := \mathbb{E}\left[\sum_{t=0}^\infty \rho^t U(X_t - \sigma(X_t))\right] = \sum_{t=0}^\infty \rho^t \mathbf{M}_\sigma^t r_\sigma(x) \qquad (5.7)$$

5.1.2 Value Iteration

The term $v_\sigma(x)$ in (5.7) gives the expected discounted reward from following the policy σ. Taking x as given, our job is to find a maximizer of $v_\sigma(x)$ over the set of policies Σ. The first technique we discuss is value iteration. To begin, define the *value function*

$$v^*(x) := \sup\{v_\sigma(x) : \sigma \in \Sigma\} \qquad (x \in S) \tag{5.8}$$

The value function satisfies a restriction known as the *Bellman equation*. Letting

$$\Gamma(x) := \{0, 1, \ldots, x \wedge M\} \qquad x \wedge M := \min\{x, M\}$$

be the set of all feasible actions (number of fish that can be frozen) when the current state is x, the Bellman equation can be written as

$$v^*(x) = \max_{a \in \Gamma(x)} \left\{ U(x - a) + \rho \sum_{z=0}^{B} v^*(a + z)\phi(z) \right\} \qquad (x \in S) \tag{5.9}$$

The idea behind (5.9)—which we later prove in some detail—is that if one knows the values of different states in terms of maximum future rewards, then the best action is found by trading off the two effects inherent in choosing an action: current reward and future reward after transitioning to a new state next period (the transition probabilities being determined by the action). The result of making this trade-off optimally is the maximum value from the current state, which is the left-hand side of (5.9).

Given $w \colon S \to \mathbb{R}$, we say that $\sigma \in \Sigma$ is *w-greedy* if

$$\sigma(x) \in \underset{a \in \Gamma(x)}{\operatorname{argmax}} \left\{ U(x - a) + \rho \sum_{z=0}^{B} w(a + z)\phi(z) \right\} \qquad (x \in S) \tag{5.10}$$

Further, a policy $\sigma \in \Sigma$ is called *optimal* if $v_\sigma = v^*$, which is to say that the value obtained from following σ is the maximum possible. A key result of chapter 10 is that a policy σ^* is optimal if and only if it is v^*-greedy. Hence computing an optimal policy is trivial if we know the value function v^*, since we need only solve (5.10) for each $x \in S$, using v^* in place of w.

So how does one solve for the value function? Equation (5.9) is a tail-chasing equation: if we know v^*, then we can substitute it into the right-hand side and find v^*. When it comes to such equations involving functions, Banach's fixed point theorem (page 53) can often be used to unravel them. Let bS be the set of functions $w \colon S \to \mathbb{R}$,[2] and define the *Bellman operator* $bS \ni v \mapsto Tv \in bS$ by

$$Tv(x) = \max_{a \in \Gamma(x)} \left\{ U(x - a) + \rho \sum_{z=0}^{B} v(a + z)\phi(z) \right\} \qquad (x \in S) \tag{5.11}$$

---

[2]This is our usual notation for the *bounded* functions from S to $\mathbb{R}$. Since S is finite, all real-valued functions on S are bounded, and bS is just the real-valued functions on S.

As will be proved in chapter 10, T is a uniform contraction of modulus ρ on (bS, d_∞), where $d_\infty(v, w) := \sup_{x \in S} |v(x) - w(x)|$. By construction, $Tv^*(x) = v^*(x)$ for all $x \in S$ (check it), so v^* is a fixed point of T. From Banach's fixed point theorem, v^* is the only fixed point of T in bS, and $d_\infty(T^n v, v^*) \to 0$ as $n \to \infty$ for any given $v \in bS$.[3]

Algorithm 5.1: Value iteration algorithm

---

pick any $v \in bS$
repeat
    compute Tv from v
    set $e = d_\infty(Tv, v)$
    set $v = Tv$
until *e is less that some tolerance*
solve for a v-greedy policy σ

This suggests the *value iteration* algorithm presented in algorithm 5.1.[4] If the tolerance is small, then the algorithm produces a function $T^n v$ that is close to v^*. Since v^*-greedy policies are optimal, and since $T^n v$ is almost equal to v^*, it seems likely that $T^n v$-greedy policies are almost optimal. This intuition is correct, and will be confirmed in §10.2.1.

The Bellman operator T is implemented in listing 5.1. In the listing, the utility function is $U(c) = c^\beta$, and the distribution ϕ is uniform on $\{0, \ldots, B\}$. The operator T is implemented as a function `T()`. It takes as input a sequence (list, etc.) `v`, which corresponds to a function v on S, and returns a list `Tv` representing the image of v under T. The outer loop steps through each $x \in S$, computing the maximum on the right-hand side of (5.11) and assigning it to `Tv[x]`. The inner loop steps through each feasible action $a \in \Gamma(x)$ to find the maximum at x.[5]

Exercise 5.1.2 Complete the implementation of algorithm 5.1 in listing 5.1, computing an (approximately) optimal policy. For the initial condition you might like to use v defined by $v(x) = U(x)$.[6]

Exercise 5.1.3 Using the code you wrote in exercise 4.3.26 (page 91), show that $(X_t)_{t \geq 0}$ is stable under the optimal policy (in particular, show numerically that $\alpha(p_{\sigma^*}) > 0$). Compute the stationary distribution.

---

[3]Recall that (bS, d_∞) is complete (see theorem 3.2.6 on page 45).

[4]Which is reminiscent of the iterative technique for computing stationary distributions explored in exercise 4.3.25 on page 91.

[5]As usual, the code is written for clarity, not speed. For large state spaces you will need to rewrite this code using libraries for fast array processing such as SciPy and NumPy, or rewrite the inner loop in C or Fortran, and then call it from Python.

[6]You will have an opportunity to check that the policy you compute is correct in the next section.

Listing 5.1 (kurtzbellman.py) Bellman operator

```python
beta, rho, B, M = 0.5, 0.9, 10, 5
S = range(B + M + 1)    # State space = 0,...,B + M
Z = range(B + 1)        # Shock space = 0,...,B

def U(c):
    "Utility function."
    return c**beta

def phi(z):
    "Probability mass function, uniform distribution."
    return 1.0 / len(Z) if 0 <= z <= B else 0

def Gamma(x):
    "The correspondence of feasible actions."
    return range(min(x, M) + 1)

def T(v):
    """An implementation of the Bellman operator.
    Parameters: v is a sequence representing a function on S.
    Returns: Tv, a list."""
    Tv = []
    for x in S:
        # Compute the value of the objective function for each
        # a in Gamma(x), and store the result in vals
        vals = []
        for a in Gamma(x):
            y = U(x - a) + rho * sum(v[a + z]*phi(z) for z in Z)
            vals.append(y)
        # Store the maximum reward for this x in the list Tv
        Tv.append(max(vals))
    return Tv
```

5.1.3   Policy Iteration

Another common technique for solving dynamic programming problems is *policy iteration*, as presented in algorithm 5.2.[7] This technique is easy to program, and is often faster than value iteration. It has the nice feature that for finite state problems the optimal policy is computed exactly (modulo numerical error) in finite time (theorem 10.2.6, page 243).

Algorithm 5.2: Policy iteration algorithm

> pick any $\sigma \in \Sigma$
> **repeat**
> > compute v_σ from σ
> > compute a v_σ-greedy policy σ'
> > set $e = \sigma - \sigma'$
> > set $\sigma = \sigma'$
> **until** $e = 0$

First, an arbitrary policy σ is chosen. Next, one computes the value v_σ of this policy. From v_σ a v_σ-greedy policy σ' is computed:

$$\sigma'(x) \in \underset{0 \leq a \leq x \wedge M}{\operatorname{argmax}} \left\{ U(x - a) + \rho \sum_{z=0}^{B} v_\sigma(a + z)\phi(z) \right\} \qquad (x \in S)$$

and e records the deviation between σ and σ'. If the policies are equal the loop terminates. Otherwise we set $\sigma = \sigma'$ and iteration continues.

The most difficult part of coding this algorithm is to compute the value of a given policy (i.e., compute v_σ from σ). Listing 5.2 gives one method for accomplishing this, based on evaluating the right-hand side of (5.6) with large T. It starts with two import statements. First we import zeros, dot and array from numpy. The first is for creating arrays of zeros, the second is for matrix multiplication, and the third is for converting other data types such as lists into NumPy arrays. Second, we import some primitives from kurtzbellman, which is listing 5.1.

Next we define the function value_of_policy(), which takes an array sigma representing a feasible policy $\sigma \in \Sigma$ as its argument, and returns a NumPy array v_sigma, which represents v_σ. At the start of the function we create a 2D array p_sigma corresponding to the stochastic kernel $p_\sigma(x, y) := \gamma(\sigma(x), y)$ defined above. From that array we create a function M_sigma() corresponding to $\mathbf{M}_\sigma$. In addition we create the array r_sigma(), which corresponds to r_σ. Finally, we step through 50 terms of $\rho^t \mathbf{M}_\sigma^t r_\sigma$, adding each one to the return value.

---

[7]Sometimes called Howard's policy improvement algorithm.

Listing 5.2 (`kurtzvsigma.py`) Approximation of v_σ

```python
from numpy import zeros, dot, array
from kurtzbellman import S, rho, phi, U   # From listing 5.1

def value_of_policy(sigma):
    "Computes the value of following policy sigma."

    # Set up the stochastic kernel p_sigma as a 2D array:
    N = len(S)
    p_sigma = zeros((N, N))
    for x in S:
        for y in S:
            p_sigma[x, y] = phi(y - sigma[x])

    # Create the right Markov operator M_sigma:
    M_sigma = lambda h: dot(p_sigma, h)

    # Set up the function r_sigma as an array:
    r_sigma = array([U(x - sigma[x]) for x in S])
    # Reshape r_sigma into a column vector:
    r_sigma.shape = (N, 1)

    # Initialize v_sigma to zero:
    v_sigma = zeros((N,1))
    # Initialize the discount factor to 1:
    discount = 1

    for i in range(50):
        v_sigma = v_sigma + discount * r_sigma
        r_sigma = M_sigma(r_sigma)
        discount = discount * rho

    return v_sigma
```

Exercise 5.1.4 Complete the implementation of algorithm 5.2. The resulting policy should be the same as the one you computed in exercise 5.1.2.

5.2 MCs and SRSs

In this section we investigate the connection between Markov chains generated by stochastic kernels on one hand, and stochastic recursive sequences (stochastic difference equations) on the other. We will see that to each Markov chain there corresponds at least one stochastic recursive sequence (in fact there are many). This provides us with new ways of analyzing and simulating Markov chains.

5.2.1 From MCs to SRSs

When we start studying processes on infinite state spaces we will often be interested in *stochastic recursive sequences* (SRSs). In the finite case, a typical SRS has the form

$$X_{t+1} = F(X_t, W_{t+1}), \quad X_0 \sim \psi \in \mathscr{P}(S), \quad F: S \times Z \to S \qquad (5.12)$$

where $(W_t)_{t\geq 1}$ is a sequence of independent shocks taking values in arbitrary set Z. As discussed in §5.1.1, the shocks W_t are to be thought of as functions on a common space Ω. At the start of time, nature selects an $\omega \in \Omega$ according to the probability $\mathbb{P}$. This provides a complete realization of the path $(W_t(\omega))_{t\geq 1}$. The value ω also determines X_0, with $\mathbb{P}\{\omega : X_0(\omega) = x_i\} = \psi(x_i)$. Given $(W_t(\omega))_{t\geq 1}$ and $X_0(\omega)$, we construct the corresponding time path $(X_t(\omega))_{t\geq 0}$ by

$$X_1(\omega) = F(X_0(\omega), W_1(\omega)), \quad X_2(\omega) = F(X_1(\omega), W_2(\omega)), \quad \text{etc.}$$

The idea that all uncertainty is realized at the start of time by a single observation ω from Ω is a convenient mathematical fiction. It's as if we could draw an entire path $(W_t(\omega))_{t\geq 1}$ by a single call to our random number generator. However, you can mentally equate the two approaches by thinking of ω as being realized at the beginning of our simulation. Generating n random numbers can be thought of as observing the first n observations of the sequence. The default behavior of most random number generators is to produce (quasi) independent sequences of shocks.

Given the SRS (5.12) we obtain a stochastic kernel p on S by

$$p(x,y) = \mathbb{P}\{F(x, W_t) = y\} := \mathbb{P}\{\omega \in \Omega : F(x, W_t(\omega)) = y\}$$

In fact, we can also go the other way, representing *any* Markov-(p, ψ) process by an SRS such as (5.12). This is useful for simulation, and for gaining a deeper understanding of the probabilistic structure of dynamics.

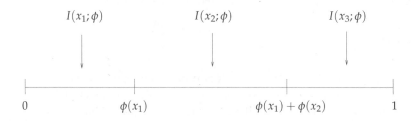

Figure 5.1 Partition $(I(x;\phi))_{x\in S}$ created by ϕ

To begin, let W be uniformly distributed on $(0,1]$. Thus, for any $a \le b \in (0,1]$, we have $\mathbb{P}\{a < W \le b\} = b - a$, which is the length of the interval $(a,b]$.[8] Given a distribution $\phi \in \mathscr{P}(S)$, let us try to construct a function $z \mapsto \tau(z;\phi)$ from $(0,1]$ to S such that $\tau(W;\phi)$ has distribution ϕ:

$$\mathbb{P}\{\tau(W;\phi) = x\} = \phi(x) \qquad (x \in S)$$

One technique is as follows: Divide the unit interval $(0,1]$ into N disjoint subintervals, one for each $x \in S$. A typical subinterval is denoted $I(x;\phi)$ and is chosen to have length $\phi(x)$. As a concrete example we could take

$$I(x_i;\phi) := (\phi(x_1) + \cdots + \phi(x_{i-1}), \ \phi(x_1) + \cdots + \phi(x_i)]$$

with $I(x_1;\phi) = (0, \phi(x_1)]$. Figure 5.1 gives the picture for $S = \{x_1, x_2, x_3\}$.

Now consider the function $z \mapsto \tau(z;\phi)$ defined by

$$\tau(z;\phi) := \sum_{x\in S} x\mathbb{1}\{z \in I(x;\phi)\} \qquad (z \in (0,1]) \tag{5.13}$$

where $\mathbb{1}\{z \in I(x;\phi)\}$ is one when $z \in I(x;\phi)$ and zero otherwise.

Exercise 5.2.1 Prove: $\forall\, x \in S$, $\tau(z;\phi) = x$ if and only if $z \in I(x;\phi)$.

The random variable $\tau(W;\phi)$ has distribution ϕ. To see this, pick any $x \in S$, and observe that the $\tau(W;\phi) = x$ precisely when $W \in I(x;\phi)$. The probability of this event is the length of the interval $I(x;\phi)$, which, by construction, is $\phi(x)$. Hence $\mathbb{P}\{\tau(W;\phi) = x\} = \phi(x)$ for all $x \in S$ as claimed. An implementation of the function $z \mapsto \tau(z;\phi)$ is given in algorithm 5.3.[9]

---

[8] The probability is the same whether inequalities are weak or strict.

[9] This algorithm was used previously in listing 4.4 (page 73).

Algorithm 5.3: The function $z \mapsto \tau(z; \phi)$

```
read in z
set a = 0
for x in S do
    if a < z ≤ a + φ(x) then return x
    a = a + φ(x)
end
```

With these results we can represent any Markov-(p, ψ) chain as an SRS. By definition, such a chain obeys $X_0 \sim \psi$ and $X_{t+1} \sim p(X_t, dy)$ for $t \geq 0$. To write this as a stochastic recursive sequence, let $(W_t)_{t \geq 0}$ be a IID and uniform on $(0, 1]$, and let

$$X_0 = \tau(W_0; \psi), \quad X_{t+1} = \tau(W_{t+1}; p(X_t, dy)) \tag{5.14}$$

The second equality can be rewritten as

$$X_{t+1} = F(X_t, W_{t+1}) \quad \text{where} \quad F(x, z) := \tau(z; p(x, dy)) \tag{5.15}$$

You should convince yourself that if W is uniform on $(0, 1]$, then $F(x, W)$ has distribution $p(x, dy)$, and that the sequence $(X_t)_{t \geq 0}$ generated by (5.14) and (5.15) obeys $X_0 \sim \psi$ and $X_{t+1} \sim p(X_t, dy)$ for $t \geq 0$.

Listing 5.3 provides some code for creating F from a given kernel p. The kernel p is represented as a sequence of sequences (see, e.g., the definition of the kernel on page 79). A call such as F = createF(p) assigns to F the function $F(x, z) = \tau(z; p(x, dy))$.

Exercise 5.2.2 Use listing 5.3 to reimplement Hamilton's Markov chain as a stochastic recursive sequence. Verify that the law of large numbers still holds by showing that $\frac{1}{n} \sum_{t=1}^{n} \mathbb{1}\{X_t = y\} \to \psi^*(y)$ as $n \to \infty$. (The right-hand side of this equality can be calculated by listing 4.6. Now calculate the left using listing 5.3 and compare.)

Incidentally, SRSs are sometimes referred to as iterated function systems. In this framework one thinks of updating the state from X_t to X_{t+1} by the random function $F_{W_{t+1}} := F(\cdot, W_{t+1})$. Although we are now dealing with "random functions," which sounds rather fancy, in practice the only change is a notational one: $X_{t+1} = F_{W_{t+1}}(X_t)$ as compared to (5.12). The main advantage is that we can now write

$$X_t = F_{W_t} \circ F_{W_{t-1}} \circ \cdots \circ F_{W_1}(X_0) = F_{W_t} \circ F_{W_{t-1}} \circ \cdots \circ F_{W_1}(\tau(W_0; \psi))$$

We see that X_t is just a fixed function of the shocks up to time t.

Listing 5.3 (p2srs.py) Creating F for kernel p

```
def createF(p):
    """Takes a kernel p on S = {0,...,N-1} and returns a
    function F(x,z) which represents it as an SRS.
    Parameters: p is a sequence of sequences, so that p[x][y]
    represents p(x,y) for x,y in S.
    Returns: A function F with arguments (x,z)."""
    S = range(len(p[0]))
    def F(x,z):
        a = 0
        for y in S:
            if a < z <= a + p[x][y]:
                return y
            a = a + p[x][y]
    return F
```

5.2.2   Application: Equilibrium Selection

In this section we consider an application where a finite state Markov chain naturally arises as an SRS. The context is equilibrium selection in games. We look at how so-called stochastically stable equilibria are identified in games with multiple Pareto ranked Nash equilibria.

The application we consider is a coordination game with N players. The players cooperate on a project that involves the use of computers. The agents choose as their individual operating system (OS) either an OS called U or a second OS called W. For this project, OS U is inherently superior. At the same time, cooperation is enhanced by the use of common systems, so W may be preferable if enough people use it.

Specifically, we assume that the individual one-period rewards for using U and W are given respectively by

$$\Pi_u(x) := \frac{x}{N}u \quad \text{and} \quad \Pi_w(x) := \frac{N-x}{N}w \qquad (0 < w < u)$$

where x is the number of players using U. Players update their choice of operating system according to current rewards. As a result of their actions the law of motion for the number of players using U is $x_{t+1} = B(x_t)$, where the function B is defined by

$$B(x) := \begin{cases} N & \text{if } \Pi_u(x) > \Pi_w(x) \\ x & \text{if } \Pi_u(x) = \Pi_w(x) \\ 0 & \text{if } \Pi_u(x) < \Pi_w(x) \end{cases} \qquad (\iff x = N(1+u/w)^{-1})$$

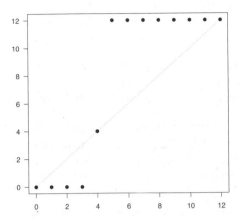

Figure 5.2 Best response dynamics

The 45 degree diagram for B is shown in figure 5.2 when $N = 12$, $u = 2$ and $w = 1$. There are three fixed points: $x = 0$, $x = x_b := N(1 + u/w)^{-1} = 4$ and $x = N$. The point x_b is the value of x such that rewards are exactly equal. That is, $\Pi_u(x_b) = \Pi_w(x_b)$.

Under these deterministic dynamics, the long-run outcome for the game is determined by the initial condition x_0, which corresponds to the number of players originally using U. Notice that a larger fraction of initial conditions lead to coordination on U, which follows from our assumption that U is inherently superior (i.e., $u > w$).

So far the dynamics are characterized by multiple equilibria and path dependence (where long-run outcomes are determined by initial conditions). Some authors have sought stronger predictions for these kinds of coordination models (in the form of unique and stable equilibria) by adding learning, or "mutation."

Suppose, for example, that after determining the choice of OS via the best response function B, players switch to the alternative OS with independent probability $\epsilon > 0$. Thus each of the $B(x)$ users of U switches to W with probability ϵ, and each of the $N - B(x)$ users of W switches to U with the same probability. Using X_t to denote the (random) number of U users at time t, the dynamics are now

$$X_{t+1} = B(X_t) + V_t^u - V_t^w \tag{5.16}$$

where V_t^u and V_t^w are independent and binomially distributed with probability ϵ and sizes $N - B(X_t)$ and $B(X_t)$ respectively.[10] Here V_t^u is the number of switches from W to U, while V_t^w is switches from U to W.

With the addition of random "mutation," uniqueness and stability of the steady state is attained:

Exercise 5.2.3 Let $p(x, y) := \mathbb{P}\{X_{t+1} = y \mid X_t = x\}$ be the stochastic kernel corresponding to the SRS (5.16), let $\mathbf{M}$ be the Markov operator, and let $S := \{0, \dots, N\}$. Argue that for any fixed $\epsilon \in (0, 1)$, the system $(\mathscr{P}(S), \mathbf{M})$ is globally stable.

Let ψ_ϵ^* be the unique stationary distribution for $\epsilon \in (0, 1)$. It has been shown (see Kandori, Mailath, and Rob 1993) that as $\epsilon \to 0$, the distribution ψ_ϵ^* concentrates on the Pareto dominant equilibrium N (i.e., $\psi_\epsilon^*(N) \to 1$ as $\epsilon \to 0$). The interpretation is that for low levels of experimentation or mutation, players rarely diverge from the most attractive equilibrium.

This concentration on N can be observed by simulation: Let $(X_t^\epsilon)_{t=0}^n$ be a time series generated for some fixed $\epsilon \in (0, 1)$. Then the law of large numbers (theorem 4.3.33, page 94) implies that for large n,

$$n^{-1} \sum_{t=1}^n \{X_t^\epsilon = N\} \cong \psi_\epsilon^*(N)$$

Figure 5.3 shows a simulation that gives $n^{-1} \sum_{t=1}^n \{X_t^\epsilon = N\}$ as ϵ ranges over the interval $[0.001, 0.1]$. The parameters are $N = 12$, $u = 2$ and $w = 1$. The series length n used in the simulation is $n = 10{,}000$. The figure shows that steady state probabilities concentrate on N as $\epsilon \to 0$.

Exercise 5.2.4 Replicate figure 5.3. Generate one time series of length $n = 10{,}000$ for each ϵ, and plot the fraction of time each series spends in state N.

5.2.3   The Coupling Method

Much of the modern theory of Markov chains is based on probabilistic methods (as opposed to analytical techniques such as fixed point theory). A prime example is *coupling*. Found in many guises, coupling is a powerful and elegant technique for studying all manner of probabilistic phenomena. It has been used to prove stability of Markov processes since the masterful work of Wolfgang Doeblin (1938).

We will use coupling to prove global stability of Markov chains for which the Dobrushin coefficient is strictly positive, without recourse to the contraction mapping argument employed in theorem 4.3.18 (page 90). Our main aim is to provide the basic

---

[10] A binomial random variable with probability p and size n counts the number of successes in n binary trials, each with independent success probability p.

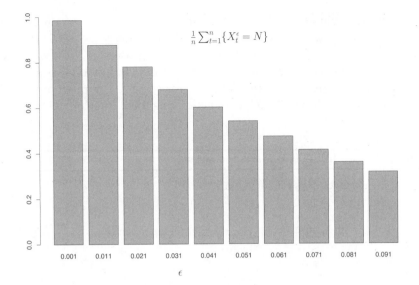

$$\frac{1}{n}\sum_{t=1}^{n}\{X_t^{\epsilon}=N\}$$

Figure 5.3 Fraction of time spent at N as $\epsilon \to 0$

feel of the coupling method. When we turn to stability of Markov chains on infinite state spaces, the intuition you have developed here will be valuable. Note, however, that the topic is technical, and those who feel they have learned enough about stability for now can move on without much loss of continuity.

To begin, consider a stochastic kernel p with $\alpha(p^t) > 0$. To simplify the argument, we are going to assume that $t = 1$. (The general case is a bit more complicated but works along the same lines.) A little thought will convince you that $\alpha(p) > 0$ is equivalent to strict positivity of

$$\epsilon := \min\left\{\sum_{y \in S} p(x,y) \cdot p(x',y) : (x,x') \in S \times S\right\} \tag{5.17}$$

The condition $\epsilon > 0$ can be understood as follows: If we run two independent chains $(X_t)_{t \geq 0}$ and $(X_t')_{t \geq 0}$, both updated with kernel p, then the kernel for the joint process $((X_t, X_t'))_{t \geq 0}$ on $S \times S$ is $p(x,y)p(x',y')$. If $X_t = x$ and $X_t' = x'$, then the probability both chains hit the same state next period (i.e., the probability that $X_{t+1} = X_{t+1}'$) is $\sum_{y \in S} p(x,y)p(x',y)$. Hence $\epsilon > 0$ means that, regardless of the current state, there is a positive ($\geq \epsilon$) probability the chains will meet next period. This in turn is associated with stability, as it suggests that initial conditions are relatively unimportant.

To make this argument more concrete, fix $\psi \in \mathscr{P}(S)$ and consider two independent Markov chains $(X_t)_{t \geq 0}$ and $(X_t^*)_{t \geq 0}$, where $(X_t)_{t \geq 0}$ is Markov-(p, ψ) and $(X_t^*)_{t \geq 0}$ is Markov-(p, ψ^*) for some stationary distribution $\psi^* \in \mathscr{P}(S)$.[11] It follows that $X_t \sim \psi M^t$ and $X_t^* \sim \psi^*$. Now consider a third process $(X_t')_{t \geq 0}$, which follows $(X_t)_{t \geq 0}$ until $\nu := \min\{t \geq 0 : X_t = X_t^*\}$, and then switches to following $(X_t^*)_{t \geq 0}$. In other words, $X_t' = X_t$ for $t \leq \nu$ and $X_t' = X_t^*$ for $t \geq \nu$. (The random variable ν is known as the coupling time.) A recipe for generating these three processes is given in algorithm 5.4.

Algorithm 5.4: Coupling two Markov chains

---

generate independent draws $X_0 \sim \psi$ and $X_0^* \sim \psi^*$
set $X_0' = X_0$
for $t \geq 0$ **do**
    draw $X_{t+1} \sim p(X_t, dy)$ and $X_{t+1}^* \sim p(X_t^*, dy)$ independently
    if $X_t' = X_t^*$ **then**
       | set $X_{t+1}' = X_{t+1}^*$
    else
       | set $X_{t+1}' = X_{t+1}$
    end
end

We claim that *the distributions of X_t and X_t' are equal for all t*, from which it follows that $X_t' \sim \psi M^t$. To verify the latter it is sufficient to show that $(X_t')_{t \geq 0}$ is Markov-(p, ψ). And indeed $(X_t')_{t \geq 0}$ *is* Markov-(p, ψ) because at time zero we have $X_0' = X_0 \sim \psi$, and subsequently $X_{t+1}' \sim p(X_t', dy)$.

That X_{t+1}' is drawn from $p(X_t', dy)$ at each $t \geq 0$ can be checked by carefully working through algorithm 5.4. Another way to verify that $X_{t+1}' \sim p(X_t', dy)$ is to cast both $(X_t)_{t \geq 0}$ and $(X_t^*)_{t \geq 0}$ as stochastic recursive sequences of the form

$$X_{t+1} = F(X_t, W_{t+1}), \quad X_0 \sim \psi, \qquad X_{t+1}^* = F(X_t^*, W_{t+1}^*), \quad X_0^* \sim \psi^*$$

where the random variables $(W_t)_{t \geq 0}$ and $(W_t^*)_{t \geq 0}$ are all independent and uniform on $(0, 1]$, and F is determined in (5.15). Now we create $(X_t')_{t \geq 0}$ by setting $X_0' = X_0$, $X_{t+1}' = F(X_t', W_{t+1})$ for $t < \nu$ and $X_{t+1}' = F(X_t', W_{t+1}^*)$ for $t \geq \nu$. By switching the source of shocks at the coupling time, X_t' changes course and starts to follow $(X_t^*)_{t \geq 0}$. Nevertheless, $(X_t')_{t \geq 0}$ is always updated by $F(\cdot, W)$ for some uniformly distributed independent W, which means that $X_{t+1}' \sim p(X_t', dy)$ at every step, and hence $X_t' \sim \psi M^t$ as claimed.

The next step of the proof uses the following *coupling inequality*.

---

[11] At least one must exist by theorem 4.3.5, page 86.

Lemma 5.2.5 *If X and Y are any random variables taking values in S and having distributions ϕ_X and ϕ_Y respectively, then*

$$\|\phi_X - \phi_Y\|_\infty := \max_{x \in S} |\phi_X(x) - \phi_Y(x)| \leq \mathbb{P}\{X \neq Y\}$$

Intuitively, if the probability that X and Y differ is small, then so is the distance between their distributions. An almost identical proof is given later in the book so we omit the proof here.[12]

Let's apply lemma 5.2.5 to X_t' and X_t^*. Since $X_t' \sim \psi M^t$ and $X_t^* \sim \psi^*$,

$$\|\psi M^t - \psi^*\|_\infty \leq \mathbb{P}\{X_t' \neq X_t^*\}$$

We wish to show that the right-hand side of this inequality goes to zero, and this is where the reason for introducing X_t' becomes clear. Not only does it have the distribution ψM^t, just as X_t does, but also we know that *if X_t' is distinct from X_t^*, then X_j and X_j^* are distinct for all $j \leq t$.* Hence $\mathbb{P}\{X_t' \neq X_t^*\} \leq \mathbb{P} \cap_{j \leq t} \{X_j \neq X_j^*\}$.[13] Therefore

$$\|\psi M^t - \psi^*\|_\infty \leq \mathbb{P} \cap_{j \leq t} \{X_j \neq X_j^*\} \tag{5.18}$$

Thus, to show that ψM^t converges to ψ^*, it is sufficient to demonstrate that the probability of X_j and X_j^* never meeting prior to t goes to zero as $t \to \infty$. And this is where positivity of ϵ in (5.17) comes in. It means that there is an ϵ chance of meeting at each time j, independent of the locations of X_{j-1} and X_{j-1}^*. Hence the probability of never meeting converges to zero. Specifically,

Proposition 5.2.6 *We have $\mathbb{P} \cap_{j \leq t} \{X_j \neq X_j^*\} \leq (1-\epsilon)^t$ for all $t \in \mathbb{N}$.*

It follows from proposition 5.2.6 and (5.18) that if ϵ is strictly positive, then $\|\psi M^t - \psi^*\|_\infty \to 0$ at a geometric rate.

Proof of proposition 5.2.6. The process $(X_t, X_t^*)_{t \geq 0}$ is a Markov chain on $S \times S$. A typical element of $S \times S$ will be denoted by (x, s). In view of independence of $(X_t)_{t \geq 0}$ and $(X_t^*)_{t \geq 0}$, the initial condition of $(X_t, X_t^*)_{t \geq 0}$ is $\psi \times \psi^*$ (i.e., $\mathbb{P}\{(X_0, X_0^*) = (x, s)\} = (\psi \times \psi^*)(x, s) := \psi(x)\psi^*(s)$), while the stochastic kernel is

$$k((x, s), (x', s')) = p(x, x')p(s, s')$$

To simplify notation let's write (x, s) as $\mathbf{x}$ so that $k((x, s), (x', s'))$ can be expressed more simply as $k(\mathbf{x}, \mathbf{x}')$, and set $D := \{(x, s) \in S \times S : x = s\}$. Evidently

$$\mathbb{P} \cap_{j \leq t} \{X_j \neq X_j^*\} = \mathbb{P} \cap_{j \leq t} \{(X_j, X_j^*) \in D^c\}$$

---

[12]See lemma 11.3.2 on page 273.

[13]If A and B are two events with $A \subset B$ (i.e., occurrence of A implies occurrence of B), then $\mathbb{P}(A) \leq \mathbb{P}(B)$. See chapter 9 for details.

In view of lemma 4.2.17 this probability is equal to

$$\sum_{x^0 \in D^c} (\psi \times \psi^*)(x^0) \sum_{x^1 \in D^c} k(x^0, x^1) \cdots \sum_{x^{t-1} \in D^c} k(x^{t-2}, x^{t-1}) \sum_{x^t \in D^c} k(x^{t-1}, x^t) \qquad (5.19)$$

Now consider the last term in this expression. We have

$$\sum_{x^t \in D^c} k(x^{t-1}, x^t) = 1 - \sum_{x^t \in D} k(x^{t-1}, x^t)$$

But from the definitions of k and D we obtain

$$\sum_{x^t \in D} k(x^{t-1}, x^t) = \sum_{(x^t, s^t) \in D} p(x^{t-1}, x^t) p(s^{t-1}, s^t) = \sum_{y \in S} p(x^{t-1}, y) p(s^{t-1}, y)$$

$$\therefore \quad \sum_{x^t \in D} k(x^{t-1}, x^t) \geq \epsilon$$

$$\therefore \quad \sum_{x^t \in D^c} k(x^{t-1}, x^t) \leq 1 - \epsilon$$

Working back through (5.19) and applying the same logic to each term shows that (5.19) is less than $(1 - \epsilon)^t$. This proves the proposition. $\qquad \square$

5.3 Commentary

Much of the early theory of dynamic programming is due to Bellman (1957). A good introduction to dynamic programming in discrete state environments can be found in Puterman (1994). An overview with applications is given in Miranda and Fackler (2002, ch. 7). Further references can be found in the commentary to chapters 6 and 10.

The representation of Markov chains as stochastic recursive sequences in §5.2.1 is loosely based on Häggström (2002, ch. 3). The application in §5.2.2 is from Kandori, Mailath, and Rob (1993). The approach to coupling in §5.2.3 is somewhat nonstandard. More information can be found in the commentary to chapter 11.

Chapter 6

Infinite State Space

In this chapter we begin working with stochastic systems on infinite state space. While a completely rigorous treatment of this area requires measure theory (chapter 7 and onward), we can build a good understanding of the key topics (dynamics, optimization, etc.) by heuristic arguments, simulation, and analogies with the finite case. Along the way we will meet some more challenging programming problems.

6.1 First Steps

In this section we study dynamics for stochastic recursive sequences (SRSs) taking values in $\mathbb{R}$. Our main interest is in tracking the evolution of probabilities over time, as represented by the marginal distributions of the process. We will also look at stationary distributions—the infinite state analogue of the stationary distributions discussed in §4.3.1—and how to calculate them.

6.1.1 Basic Models and Simulation

Our basic model is as follows: Let the state space S be a subset of $\mathbb{R}$ and let $Z \subset \mathbb{R}$. Let $F \colon S \times Z \to S$ be a given function, and consider the SRS

$$X_{t+1} = F(X_t, W_{t+1}), \quad X_0 \sim \psi, \quad (W_t)_{t \geq 1} \overset{\text{IID}}{\sim} \phi \tag{6.1}$$

Here $(W_t)_{t \geq 1} \overset{\text{IID}}{\sim} \phi$ means that $(W_t)_{t \geq 1}$ is an IID sequence of shocks with *cumulative distribution function* ϕ. In other words, $\mathbb{P}\{W_t \leq z\} = \phi(z)$ for all $z \in Z$. Likewise ψ is the cumulative distribution function of X_0, and X_0 is independent of $(W_t)_{t \geq 1}$. Note that X_t and W_{t+1} are independent, since X_t depends only on the initial condition and the shocks $W_1, \ldots, W_t$, all of which are independent of W_{t+1}.

Example 6.1.1 Consider a stochastic version of the Solow–Swan growth model, where output is a function f of capital k and a real-valued shock W. The sequence of productivity shocks $(W_t)_{t \geq 1}$ is $\overset{\text{IID}}{\sim} \phi$. Capital at time $t + 1$ is equal to that fraction s of output that was saved last period, plus undepreciated capital, giving law of motion

$$k_{t+1} = F(k_t, W_{t+1}) := s f(k_t, W_{t+1}) + (1 - \delta)k_t \tag{6.2}$$

Consumption is given by $c_t = (1 - s)f(k_t, W_{t+1})$. The production function satisfies $f \colon \mathbb{R}_+^2 \to \mathbb{R}_+$ and $f(k, z) > 0$ whenever $k > 0$ and $z > 0$. For the state space we can choose either $S_0 = \mathbb{R}_+$ or $S = (0, \infty)$, while $Z := (0, \infty)$.

Exercise 6.1.2 Show that if $k \in S_0$ (resp., S) and $z \in Z$, then next period's stock $F(k, z)$ is in S_0 (resp., S).

Example 6.1.3 Let $Z = S = \mathbb{R}$, and consider the smooth transition threshold autoregression (STAR) model

$$X_{t+1} = g(X_t) + W_{t+1}, \quad (W_t)_{t \geq 1} \overset{\text{IID}}{\sim} \phi \tag{6.3}$$

$$g(x) := (\alpha_0 + \alpha_1 x)(1 - G(x)) + (\beta_0 + \beta_1 x)G(x)$$

Here $G \colon S \to [0, 1]$ is a smooth transition function, such as the logistic function, satisfying $G' > 0$, $\lim_{x \to -\infty} G(x) = 0$ and $\lim_{x \to \infty} G(x) = 1$.

Code for simulating time series from an arbitrary SRS is given in listing 6.1. The listing defines a class SRS implementing the canonical SRS in (6.1).[1] The behavior of the class is similar to that of the class MC in listing 4.4 (page 73), and the methods are explained in the doc strings.

Listing 6.2 provides an example of usage. The code creates an instance of SRS, corresponding to the Solow model (6.2) when $f(k, W) = k^\alpha W$ and $\ln W_t \sim N(0, \sigma^2)$. The two sample paths generated by this code are shown in figure 6.1.

Exercise 6.1.4 Repeat exercise 5.2.2 on page 109 using the class SRS.

Returning to the SRS (6.1), let's consider the distribution of X_t for arbitrary $t \in \mathbb{N}$. This distribution will be denoted by ψ_t, and you can think of it for now as a cumulative distribution function (i.e., $\psi_t(x)$ is the probability that $X_t \leq x$). It is also called the marginal distribution of X_t; conceptually it is equivalent to its discrete state namesake that we met in §4.2.2.

In order to investigate ψ_t via simulation, we need to sample from this distribution. The simplest technique is this: First draw $X_0 \sim \psi$ and generate a sample path stopping at time t. Now repeat the exercise, but with a new set of draws $X_0, W_1, \ldots, W_t$,

---

[1] The benefit of designing an abstract class for SRSs is code reuse: The class can be used to study any system. Readers are encouraged to add functionality to the class as they do the exercises in the chapter.

Listing 6.1 (`srs.py`) Simulation of SRSs

```
class SRS:

    def __init__(self, F=None, phi=None, X=None):
        """Represents X_{t+1} = F(X_t, W_{t+1}); W ~ phi.
        Parameters: F and phi are functions, where phi()
        returns a draw from phi. X is a number representing
        the initial condition."""
        self.F, self.phi, self.X = F, phi, X

    def update(self):
        "Update the state according to X = F(X, W)."
        self.X = self.F(self.X, self.phi())

    def sample_path(self, n):
        "Generate path of length n from current state."
        path = []
        for i in range(n):
            path.append(self.X)
            self.update()
        return path
```

Listing 6.2 (`testsrs.py`) Example application

```
from srs import SRS                    # Import from listing 6.1
from random import lognormvariate

alpha, sigma, s, delta = 0.5, 0.2, 0.5, 0.1
# Define F(k, z) = s k^alpha z + (1 - delta) k
F = lambda k, z: s * (k**alpha) * z + (1 - delta) * k
lognorm = lambda: lognormvariate(0, sigma)

solow_srs = SRS(F=F, phi=lognorm, X=1.0)
P1 = solow_srs.sample_path(500)        # Generate path from X = 1
solow_srs.X = 60                       # Reset the current state
P2 = solow_srs.sample_path(500)        # Generate path from X = 60
```

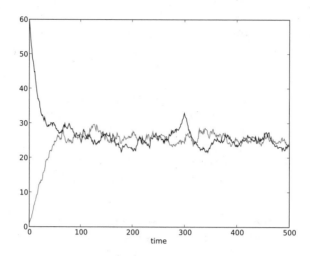

Figure 6.1 Time series plot

leading to a new draw of X_t that is independent of the first. If we do this n times, we get n independent samples $X_t^1, \ldots, X_t^n$ from the target distribution ψ_t. Algorithm 6.1 contains the pseudocode for this operation. Figure 6.2 is a visualization of the algorithm after 3 iterations of the outer loop.

Exercise 6.1.5 Investigate the mean of ψ_t for the Solow–Swan model when $f(k, W) = k^\alpha W$ and $\ln W_t \sim N(0, \sigma^2)$. Carry out a simulation with $k_0 = 1$, $t = 20$, $\delta = 0.1$, $s = 1/2$, $\sigma^2 = 0.2$ and $\alpha = 0.3$. Draw $n = 1000$ samples. Compute $\mathbb{E}k_t$ using the statistic $n^{-1} \sum_{i=1}^n k_t^i$. Explain how theorem 4.3.31 on page 94 justifies your statistic.

Exercise 6.1.6 Repeat exercise 6.1.5, but now setting $s = 3/4$. How does your estimate change? Interpret.

Exercise 6.1.7 Repeat exercise 6.1.5, but now set $k_0 = 5$, $k_0 = 10$, and $k_0 = 20$. To the extent that you can, interpret your results.

Exercise 6.1.8 Repeat exercise 6.1.5, but now set $t = 50$, $t = 100$, and $t = 200$. What happens to your estimates? Interpret.

Exercise 6.1.9 Repeat exercise 6.1.7, but with $t = 200$ instead of $t = 20$. Try to interpret your results.

Algorithm 6.1: Draws from the marginal distribution

---

for *i in 1 to n* **do**
  draw X from the initial condition ψ
  for *j in 1 to t* **do**
    draw W from the shock distribution ϕ
    set $X = F(X, W)$
  end
  set $X_t^i = X$
end
return $(X_t^1, \ldots, X_t^n)$

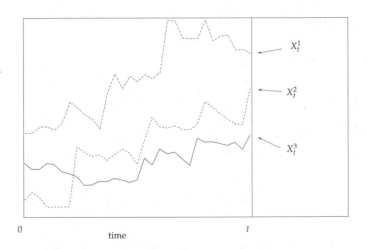

Figure 6.2 Sampling from the marginal distribution

Recall that if $X_1, \ldots, X_n$ is a sample of IID random variables, then the *sample mean* is defined as $\bar{X}_n := \frac{1}{n} \sum_{i=1}^{n} X_i$, while the *sample variance* is

$$\hat{\sigma}_n^2 := \frac{1}{n-1} \sum_{i=1}^{n} (X_i - \bar{X}_n)^2$$

Assuming that the second moment of X_i is finite, the central limit theorem and a convergence result often referred to as Slutsky's theorem give

$$\frac{\sqrt{n}(\bar{X}_n - \mathbb{E}X_1)}{\hat{\sigma}_n} \xrightarrow{d} N(0,1) \text{ as } n \to \infty, \text{ where } \hat{\sigma}_n := \sqrt{\hat{\sigma}_n^2}$$

Exercise 6.1.10 Based on this fact, construct a 95% confidence interval for your estimate of $\mathbb{E}k_t$. (Use the parameters from exercise 6.1.5.)

Exercise 6.1.11 Consider the same model as in exercise 6.1.5, but now set $\delta = 1$. The classical golden rule optimization problem is to choose the savings rate s in the Solow–Swan model to maximize steady state consumption. Let's consider the stochastic analogue. The simplest criterion is to maximize *expected* steady state consumption. For this model, by $t = 100$ the distribution of c_j varies little for $j \geq t$ (we'll learn more about this later on). As such, let's consider $\mathbb{E}c_{100}$ as expected steady state consumption. Compute $n = 5{,}000$ observations of c_{100}, and take the sample average to obtain an approximation of the expectation. Repeat for s in a grid of values in $(0, 1)$. Plot the function, and report the maximizer.

6.1.2   Distribution Dynamics

While the mean conveys some information about the random variable k_t, at times we wish to know about the entire (cumulative) distribution ψ_t. How might one go about computing ψ_t by simulation?

The standard method is with the *empirical distribution function*, which, for independent samples $(X_i)_{i=1}^{n}$ of random variable $X \in \mathbb{R}$, is given by

$$F_n(x) := \frac{1}{n} \sum_{i=1}^{n} \mathbb{1}\{X_i \leq x\} \qquad (x \in \mathbb{R}) \tag{6.4}$$

Thus $F_n(x)$ is the fraction of the sample that falls below x. The LLN (theorem 4.3.31, page 94) can be used to show that if X has cumulative distribution F, then $F_n(x) \to F(x)$ with probability one for each x as $n \to \infty$.[2] These results formalize the fundamental idea that empirical frequencies converge to probabilities when the draws are independent.

---

[2]Later we will cover how to do these kinds of proofs. In this case much more can be proved—interested readers should refer to the Glivenko–Cantelli theorem.

Listing 6.3 (ecdf.py) Empirical distribution function

```
class ECDF:

    def __init__(self, observations):
        self.observations = observations

    def __call__(self, x):
        counter = 0.0
        for obs in self.observations:
            if obs <= x:
                counter += 1
        return counter / len(self.observations)
```

An implementation of the empirical distribution function is given in listing 6.3, based on a class called ECDF.[3] The class is initialized with samples $(X_t)_{t=1}^n$ stored in a sequence (list or tuple) called observations, and an instance is created with a call such as F = ECDF(data). The method __call__ evaluates $F_n(x)$ for a given value of x. (Recall from §2.2.3 that __call__ is a special method that makes an instance F *callable*, so we can evaluate $F_n(x)$ using the simple syntax F(x).) Here is an example:

```
from ecdf import ECDF              # Import from listing 6.3
from random import uniform
samples = [uniform(0, 1) for i in range(10)]
F = ECDF(samples)
F(0.5)   # Returned 0.29
F.observations = [uniform(0, 1) for i in range(1000)]
F(0.5)   # Returned 0.479
```

Figure 6.3 gives four plots of the empirical distribution function corresponding to the time t distribution of the Solow–Swan model. The plots are for $n = 4, n = 25, n = 100$, and $n = 5,000$. The parameters are $k_0 = 1, t = 20, \delta = 0.1, s = 1/2, \sigma^2 = 0.2$, and $\alpha = 0.3$.

Exercise 6.1.12 Add a method to the ECDF class that uses Matplotlib to plot the empirical distribution over a specified interval. Replicate the four graphs in figure 6.3 (modulo randomness).

Consider now a variation of our growth model with additional nonlinearities. In

---

[3] As usual, the code is written for clarity rather than speed. See the text home page for more optimized solutions.

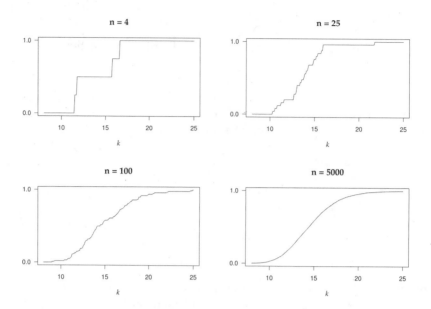

Figure 6.3 Empirical distribution functions

example 4.1.8 (page 57) we looked at a model with "threshold" nonconvexities. A stochastic version is

$$k_{t+1} = sA(k_t)k_t^{\alpha}W_{t+1} + (1-\delta)k_t \tag{6.5}$$

where the shock is assumed to be lognormally distributed (and independent), and A is the step function

$$A(k) = A_1 \mathbb{1}\{0 < k < k_b\} + A_2 \mathbb{1}\{k_b \leq k < \infty\} = \begin{cases} A_1 & \text{if } 0 < k < k_b \\ A_2 & \text{if } k_b \leq k < \infty \end{cases}$$

with $k_b \in S = (0, \infty)$ interpreted as the threshold, and $0 < A_1 < A_2$.

Figures 6.4 and 6.5 each show two time series generated for this model, with initial conditions $k_0 = 1$ and $k_0 = 80$. The parameters for figure 6.4 are set at $\alpha = 0.5$, $s = 0.25$, $A_1 = 15$, $A_2 = 25$, $\sigma^2 = 0.02$, and $k_b = 21.6$, while for figure 6.5, $k_b = 24.1$. Notice how initial conditions tend to persist, although time series occasionally cross the threshold k_b—what might be referred to in physics as a "phase transition." Informally, the state variable moves from one locally attracting region of the state space to another.

Exercise 6.1.13 Compute the empirical distribution functions at $t = 100$ for the two

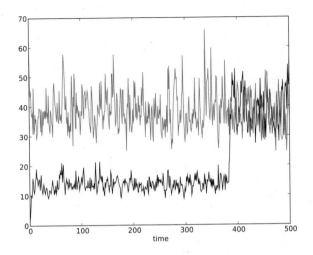

Figure 6.4 Persistence in time series

sets of parameters used in figures 6.4 and 6.5. Plot the functions and interpret their shapes.

Aside from computing the distributions, another interesting question is: How long do we expect it to take on average for the transition (crossing of the threshold k_b) to occur for an economy with initial condition $k_0 = 1$? More mathematically, what is the expectation of $\tau := \inf\{t \geq 0 : k_t > k_b\}$ when regarded as a random variable on $\mathbb{N}$? (Here τ is usually called the *first passage time* of $(k_t)_{t \geq 0}$ to (k_b, ∞).)

Exercise 6.1.14 Using $\alpha = 0.5$, $s = 0.25$, $A_1 = 15$, $A_2 = 25$, $\sigma^2 = 0.02$, $k_0 = 1$ and $k_b = 21.6$, compute an approximate expectation of τ by sample mean. (Set $n = 5,000$.) Do the same for $k_b = 24.1$, which corresponds to figure 6.5. How does your answer change? Interpret.

6.1.3 Density Dynamics

Now let's look more deeply at distribution dynamics for SRSs, with an emphasis on density dynamics. In reading this section, you should be aware that all densities create distributions but not all distributions are created by densities. If f is a density function on $\mathbb{R}$, then $F(x) := \int_{-\infty}^{x} f(u)du$ is a cumulative distribution function. However, if F is a cumulative distribution function with jumps—corresponding to positive probability

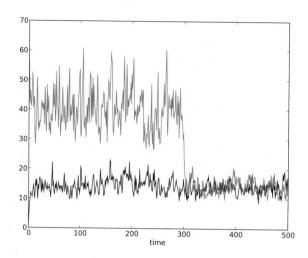

Figure 6.5 Persistence in time series

mass on individual points—then there exists no density f with $F(x) := \int_{-\infty}^{x} f(u)du$ for all $x \in \mathbb{R}$. (More about this later on.)

For the sake of concreteness, let's focus on a model of the form

$$X_{t+1} = g(X_t) + W_{t+1}, \quad X_0 \sim \psi, \quad (W_t)_{t \geq 1} \overset{\text{IID}}{\sim} \phi \tag{6.6}$$

where $Z = S = \mathbb{R}$, and both ψ and ϕ are *density* functions on $\mathbb{R}$. For this model, the distribution of X_t can be represented by a density ψ_t for any $t \geq 1$, and ψ_t and ψ_{t+1} are linked by the recursion

$$\psi_{t+1}(y) = \int p(x,y)\psi_t(x)dx, \quad \text{where } p(x,y) := \phi(y - g(x)) \tag{6.7}$$

Here p is called the *stochastic density kernel* corresponding to (6.3). It represents the distribution of $X_{t+1} = g(X_t) + W_{t+1}$ given $X_t = x$ (see below). The left-hand side of (6.7) is a continuous state version of (4.10) on page 75. It links the marginal densities of the process from one period to the next. In fact it defines the whole sequence of densities $(\psi_t)_{t \geq 1}$ for the process once an initial condition is given.

Let's try to understand why (6.7) holds, leaving fully rigorous arguments until later. First we need the following lemma.

Lemma 6.1.15 *If $W \sim \phi$, then $Y := g(x) + W$ has density $\phi(y - g(x))dy$.*[4]

---

[4]Here the symbol dy indicates that $\phi(y - g(x))$ is a density in y rather than in x.

Proof. Let F be the cumulative distribution function (cdf) of Y, and let Φ be the cdf corresponding to ϕ (i.e., $\Phi' = \phi$). We have

$$F(y) = \mathbb{P}\{g(x) + W \leq y\} = \mathbb{P}\{W \leq y - g(x)\} = \Phi(y - g(x))$$

The density of Y is $F'(y) = \phi(y - g(x))$ as claimed. $\square$

Returning to (6.7), recall that if X and Y are random variables with joint density $p_{X,Y}(x, y)$, then their marginal densities satisfy

$$p_X(x) = \int p_{X,Y}(x, y) dy, \quad p_Y(y) = \int p_{X,Y}(x, y) dx$$

Moreover the conditional density $p_{Y|X}(x, y)$ of Y given $X = x$ is given by

$$p_{Y|X}(x, y) = \frac{p_{X,Y}(x, y)}{p_X(x)} \qquad (x, y \in S)$$

Some simple manipulations now yield the expression

$$p_Y(y) = \int p_{Y|X}(x, y) p_X(x) dx \qquad (y \in S)$$

We have almost established (6.7). Letting $X_{t+1} = Y$ and $X_t = X$, we have

$$\psi_{t+1}(y) = \int p_{X_{t+1}|X_t}(x, y) \psi_t(x) dx \qquad (y \in S)$$

The function $p_{X_{t+1}|X_t}(x, y)$ is the density of $g(X_t) + W_{t+1}$ given $X_t = x$, or, more simply, the density of $g(x) + W_{t+1}$. By lemma 6.1.15, this is $\phi(y - g(x)) =: p(x, y)$, confirming (6.7).

Now let's look at the dynamics implied by the law of motion (6.7). The initial condition is $\psi_0 = \psi$, which is the density of X_0 (regarded as given). From this initial condition, (6.7) defines the entire sequence $(\psi_t)_{t \geq 0}$. There are a couple of ways that we can go about computing elements of this sequence. One is numerical integration. For example, ψ_1 could be calculated by evaluating $\psi_1(y) = \int p(x, y) \psi(x) dx$ at each $y \in S$. Come to think of it, though, this is impossible: there is an infinity of such y. Instead, we would have to evaluate on a finite grid, use our results to form an approximation $\hat{\psi}_1$ of ψ_1, then do the same to obtain $\hat{\psi}_2$, and so on.

Actually this process is not very efficient, and it is difficult to obtain a measure of accuracy. So let's consider some other approaches. Say that we wish to compute ψ_t, where t is a fixed point in time. Since we know how to compute empirical distribution functions by simulation, we could generate n observations of X_t (see algorithm 6.1), compute the empirical distribution function F_t^n, and differentiate F_t^n to obtain an approximation ψ_t^n to ψ_t.

It turns out that this is not a good plan either. The reason is that F_t^n is not differentiable everywhere on S. And although it is differentiable at many points in S, at those points the derivative is zero. So the derivative of F_t^n contains *no* information about ψ_t.[5]

Another plan would be to generate observations of X_t and histogram them. This is a reasonable and common way to proceed—but not without its flaws. The main problem is that the histogram converges rather slowly to its target ψ_t. This is because histograms have no notion of neighborhood, and do not make use of all knowledge we have at hand. For example, if the number of bins is large, then it is often the case that no points from the sample will fall in certain bins. This may happen even for bins close to the mean of the distribution. The result is a "spiky" histogram, even when the true density is smooth.[6]

We can include the prior information that the density of ψ_t is relatively smooth using *Parzen windows*, or nonparametric kernel density estimates. The kernel density estimate f_n of unknown density f from observations $(Y_i)_{i=1}^n$ is defined as

$$f_n(x) := \frac{1}{n \cdot \delta_n} \sum_{i=1}^n K\left(\frac{x - Y_i}{\delta_n}\right) \qquad (x \in \mathbb{R}) \tag{6.8}$$

where K is some density on $\mathbb{R}$, and δ_n is either a parameter or a function of the data, usually referred to as the *bandwidth*.

Essentially, f_n is a collection of n "bumps," one centered on each data point Y_i. These are then summed and normalized to create a density.

Exercise 6.1.16 Using a suitable change of variables in the integral, show that f_n is a density for every n.

The bandwidth parameter plays a role similar to the number of bins used in the histogram: A high value means that the densities we place on each data point are flat with large tails. A low value means they are concentrated around each data point, and f_n is spiky.

Exercise 6.1.17 Implement the nonparametric kernel density estimator (6.8) using a standard normal density for K. Base your code on the empirical distribution function class in listing 6.3. Generate a sample of 100 observations from the standard normal distribution and plot the density esimate for bandwidth values 0.01, 0.1, and 0.5.

Although the nonparametric kernel density estimator produces good results in a broad range of environments, it turns out that for the problem at hand there is a better

---

[5]Readers familiar with the theory of ill-posed problems will have a feel for what is going on here. The density computation problem is ill-posed!

[6]We get ψ_t by integrating ψ_{t-1} with respect to a kernel $\phi(y - g(x))$, and functions produced in this way are usually smooth rather than spiky.

way: The *look-ahead estimator* ψ_t^n of ψ_t is defined by generating n independent draws $(X_{t-1}^1, \ldots, X_{t-1}^n)$ of X_{t-1} and then setting

$$\psi_t^n(y) := \frac{1}{n} \sum_{i=1}^n p(X_{t-1}^i, y) \qquad (y \in \mathbb{R}) \tag{6.9}$$

where $p(x, y) = \phi(y - g(x))$.[7] This estimator has excellent asymptotic and finite sample properties. While we won't go into them too deeply, note that

Lemma 6.1.18 *The look-ahead estimator ψ_t^n is pointwise unbiased for ψ_t, in the sense that $\mathbb{E}\psi_t^n(y) = \psi_t(y)$ for every $y \in S$. Moreover $\psi_t^n(y) \to \psi_t(y)$ as $n \to \infty$ with probability one.*

Proof. Fix $y \in S$, and consider the random variable $Y := p(X_{t-1}, y)$. The look-ahead estimator $\frac{1}{n} \sum_{i=1}^n p(X_{t-1}^i, y)$ is the sample mean of IID copies of Y, while the mean is

$$\mathbb{E}Y = \mathbb{E}p(X_{t-1}^i, y) = \int p(x, y)\psi_{t-1}(x)dx = \psi_t(y)$$

(The last equality is due to (6.7)). The desired result now follows from the fact that the sample mean of an IID sequence of random variables is an unbiased and consistent estimator of the mean.[8] □

Figure 6.6 shows a sequence of densities from the STAR model (6.3), computed using the look-ahead estimator with 1,000 observations per density. The transition function G is the cumulative distribution function for the standard normal distribution, $\alpha_0 = 1$, $\alpha_1 = 0.4$, $\beta_0 = 10$, and $\beta_1 = 0.8$. The density ϕ is standard normal.

Exercise 6.1.19 Implement the look-ahead estimator for the STAR model using the same parameters used for figure 6.6. Replicate the figure (modulo the influence of randomness).

6.1.4 Stationary Densities: First Pass

The sequence of densities $(\psi_t)_{t \geq 0}$ in figure 6.6 appears to be converging.[9] Indeed it can be shown (see chapter 8) that there is a limiting distribution ψ^* to which $(\psi_t)_{t \geq 0}$ is converging, and that the limit ψ^* is independent of the initial condition ψ_0. The density ψ^* is called a stationary density, and it satisfies

$$\psi^*(y) = \int p(x, y)\psi^*(x)dx \qquad (y \in \mathbb{R}) \tag{6.10}$$

---

[7] The independent draws $(X_{t-1}^1, \ldots, X_{t-1}^n)$ can be obtained via algorithm 6.1 on page 121.

[8] If you don't know the proof of this fact then try to do it yourself. Consistency follows from the law of large numbers (theorem 4.3.31 on page 94).

[9] See figure 6.12 on page 142 for another sequence of densities converging to a limit.

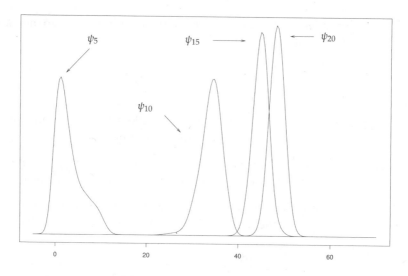

Figure 6.6 Density sequence

More generally, a density ψ^* on $\mathbb{R}$ is called *stationary* for the SRS (6.6) if (6.10) holds, where the density kernel p satisfies $p(x, y) = \phi(y - g(x))$. The SRS is called globally stable if there exists one and only one such density on $\mathbb{R}$, and the sequence of marginal distributions $(\psi_t)_{t \geq 0}$ converges to it as $t \to \infty$. (A more formal definition is given in chapter 8.)

You will recall that in the finite case a distribution ψ^* is called stationary if $\psi^* = \psi^* \mathbf{M}$, or equivalently, $\psi^*(y) = \sum_{x \in S} p(x, y) \psi^*(x)$ for all $y \in S$. The expression (6.10) simply replaces the sum with an integral, and the basic idea is the same: if the current marginal density is stationary, then updating to the next period leaves probabilities unchanged. Note, however, that when the state space is infinite a stationary density may fail to exist. You will be asked to give an example in exercise 8.2.2.

Recall that in the finite state case, when the stochastic kernel p is globally stable, each Markov chain generated by the kernel satisfies a law of large numbers (theorem 4.3.33, page 94). Here we have an analogous result. As shown in theorem 8.2.15 (page 206), given global stability and a function h such that $\int |h(x)| \psi^*(x) dx$ is finite, we have

$$\frac{1}{n} \sum_{t=1}^{n} h(X_t) \to \int h(x) \psi^*(x) dx \quad \text{as } n \to \infty \tag{6.11}$$

with probability one, where $(X_t)_{t \geq 0}$ is a time series generated by the model.

Exercise 6.1.20 Consider the STAR model (6.3) with $\alpha_0 = \beta_0 = 0$ and $\alpha_1 = \beta_1 = a$, where a is a constant with $|a| < 1$. Suppose that ϕ is standard normal. We will see later that this is a stable parameter configuration, and $\psi^* = N(0, 1/(1 - a^2))$ is stationary for this kernel. From (6.11) we have

$$\frac{1}{n} \sum_{t=1}^{n} X_t^2 \cong \frac{1}{1 - \alpha^2} \quad \text{for large } n$$

Write a simulation that compares these two expressions for large n.

The LLN gives us a method to investigate the steady state distribution ψ^* for globally stable systems. For example, we can form the empirical distribution function

$$F^n(x) := \frac{1}{n} \sum_{t=1}^{n} \mathbb{1}\{X_t \leq x\} = \frac{1}{n} \sum_{t=1}^{n} \mathbb{1}_{(-\infty, x]}(X_t) \qquad (x \in \mathbb{R})$$

where $\mathbb{1}_{(-\infty, x]}(y) = 1$ if $y \leq x$ and zero otherwise and $(X_t)_{t \geq 0}$ is a simulated time series generated by the model. The empirical distribution function was discussed previously in §6.1.2. We will see in what follows that $\int \mathbb{1}_{(-\infty, x]}(y) \psi^*(y) dy$ is the probability that a draw from ψ^* falls below x. In other words, $F(x) := \int \mathbb{1}_{(-\infty, x]}(y) \psi^*(y) dy$ is the cumulative distribution function associated with ψ^*. Setting $h = \mathbb{1}_{(-\infty, x]}$ in (6.11), we then have $F^n(x) \to F(x)$ with probability one, $\forall x \in \mathbb{R}$, and the empirical distribution function is consistent for F.

Exercise 6.1.21 Use the empirical distribution function to compute an estimate of F for (6.3) under the same parameters used in figure 6.6.

There is, however, a more powerful technique for evaluating ψ^* when global stability holds. Taking our simulated time series $(X_t)_{t=1}^{n}$, define

$$\psi_n^*(y) := \frac{1}{n} \sum_{t=1}^{n} p(X_t, y) \qquad (y \in \mathbb{R}) \tag{6.12}$$

This expression is almost identical to the look-ahead estimator developed in §6.1.3 (see (6.9) on page 129), with the difference being that the random samples are now a single time series rather than repeated draws at a fixed point in time. To study the properties of ψ_n^*, observe that for any fixed $y \in S$, the LLN (6.11) gives us

$$\psi_n^*(y) := \frac{1}{n} \sum_{t=1}^{n} p(X_t, y) \to \int p(x, y) \psi^*(x) dx = \psi^*(y)$$

where the last equality is by (6.10). Thus $\psi_n^*(y)$ is consistent for $\psi^*(y)$.

In fact much stronger results are true, and ψ_n^* is an excellent estimator for ψ^* (see, e.g., Stachurski and Martin 2008). The reason is that while estimators such as F^n use

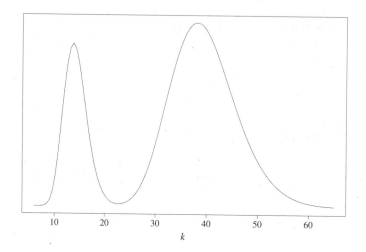

Figure 6.7 Look-ahead estimator

only the information contained in the sampled time series (X_t), the look-ahead estimator ψ_n^* also incorporates the stochastic kernel p, which encodes the entire dynamic structure of the model.

Exercise 6.1.22 Use the look-ahead estimator (6.12) to compute an estimate of ψ^* for (6.3) under the same parameters used in figure 6.6.

Here is a second application. Consider the nonconvex growth model in (6.5) on page 124 with $\delta = 1$. We will prove below that the stochastic density kernel for this model is

$$p(x,y) = \phi\left(\frac{y}{sA(x)x^\alpha}\right)\frac{1}{sA(x)x^\alpha} \qquad (x,y > 0) \tag{6.13}$$

and that the model is globally stable. From stability we obtain the LLN (6.11), and hence the look-ahead estimator (6.12) is consistent for the unique stationary density ψ^*. Figure 6.7 shows a realization of ψ_n^* when A is the step function

$$A(k) := A_1\mathbb{1}\{k \leq k_b\} + A_2\mathbb{1}\{k > k_b\} \qquad (k > 0)$$

and $W_t = e_t^\xi$, where $\xi_t \sim N(0,\sigma^2)$. The parameters are $\alpha = 0.5$, $s = 0.25$, $A_1 = 15$, $A_2 = 25$, $\sigma^2 = 0.02$, $k_b = 22.81$, and $k_0 = k_b$.

Exercise 6.1.23 Replicate figure 6.7. Sample a time series $(k_t)_{t\geq0}$ from the model, and implement ψ_n^* with $(k_t)_{t\geq0}$ and the kernel in (6.13).[10]

---

[10]You will require a value of n around 100,000 to get a reasonable estimate, and even then some variation

6.2   Optimal Growth, Infinite State

Let's now look at a simple optimal growth model on an infinite state space. We will compute the optimal policy for the model numerically using value iteration and policy iteration. We also study via simulation the dynamics of the model under that policy.

6.2.1   Optimization

Consider again the optimal growth model discussed in chapter 1. At time t an agent receives income y_t, which is split into consumption c_t and savings k_t. Given k_t, output at $t+1$ is $y_{t+1} = f(k_t, W_{t+1})$, where $(W_t)_{t \geq 1}$ is IID and takes values in $Z := (0, \infty)$ according to density ϕ. The agent's behavior is specified by a policy function σ, which is a map from $S := \mathbb{R}_+$ to $\mathbb{R}$ satisfying $0 \leq \sigma(y) \leq y$ for all $y \in S$. The value $\sigma(y)$ should be interpreted as the agent's choice of savings when income $= y$, while $0 \leq \sigma(y) \leq y$ is a feasibility constraint ensuring that savings is nonnegative and does not exceed income. The set of all such policies will be denoted by Σ.

   As with the finite state case, choice of a policy function $\sigma \in \Sigma$ also determines an SRS for the state variable, given by

$$y_{t+1} = f(\sigma(y_t), W_{t+1}), \quad (W_t)_{t \geq 1} \overset{\text{IID}}{\sim} \phi, \quad y_0 = y \tag{6.14}$$

where y is initial income. Letting U be the agent's utility function and $\rho \in (0,1)$ be the discount factor, the agent's decision problem is

$$\max_{\sigma \in \Sigma} v_\sigma(y), \quad \text{where} \quad v_\sigma(y) := \mathbb{E} \left[\sum_{t=0}^{\infty} \rho^t U(y_t - \sigma(y_t)) \right] \tag{6.15}$$

Here $v_\sigma(y)$ is the expected discounted value of following policy σ when initial income is $y_0 = y$. For now we assume that $U \colon \mathbb{R}_+ \to \mathbb{R}_+$ is bounded and continuous, and that $f \colon \mathbb{R}_+ \times Z \to \mathbb{R}_+$ is continuous.

   As discussed in the finite case—see §5.1.1, and in particular the discussion surrounding (5.4) on page 100—a rigorous definition of the expectation in (6.15) requires measure theory. The details are deferred until chapter 10. Among other things, we will see that the expectation can be passed through the sum to obtain

$$v_\sigma(y) = \sum_{t=0}^{\infty} \rho^t \mathbb{E} U(y_t - \sigma(y_t)) \qquad (y \in S = \mathbb{R}_+) \tag{6.16}$$

---

will be observable over different realizations. This is due to the nonlinearity in the model and resulting slow convergence.

This expression is simpler to interpret, with each term $\mathbb{E}U(y_t - \sigma(y_t))$ defined in terms of integrals over $\mathbb{R}$. Specifically, we integrate the function $y \mapsto U(y - \sigma(y))$ with respect to the marginal distribution ψ_t of y_t, where y_t is defined recursively in (6.14).

Given v_σ in (6.16), we can define the value function v^* in exactly the same way as (5.8) on page 102: $v^*(y) := \sup\{v_\sigma(y) : \sigma \in \Sigma\}$.[11] Just as in §5.1.2, the value function satisfies a Bellman equation: Letting $\Gamma(y) := [0, y]$ be the feasible savings choices when income is y, we have

$$v^*(y) = \max_{k \in \Gamma(y)} \left\{ U(y - k) + \rho \int v^*(f(k, z))\phi(z)dz \right\} \qquad (y \in S) \qquad (6.17)$$

The intuition behind (6.17) is similar to that for the finite state Bellman equation on page 102, and won't be repeated here. A proof that v^* satisfies (6.17) will be provided via theorem 10.1.11 on page 234. In the same theorem it is shown that v^* is continuous.

Recall that bcS is the set of continuous bounded real-valued functions on S. Given a $w \in bcS$, we say that $\sigma \in \Sigma$ is w-greedy if

$$\sigma(y) \in \operatorname*{argmax}_{k \in \Gamma(y)} \left\{ U(y - k) + \rho \int w(f(k, z))\phi(z)dz \right\} \qquad (y \in S) \qquad (6.18)$$

Later in the text we will see that continuity of w implies continuity of the objective function in (6.18), and since $\Gamma(y)$ is compact, the existence of a maximizer $\sigma(y)$ for each y is guaranteed by theorem 3.2.22 (page 49).

We will also prove that a policy σ^* is optimal in terms of maximizing expected discounted rewards if and only if it is v^*-greedy (theorem 10.1.11). In view of continuity of v^* and the previous comment regarding existence of maximizers, this result shows that at least one optimal policy exists. Moreover we can compute σ^* by first solving for v^* and then obtaining σ^* as the maximizer in (6.18) with v^* in place of w.

In order to compute v^*, we define the Bellman operator T, which maps $w \in bcS$ into $Tw \in bcS$ via

$$Tw(y) = \max_{k \in \Gamma(y)} \left\{ U(y - k) + \rho \int w(f(k, z))\phi(z)dz \right\} \qquad (y \in S) \qquad (6.19)$$

We prove in chapter 10 that T is a uniform contraction of modulus ρ on the metric space (bcS, d_∞), where $d_\infty(v, w) := \sup_{y \in S} |v(y) - w(y)|$. In view of Banach's fixed point theorem (page 53), T then has a unique fixed point $\bar{v} \in bcS$, and $T^n v \to \bar{v}$ in d_∞ as $n \to \infty$ for all $v \in bcS$. Moreover it is immediate from the definition of T and the Bellman equation that $Tv^*(y) = v^*(y)$ for all $y \in S$, so $\bar{v} = v^*$. We conclude that all trajectories of the dynamical system (bcS, T) converge to v^*.

---

[11]From boundedness of U it can be shown that this supremum is taken over a bounded set, and hence v^* is well defined. See exercise 10.1.8 on page 233. We treat unbounded rewards in §12.2.

These observations suggest that to solve for an optimal policy we can use the value iteration technique presented in algorithm 5.1 on page 103, replacing bS with bcS for the set from which the initial condition is chosen. The algorithm returns a v-greedy policy σ, computed from a function $v \in bcS$ that is close to v^*. If v is close to v^*, then v-greedy policies are "almost optimal." See §10.2.1 for details.

6.2.2   Fitted Value Iteration

Let's turn to numerical techniques. With regard to value iteration, the fact that the state space is infinite means that implementing the sequence of functions generated by the algorithm on a computer is problematic. Essentially, the issue is that if w is an arbitrary element of bcS, then to store w in memory we need to store the values $w(y)$ for every $y \in S$. For infinite S this is not generally possible.

At the same time, some functions from S to $\mathbb{R}$ can be stored on a computer. For example, if w is a polynomial function such as $w(y) = \sum_{i=0}^{n-1} a_i y^i$, then to store w in memory, we need only store the n coefficients $(a_i)_{i=0}^{n-1}$ and the instructions for obtaining $w(y)$ from these coefficients. Functions that can be recorded in this way (i.e., with a finite number of parameters) are said to have *finite parametric representation*.

Unfortunately, iterates of the Bellman operator do not naturally present themselves in finite parametric form. To get Tv from v, we need to solve a maximization problem at each y and record the result. Again, this is not possible when S is infinite. A common kludge is discretization, where S is replaced with a grid of size k, and the original model with a "similar" model that evolves on the grid. This is rarely the best way to treat continuous state problems, since a great deal of useful information is discarded, and there is little in the way of theory guaranteeing that the limiting policy converges to the optimal policy as $k \to \infty$.[12]

Another approach is fitted value iteration, as described in algorithm 6.2. Here $\mathscr{F}$ is a class of functions with finite parametric representation. The map $v \mapsto w$ defined by the first two lines of the loop is, in effect, an approximate Bellman operator $\hat{T}$, and fitted value iteration is equivalent to iteration with $\hat{T}$ in place of T. A detailed theoretical treatment of this algorithm is given in §10.2.3. At this stage let us try to grasp the key ideas, and then look at implementation.

The first thing to consider is the particular approximation scheme to be used in the step that sends Tv into $w \in \mathscr{F}$. A number of schemes have been used in economic modeling, from Chebychev polynomials to splines and neural nets. In choosing the best method we need to consider how the scheme interacts with the iteration process

---

[12]One reason is that the resulting policy is not an element of the original policy space Σ, making it difficult to discuss the error induced by approximation. As an aside, some studies actually treat discrete state problem using continuous approximations in order to reduce the number of parameters needed to store the value function.

Algorithm 6.2: Fitted value iteration

---

initialize $v \in bcS$
repeat

> sample the function Tv at finite set of grid points $(y_i)_{i=1}^k$
> use the samples to construct an approximation $w \in \mathscr{F}$ of Tv
> set $e = d_\infty(v, w)$
> set $v = w$

until *e is less that some tolerance*
solve for a v-greedy policy σ

---

used to compute the fixed point v^*. A scheme that approximates individual functions well with respect to some given criterion does not always guarantee good dynamic properties for the sequence $(\hat{T}^n v)_{n \geq 1}$.

To try to pin down a suitable technique for approximation, let's decompose $\hat{T}$ into the action of two operators L and T. First T is applied to v—in practice Tv is evaluated only at finitely many points—and then an approximation operator L sends the result into $w = \hat{T}v \in \mathscr{F}$. Thus, $\hat{T} = L \circ T$. Figure 6.8 illustrates iteration of $\hat{T}$.

We aim to choose L such that (1) the sequence $(\hat{T}^n v)_{n \geq 1}$ converges, and (2) the collection of functions $\mathscr{F}$ is sufficiently rich that the limit of this sequence (which lives in $\mathscr{F}$) can be close to the fixed point v^* of T (which lives in bcS).[13] The richness of $\mathscr{F}$ depends on the choice of the approximation scheme and the number of grid points k in algorithm 6.2. In the formal results presented in §10.2.3, we will see that the approximation error depends on $d_\infty(Lv^*, v^*)$, which indicates how well v^* can be approximated by an element of $\mathscr{F}$.

Returning to point (1), any serious attempt at theory requires that the sequence $(\hat{T}^n v)_{n \geq 1}$ converges in some sense as $n \to \infty$. In this connection, note the following result.

Exercise 6.2.1 Let M and N be operators sending metric space (U, d) into itself. Show that if N is a uniform contraction with modulus ρ and M is nonexpansive, then $M \circ N$ is a uniform contraction with modulus ρ.

As T is a uniform contraction on (bcS, d_∞), we see that $\hat{T}$ is uniformly contracting whenever L is nonexpansive on (bcS, d_∞). While for some common approximation architectures this fails, it does hold for a number of useful schemes. When attention is restricted to these schemes the sequence $(\hat{T}^n v)_{n \geq 1}$ is convergent by Banach's fixed point theorem, and we can provide a detailed analysis of the algorithm.

---

[13]More correctly, the limit of the sequence lives in cl $\mathscr{F} \subset bcS$.

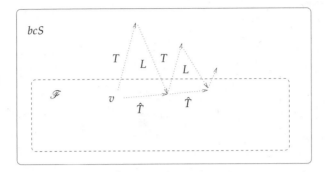

Figure 6.8 The map $\hat{T} := L \circ T$

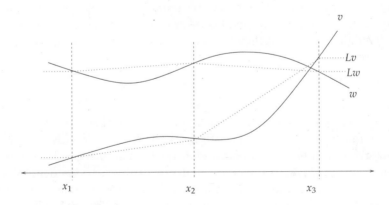

Figure 6.9 Approximation via linear interpolation

Listing 6.4 (`lininterp.py`) An interpolation class

```python
from scipy import interp

class LinInterp:
    "Provides linear interpolation in one dimension."

    def __init__(self, X, Y):
        """Parameters: X and Y are sequences or arrays
        containing the (x,y) interpolation points.
        """
        self.X, self.Y = X, Y

    def __call__(self, z):
        """Parameters: z is a number, sequence or array.
        This method makes an instance f of LinInterp callable,
        so f(z) returns the interpolation value(s) at z.
        """
        return interp(z, self.X, self.Y)
```

Let's move on to implementation, deferring further theory until §10.2.3. The approximation scheme we will use is piecewise linear interpolation, as shown in figure 6.9. (Outside the set of grid points, the approximations are constant.) With reference to the figure, it is not difficult to see that for any $v, w \in bcS$, and any x in the domain, we have

$$|Lv(x) - Lw(x)| \leq \sup_{1 \leq i \leq k} |v(x_i) - w(x_i)| \leq \|v - w\|_\infty$$

Taking the supremum over $x \in S$, we see that L is nonexpansive on (bcS, d_∞).

Consider the class `LinInterp` in listing 6.4. This class provides an interface to SciPy's `interp()` function, which implements linear interpolation. The latter is called using syntax `interp(Z, X, Y)`, where X and Y are arrays of equal length providing the (x, y) interpolation points, and Z is a number or array of point(s) where the interpolant is to be evaluated. `LinInterp` instances store the interpolation points X and Y internally, so that evaluations of the interpolant can be obtained without having to pass the interpolation points each time.[14]

The main part of the code is in listing 6.5. The utility function is set to $U(c) = 1 - e^{-\theta c}$, where the risk aversion parameter θ determines the curvature of U. The

---

[14]For discussion of the `__call__` method see §2.2.3.

Listing 6.5 (fvi.py) Fitted value iteration

```
from scipy import linspace, mean, exp, randn
from scipy.optimize import fminbound
from lininterp import LinInterp        # From listing 6.4

theta, alpha, rho = 0.5, 0.8, 0.9      # Parameters
def U(c): return 1 - exp(- theta * c)  # Utility
def f(k, z): return (k**alpha) * z     # Production
W = exp(randn(1000))                   # Draws of shock

gridmax, gridsize = 8, 150
grid = linspace(0, gridmax**1e-1, gridsize)**10

def maximum(h, a, b):
    return float(h(fminbound(lambda x: -h(x), a, b)))

def bellman(w):
    """The approximate Bellman operator.
    Parameters: w is a vectorized function (i.e., a
    callable object which acts pointwise on arrays).
    Returns: An instance of LinInterp.
    """
    vals = []
    for y in grid:
        h = lambda k: U(y - k) + rho * mean(w(f(k,W)))
        vals.append(maximum(h, 0, y))
    return LinInterp(grid, vals)
```

production function is $f(k, z) = k^\alpha z$. The shock is assumed to be lognormal; $W = e^\xi$, where ξ is standard normal.

Stepping through the logic of listing 6.5, the grid is formed by a call to SciPy's linspace() function, which returns an evenly spaced sequence on $[0, 8^{0.1}]$. Algebraic operations on SciPy (NumPy) arrays are performed elementwise, so appending **10 to the end of the line raises each element of the grid to the power of 10. The overall effect is to create a grid on $[0, 8]$ such that most grid points are close to zero. This is desirable since most the value function's curvature is close to zero, and more curvature requires closer grid points to achieve the same level of error.

Next we define the function maximum() that takes a function h and two points a and

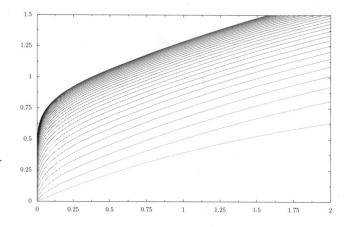

Figure 6.10 FVI algorithm iterates

b, and returns the maximum of h on the interval $[a, b]$. As defined here, maximum() is just a wrapper for SciPy's fminbound() function. The latter computes minimizers (not minima), and our wrapper uses this functionality to compute maxima. Specifically, we exploit the fact that the maximizer of h on $[a, b]$ is the minimizer x^* of $-h$ on $[a, b]$. Hence $h(x^*)$ is the maximum.

The function bellman() is the approximate Bellman operator $\hat{T} = L \circ T$, taking a function w and returning a new function $\hat{T}w$. The **for** loop steps through each grid point y_i, computing $Tw(y_i)$ as defined in (6.19) and recording this value in vals. Note how the expression mean(w(f(k,W))) is used to approximate the expection in $\int w(f(k, z))\phi(z)dz$. We are using the fact that w is vectorized (i.e., acts elementwise on NumPy arrays), so w(f(k,W)) is an array produced by applying $w(f(k, \cdot))$ to each element of the shock array W. Vectorized operations are typically faster than **for** loops.[15]

After collecting the values $Tw(y_i)$ in the **for** loop, the last line of the function definition returns an instance of LinInterp, which provides linear interpolation between the set of points $(y_i, Tw(y_i))$. This object corresponds to $\hat{T}w = L(Tw)$. Figure 6.10 shows convergence of the sequence of iterates, starting from initial condition U.

We can now compute a greedy policy σ from the last of these iterates, as shown in figure 6.11; along with the function $y \mapsto f(\sigma(y), m)$, where m is the mean of the shock. If the shock is always at its mean, then from every positive initial condition the income process $(y_t)_{t \geq 0}$ would converge to the unique fixed point at $\cong 1.25$. Notice that when income is low, the agent invests all available income.

Of course, what happens when the shock is at its mean gives us little feel for the

---

[15]Alternatively, the integral can be computed using a numerical integration routine.

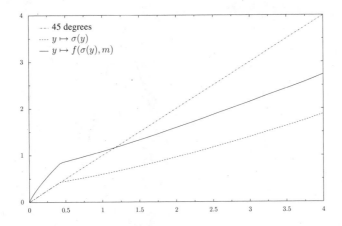

Figure 6.11 Approximate optimal policy

true dynamics. To generate the sequence of densities corresponding to the process $y_{t+1} = f(\sigma(y_t), W_{t+1})$, we can use the look-ahead estimator (6.9) on page 129. To use the theory developed there it is easiest to operate in logs. Given our parameterization of f, taking logs of both sides of $y_{t+1} = f(\sigma(y_t), W_{t+1}) = \sigma(y_t)^\alpha W_{t+1}$ yields

$$x_{t+1} = \alpha \ln \sigma(\exp(x_t)) + w_{t+1} := g(x_t) + w_{t+1}$$

where $x_t := \ln y_t$, $w_t := \ln W_t$ and $g(x) := \alpha \ln \sigma(\exp(x))$. This SRS is of the form (6.6) on page 126, and the look-ahead estimator of the density ψ_t of x_t can be computed via

$$\psi_t^n(y) := \frac{1}{n} \sum_{i=1}^{n} p(x_{t-1}^i, y) \qquad (y \in \mathbb{R}) \qquad (6.20)$$

where $p(x, y) = \phi(y - g(x))$ and $(x_{t-1}^i)_{i=1}^n$ is n independent draws of x_{t-1} starting from a given initial condition x_0. Here ϕ is the density of w_t—in this case it's $N(0, 1)$. Figure 6.12 shows the densities ψ_1 to ψ_{15} starting at $x_0 \equiv -7.5$ and using $n = 1,000$.

Exercise 6.2.2 Using your preferred plotting utility and extending the code in listing 6.5, replicate figures 6.10 through 6.12. Starting from initial condition U, about 30 iterates of $\hat{T}$ produces a good approximation of v^*. When you compute the densities via the look-ahead estimator, try experimenting with different initial conditions. Observe how the densities always converge to the same limit.

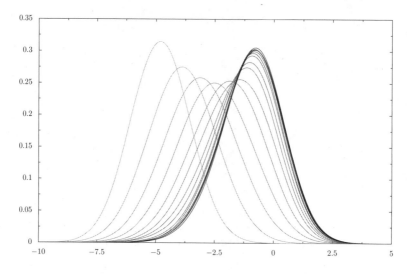

Figure 6.12 Densities of the income process

6.2.3 Policy Iteration

Recall that in §5.1.3 we solved the finite state problem using a second algorithm, called policy iteration. (See in particular algorithm 5.2 on page 105.) We can do the same thing here, although we will need to use approximation techniques similar to those we used for fitted value iteration. The basic idea is presented in algorithm 6.3. (In practice, the functions v_σ and σ will have to be approximated at each step.)

Algorithm 6.3: Policy iteration algorithm

---

pick any $\sigma \in \Sigma$
repeat
  compute v_σ from σ
  solve for a v_σ-greedy policy σ'
  set $\sigma = \sigma'$
until *until some stopping condition is satisfied*

The theory behind policy iteration is presented in §10.2.2. In this section our interest will be in implementation. When considering implementation the most difficult part is to compute v_σ from σ (i.e., to calculate the value of a given policy). Let's think about how we might do this given the definition $v_\sigma(y) = \sum_{t=0}^{\infty} \rho^t \mathbb{E} U(y_t - \sigma(y_t))$ in (6.16).

Fix y and consider the evaluation of $v_\sigma(y)$. We need to sum the terms $\mathbb{E}U(y_t - \sigma(y_t))$ discounted by ρ^t, at least for all $t \le T$ where T is large. How should we evaluate $\mathbb{E}U(y_j - \sigma(y_j))$ for fixed j? One idea would be to use Monte Carlo. The initial condition is y, and from there the process $(y_t)_{t \ge 0}$ follows the SRS in (6.14). We can generate n independent observations $y_j^1, \ldots, y_j^n$ of y_j using a version of algorithm 6.1 (page 121). The mean $n^{-1} \sum_{i=1}^n U(y_j^i - \sigma(y_j^i))$ is close to $\mathbb{E}U(y_j - \sigma(y_j))$ for large n.[16]

While Monte Carlo methods can be useful for high-dimensional problems, this approach is not the easiest to implement. The reason is that even if we obtain good approximations to $\mathbb{E}U(y_t - \sigma(y_t))$ for each t and then sum them (discounting by ρ^t) to obtain $v_\sigma(y)$, we have only obtained an evaluation of the function v_σ at a single point y. But for algorithm 6.3 we need to evaluate v_σ at every point y (or at least on a grid of points so we can approximate v_σ).

Rather than us going down this path, consider the following iterative technique. For given $\sigma \in \Sigma$, define the operator T_σ that sends $w \in bcS$ into $Tw \in bcS$ by

$$T_\sigma w(y) = U(y - \sigma(y)) + \rho \int w(f(\sigma(y), z))\phi(z)dz \qquad (y \in S) \qquad (6.21)$$

For us the pertinent features of T_σ are summarized in the following result.

Lemma 6.2.3 For each $\sigma \in \Sigma$, the operator T_σ is a uniform contraction on (bcS, d_∞), and the unique fixed point of T_σ in bcS is v_σ.

The proof is not overly difficult but does involve some measure theory. As such we defer it until §10.1.3. What is important for us now is that from any initial guess $v \in bcS$ we have $T_\sigma^n v \to v_\sigma$ so, by iterating with T_σ, we can obtain an approximation to v_σ. In doing so we need to approximate at each iteration, just as we did for fitted value iteration (algorithm 6.2, but with T_σ in place of T). Listing 6.6 provides an implementation of this scheme, along with some additional routines.

The first line of the listing imports everything from the module fvi, which is the file in listing 6.5. This provides the primitives of the model and the grid. Next some functions are imported from scipy. The first function maximizer() is similar to maximum() in listing 6.5 but returns maximizers rather than maxima.

The function T takes functions sigma and w, representing σ and w respectively, and returns $T_\sigma w$ as an instance of the class LinInterp defined in listing 6.4. The code is relatively self-explanatory, as is that for the function get_greedy(), which takes as its argument a function w and returns a w-greedy policy as an instance of LinInterp.

The function get_value is used to calculate v_σ from σ. It takes as arguments the functions sigma and v, which represent σ and a guess for v_σ. We then iterate on this

---

[16] An alternative Monte Carlo strategy would be to use the densities computed by the look-ahead estimator (see figure 6.12). Numerical integration of $y \mapsto U(y - \sigma(y))$ with respect to the density of y_j then gives another approximation to $\mathbb{E}U(y_j - \sigma(y_j))$.

Listing 6.6 (fpi.py) Fitted policy iteration

```python
from fvi import *    # Import all definitions from listing 6.5
from scipy import absolute as abs

def maximizer(h, a, b):
    return float(fminbound(lambda x: -h(x), a, b))

def T(sigma, w):
    "Implements the operator L T_sigma."
    vals = []
    for y in grid:
        Tw_y = U(y - sigma(y)) + rho * mean(w(f(sigma(y), W)))
        vals.append(Tw_y)
    return LinInterp(grid, vals)

def get_greedy(w):
    "Computes a w-greedy policy."
    vals = []
    for y in grid:
        h = lambda k: U(y - k) + rho * mean(w(f(k, W)))
        vals.append(maximizer(h, 0, y))
    return LinInterp(grid, vals)

def get_value(sigma, v):
    """Computes an approximation to v_sigma, the value
    of following policy sigma. Function v is a guess.
    """
    tol = 1e-2              # Error tolerance
    while 1:
        new_v = T(sigma, v)
        err = max(abs(new_v(grid) - v(grid)))
        if err < tol:
            return new_v
        v = new_v
```

guess with T_σ (actually $L \circ T_\sigma$ where L is the linear interpolation operator) to obtain an approximation to v_σ. Iteration proceeds until the maximum distance over the grid points between the new and previous iterate falls below some tolerance.

Exercise 6.2.4 Compute an approximate optimal policy from listing 6.6. Check that this policy is similar to the one you computed in exercise 6.2.2.[17]

6.3 Stochastic Speculative Price

This section applies some of the ideas we have developed to the study of prices in a commodity market with consumers and speculators. After specifying and solving the model, we will also investigate how the solution can be obtained using the optimal growth model of §6.2. In this way, we will see that the optimal growth model, which at first pass seems rather limited, can in fact be applied to the study of decentralized economies with a large number of agents.

6.3.1 The Model

Consider a market for a single commodity, whose price is given at t by p_t. The "harvest" of the commodity at time t is W_t. We assume that the sequence $(W_t)_{t \geq 1}$ is IID with common density function ϕ. The harvests take values in $S := [a, \infty)$, where $a > 0$. The commodity is purchased by "consumers" and "speculators." We assume that consumers generate demand quantity $D(p)$ corresponding to price p. Regarding the inverse demand function $D^{-1} =: P$ we assume that

Assumption 6.3.1 The function $P: (0, \infty) \to (0, \infty)$ exists, is strictly decreasing and continuous, and satisfies $P(x) \uparrow \infty$ as $x \downarrow 0$.

Speculators can store the commodity between periods, with I_t units purchased in the current period yielding αI_t units in the next, $\alpha \in (0, 1)$. For simplicity, the risk free interest rate is taken to be zero, so expected profit on I_t units is

$$\mathbb{E}_t p_{t+1} \cdot \alpha I_t - p_t I_t = (\alpha \mathbb{E}_t p_{t+1} - p_t) I_t$$

Here $\mathbb{E}_t p_{t+1}$ is the expectation of p_{t+1} taken at time t. Speculators are assumed to be risk neutral. Nonexistence of arbitrage requires that

$$\alpha \mathbb{E}_t p_{t+1} - p_t \leq 0 \tag{6.22}$$

---

[17]Hint: Suppose that σ_n is the policy computed at the n-th iteration. A good initial condition for the guess of v_{σ_n} in get_value() is $v_{\sigma_{n-1}}$.

Profit maximization gives the additional condition

$$\alpha \mathbb{E}_t p_{t+1} - p_t < 0 \text{ implies } I_t = 0 \tag{6.23}$$

We also require that the market clears in each period. Supply X_t is the sum $\alpha I_{t-1} + W_t$ of carryover by speculators and the current harvest, while demand is $D(p_t) + I_t$ (i.e., purchases by consumers and speculators). The market equilibrium condition is therefore

$$\alpha I_{t-1} + W_t =: X_t = D(p_t) + I_t \tag{6.24}$$

The initial condition $X_0 \in S$ is treated as given.

Now to find an equilibrium. Constructing a system $(I_t, p_t, X_t)_{t \geq 0}$ for investment, prices, and supply that satisfies (6.22)–(6.24) is not trivial. Our path of attack will be to *seek a system of prices that depend only on the current state*. In other words, we take a function $p \colon S \to (0, \infty)$ and set $p_t = p(X_t)$ for every t. The vector $(I_t, p_t, X_t)_{t \geq 0}$ then evolves as

$$p_t = p(X_t), \quad I_t = X_t - D(p_t), \quad X_{t+1} = \alpha I_t + W_{t+1} \tag{6.25}$$

For given X_0 and exogenous process $(W_t)_{t \geq 1}$, the system (6.25) determines the time path for $(I_t, p_t, X_t)_{t \geq 0}$ as a sequence of random variables. We seek a p such that (6.22) and (6.23) hold for the corresponding system (6.25).[18]

To this end, suppose that there exists a particular function $p^* \colon S \to (0, \infty)$ satisfying

$$p^*(x) = \max \left\{ \alpha \int p^*(\alpha I(x) + z) \phi(z) dz, P(x) \right\} \quad (x \in S) \tag{6.26}$$

where

$$I(x) := x - D(p^*(x)) \quad (x \in S) \tag{6.27}$$

It turns out that such a p^* will suffice, in the sense that (6.22) and (6.23) hold for the corresponding system (6.25). To see this, observe first that[19]

$$\mathbb{E}_t p_{t+1} = \mathbb{E}_t p^*(X_{t+1}) = \mathbb{E}_t p^*(\alpha I(X_t) + W_{t+1}) = \int p^*(\alpha I(X_t) + z) \phi(z) dz$$

Thus (6.22) requires that

$$\alpha \int p^*(\alpha I(X_t) + z) \phi(z) dz \leq p^*(X_t)$$

---

[18]Given (6.25) we have $X_t = I_t + D(p_t)$, so (6.24) automatically holds.

[19]If the manipulations here are not obvious don't be concerned—we will treat random variables in detail later on. The last inequality uses the fact that if U and V are independent and V has density ϕ then the expectation of $h(U, V)$ given U is $\int h(U, z) \phi(z) dz$.

This inequality is immediate from (6.26). Second, regarding (6.23), suppose that

$$\alpha \int p^*(\alpha I(X_t) + z)\phi(z)dz < p^*(X_t)$$

Then by (6.26) we have $p^*(X_t) = P(X_t)$, whence $D(p^*(X_t)) = X_t$, and $I_t = I(X_t) = 0$. (Why?) In conclusion, both (6.22) and (6.23) hold, and the system $(I_t, p_t, X_t)_{t\geq 0}$ is an equilibrium.

The only issue remaining is whether there does in fact exist a function $p^* \colon S \to (0, \infty)$ satisfying (6.26). This is not obvious, but can be answered in the affirmative by harnessing the power of Banach's fixed point theorem. To begin, let $\mathscr{C}$ denote the set of decreasing (i.e., nonincreasing) continuous functions $p \colon S \to \mathbb{R}$ with $p \geq P$ pointwise on S.

Exercise 6.3.2 Show that $\mathscr{C} \subset bcS$.

Exercise 6.3.3 Show that if $(h_n) \subset \mathscr{C}$ and $d_\infty(h_n, h) \to 0$ for some function $h \in bcS$, then h is decreasing and dominates P.

Lemma 6.3.4 *The metric space* $(\mathscr{C}, d_\infty)$ *is complete.*

Proof. By theorem 3.2.3 on page 45 (closed subsets of complete spaces are complete) and the completeness of bcS (theorem 3.2.9 on page 46) we need only show that $\mathscr{C}$ is closed as a subset of bcS. This follows from exercise 6.3.3. □

As $\mathscr{C}$ is complete, it provides a suitable space in which we can introduce an operator from $\mathscr{C}$ to $\mathscr{C}$ and—appealing to Banach's fixed point theorem—show the existence of an equilibrium. The idea is to construct the operator such that (1) any fixed point satisfies (6.26), and (2) the operator is uniformly contracting on $\mathscr{C}$. The existence of an operator satisfying conditions (1) and (2) proves the existence of a solution to (6.26).

So let p be a given element of $\mathscr{C}$, and consider the new function on S constructed by associating to each $x \in S$ the real number r satisfying

$$r = \max\left\{\alpha \int p(\alpha(x - D(r)) + z)\phi(z)dz, \ P(x)\right\} \tag{6.28}$$

We denote the new function by Tp, where $Tp(x)$ is the r that solves (6.28), and regard T as an operator sending elements of $\mathscr{C}$ into new functions on S. It is referred to below as the pricing functional operator.

Theorem 6.3.5 *The following results hold:*

1. *The pricing functional operator T is well-defined, in the sense that $Tp(x)$ is a uniquely defined real number for every $p \in \mathscr{C}$ and $x \in S$. Moreover*

$$P(x) \leq Tp(x) \leq v(x) := \max\left\{\alpha \int p(z)\phi(z)dz, P(x)\right\} \qquad (x \in S)$$

2. *T maps $\mathscr{C}$ into itself. That is, $T(\mathscr{C}) \subset \mathscr{C}$.*

The proof is only sketched. You might like to come back after reading up on measure theory and fill out the details. To start, let $p \in \mathscr{C}$ and $x \in S$. Define

$$h_x(r) := \max\left\{ \alpha \int p(\alpha(x - D(r)) + z)\phi(z)dz, \ P(x) \right\} \qquad (P(x) \leq r \leq v(x))$$

Although we skip the proof, this function is continuous and decreasing on the interval $[P(x), v(x)]$. To establish the claim that there is a unique $r \in [P(x), v(x)]$ satisfying (6.28), we must show that h_x has a unique fixed point in this set. As h_x is decreasing, uniqueness is trivial.[20] Regarding existence, it suffices to show that there exist numbers $r_1 \leq r_2$ in $[P(x), v(x)]$ with

$$r_1 \leq h_x(r_1) \quad \text{and} \quad h_x(r_2) \leq r_2$$

Why does this suffice? The reason is that if either holds with equality then we are done, and if both hold inequalities are strict, then we can appeal to continuity of h_x and the intermediate value theorem (page 335).[21]

A suitable value for r_1 is $P(x)$. (Why?) For r_2 we can use $v(x)$, as

$$h_x(r_2) = \max\left\{ \alpha \int p(\alpha(x - D(r_2)) + z)\phi(z)dz, \ P(x) \right\}$$

$$\leq \max\left\{ \alpha \int p(z)\phi(z)dz, \ P(x) \right\} = v(x) = r_2$$

The claim is now established, and with it part 1 of the theorem.

To prove part 2, we must show that Tp (1) dominates P, (2) is decreasing on S, and (3) is continuous on S. Of these, (1) is implied by previous results, while (2) and (3) hold but proofs are omitted—we won't cover the necessary integration theory until chapter 7.[22]

Exercise 6.3.6 Verify that if p^* is a fixed point of T, then it solves (6.26).

Theorem 6.3.7 *The operator T is a uniform contraction of modulus α on $(\mathscr{C}, d_\infty)$.*

It follows from theorem 6.3.7 that there exists a unique $p^* \in \mathscr{C}$ with $Tp^* = p^*$. In view of exercise 6.3.6, p^* satisfies (6.26), and we have solved our existence problem. Thus it only remains to confirm theorem 6.3.7, which can be proved using Blackwell's sufficient condition for a uniform contraction. To state the latter, consider the metric space (M, d_∞), where M is a subset of bU, the bounded real-valued functions on arbitrary set U.

---

[20]This was discussed in exercise 3.2.31 on page 51.

[21]Can you see why? Apply the theorem to $g(r) = r - h(r)$.

[22]The proof of (3) uses theorem B.1.4 on page 341.

Theorem 6.3.8 (Blackwell) *Let M be a subset of bU with the property that $u \in M$ and $\gamma \in \mathbb{R}_+$ implies $u + \gamma \mathbb{1}_U \in M$. If $T: M \to M$ is monotone and*

$$\exists \lambda \in [0,1) \quad \text{s.t.} \quad T(u + \gamma \mathbb{1}_U) \leq Tu + \lambda \gamma \mathbb{1}_U \quad \forall u \in M \text{ and } \gamma \in \mathbb{R}_+ \qquad (6.29)$$

then T is uniformly contracting on (M, d_∞) with modulus λ.

Monotonicity means that if $u, v \in M$ and $u \leq v$, then $Tu \leq Tv$, where all inequalities are pointwise on U. The proof of theorem 6.3.8 is given in on page 344, and we now return to the proof of theorem 6.3.7.

Exercise 6.3.9 Let h_1 and h_2 be decreasing functions on S with (necessarily unique) fixed points x_1 and x_2. Show that if $h_1 \leq h_2$, then $x_1 \leq x_2$.

Using this exercise it is easy to see that T is a monotone operator on $\mathscr{C}$: Pick any $p, q \in \mathscr{C}$ with $p \leq q$, and any $x \in S$. Let $r \mapsto h_p(r)$ be defined by

$$h_p(r) := \max \left\{ \alpha \int p(\alpha(x - D(r)) + z)\phi(z)dz, \ P(x) \right\}$$

and let $h_q(r)$ be defined analogously. Clearly, $Tp(x)$ is the fixed point of $r \mapsto h_p(r)$, as is $Tq(x)$ the fixed point of $r \mapsto h_q(r)$. Since $h_p(r) \leq h_q(r)$ for all r, it must be that $Tp(x) \leq Tq(x)$. As x was arbitrary we have $Tp \leq Tq$.

To apply Blackwell's condition, we need to show in addition that if $p \in \mathscr{C}$ and $\gamma \in \mathbb{R}_+$, then (1) $p + \gamma \mathbb{1}_S \in \mathscr{C}$, and (2) there exists a $\lambda < 1$ independent of p and γ and having the property

$$T(p + \gamma \mathbb{1}_S) \leq Tp + \lambda \gamma \mathbb{1}_S \qquad (6.30)$$

Statement (1) is obviously true. Regarding statement (2), we make use of the following easy lemma:

Lemma 6.3.10 *Let a, b, and c be real numbers with $b \geq 0$. We have*

$$\max\{a + b, c\} \leq \max\{a, c\} + b$$

If you're not sure how to prove these kinds of inequalities, then here is how they are done: Observe that both

$$a + b \leq \max\{a, c\} + b \quad \text{and} \quad c \leq \max\{a, c\} + b$$

$$\therefore \quad \max\{a + b, c\} \leq \max\{a, c\} + b$$

To continue, let p and γ be as above, and let $q := p + \gamma \mathbb{1}_S$. Pick any $x \in S$. Let r_p stand for $Tp(x)$ and let r_q stand for $Tq(x)$. We have

$$
\begin{aligned}
r_q &= \max \left\{ \alpha \int q(\alpha(x - D(r_q)) + z)\phi(z)dz, \ P(x) \right\} \\
&\leq \max \left\{ \alpha \int q(\alpha(x - D(r_p)) + z)\phi(z)dz, \ P(x) \right\} \\
&= \max \left\{ \alpha \int p(\alpha(x - D(r_p)) + z)\phi(z)dz + \alpha\gamma, \ P(x) \right\} \\
&\leq \max \left\{ \alpha \int p(\alpha(x - D(r_p)) + z)\phi(z)dz, \ P(x) \right\} + \alpha\gamma \\
&= r_p + \alpha\gamma
\end{aligned}
$$

Here the first inequality follows from the fact that $r_p \leq r_q$ (since $p \leq q$ and T is monotone), and the second from lemma 6.3.10.

We have show that $T(p + \gamma \mathbb{1}_S)(x) \leq Tp(x) + \alpha\gamma$. Since x is arbitrary and $\alpha < 1$, the inequality (6.30) is established with $\lambda := \alpha$.

6.3.2 Numerical Solution

In this section we compute the rational expectations pricing functional p^* numerically via Banach's fixed point theorem. To start, recall that in §6.3.1 we established the existence of a function $p^*: S \to (0, \infty)$ in $\mathscr{C}$ satisfying (6.26). In the proof, p^* was shown to be the fixed point of the pricing operator $T: \mathscr{C} \ni p \mapsto Tp \in \mathscr{C}$. In view of Banach's theorem we have $d_\infty(T^n p, p^*) \to 0$ as $n \to \infty$ for any $p \in \mathscr{C}$, so a natural approach to computing p^* is by iterating on an arbitrary element p of $\mathscr{C}$ (such as P). In doing so, we will need to approximate the iterates $T^n p$ at each step, just as for fitted value iteration (algorithm 6.2, page 136). As before we use linear interpolation, which is nonexpansive with respect to d_∞.

So suppose that $p \in \mathscr{C}$ and $x \in S$ are fixed, and consider the problem of obtaining $Tp(x)$, which, by definition, is the unique $r \in [P(x), v(x)]$ such that (6.28) holds. Regarding this r,

Exercise 6.3.11 Show that $r = P(x)$ whenever $\alpha \int p(z)\phi(z)dz \leq P(x)$.

Exercise 6.3.12 Show that if $\alpha \int p(z)\phi(z)dz > P(x)$, then r satisfies

$$
r = \alpha \int p(\alpha(x - D(r)) + z)\phi(z)dz
$$

Together, exercises 6.3.11 and 6.3.12 suggest the method for finding r presented in algorithm 6.4, which returns $Tp(x)$ given p and x. An implementation of the algorithm is given in listing 6.7. The demand curve D is set to $1/x$, while for the shock we

---

Algorithm 6.4: Computing $Tp(x)$

---

evaluate $y = \alpha \int p(z)\phi(z)dz$
if $y \leq P(x)$ **then** return $P(x)$
else
  define $h(r) = \alpha \int p(\alpha(x - D(r)) + z)\phi(z)dz$
  return the fixed point of h in $[P(x), y]$
end

---

assume that $W_t = a + cB_t$, where B_t is beta with shape parameters $(5, 5)$. The function
`fixed_point()` computes fixed points inside a specified interval using SciPy's `brentq`
root-finding algorithm. The function `T()` implements algorithm 6.4.[23]

Once we can evaluate $Tp(x)$ for each p and x, we can proceed with the iteration
algorithm, as shown in algorithm 6.5. A sequence of iterates starting at P is displayed
in figure 6.13.

Algorithm 6.5: Computing the pricing function

---

set $p = P$
repeat
  sample Tp at finite set of grid points $(x_i)_{i=1}^k$
  use samples to construct linear interpolant q of Tp
  set $p = q$
until *a suitable stopping rule is satisfied*

Exercise 6.3.13 Implement algorithm 6.5 and replicate figure 6.13.[24]

Given p^*, we have a dynamic system for quantities defined by

$$X_{t+1} = \alpha I(X_t) + W_{t+1}, \qquad (W_t)_{t \geq 1} \overset{\text{IID}}{\sim} \phi \qquad (6.31)$$

where $I(x) := x - D(p^*(x))$. As shown in §6.1.3, the distribution ψ_t of X_t is a density
for each $t \geq 1$, and the densities satisfy

$$\psi_{t+1}(y) = \int p(x, y)\psi_t(x)dx \qquad (y \in S)$$

---

[23] Although the `else` statement from algorithm 6.4 is omitted in the definition of `T()`, it is unnecessary
because the last two lines in the definition of `T()` are only executed if the statement `y <= P(x)` is false. Note
also that the function `p()` passed to `T()` must be vectorized.

[24] Hint: You can use the LinInterp class in listing 6.4 (page 138) for interpolation.

Listing 6.7 (cpdynam.py) Computing $Tp(x)$

```python
from scipy import mean
from scipy.stats import beta
from scipy.optimize import brentq

alpha, a, c = 0.8, 5.0, 2.0
W = beta(5, 5).rvs(1000) * c + a      # Shock observations
D = P = lambda x: 1.0 / x

def fix_point(h, lower, upper):
    """Computes the fixed point of h on [upper, lower]
    using SciPy's brentq routine, which finds the
    zeros (roots) of a univariate function.
    Parameters: h is a function and lower and upper are
    numbers (floats or integers).  """
    return brentq(lambda x: x - h(x), lower, upper)

def T(p, x):
    """Computes Tp(x), where T is the pricing functional
    operator.
    Parameters: p is a vectorized function (i.e., acts
    pointwise on arrays) and x is a number.  """
    y = alpha * mean(p(W))
    if y <= P(x):
        return P(x)
    h = lambda r: alpha * mean(p(alpha*(x - D(r)) + W))
    return fix_point(h, P(x), y)
```

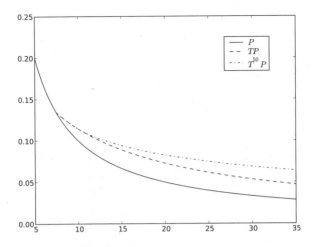

Figure 6.13 The trajectory $T^n P$

where $p(x,y) = \phi(y - \alpha I(x))$, and ϕ is the density of the harvest W_t.[25]

Exercise 6.3.14 We will see later that the process (6.31) is stable, with a unique stationary density ψ^*, and that the look-ahead estimator given in (6.12) on page 131 can be used to estimate it. Using this estimator, show graphically that for these particular parameter values, speculators do not affect long-run probabilities for the state, in the sense that $\psi^* \cong \phi$.[26]

6.3.3 Equilibria and Optima

In §6.3.1 we used Banach's fixed point theorem to show the existence of a pricing functional p^* such that the resulting system for prices and quantities was a competitive equilibrium. There is another way we can obtain the same result using dynamic programming and the optimal growth model. Solving the problem this way illustrates one of the many fascinating links between decentralized equilibria and optimality.

Before getting started we are going to complicate the commodity pricing model slightly by removing the assumption that the interest rate is zero. With a positive and constant interest rate r, next period returns must be discounted by $\rho := 1/(1 + r)$. As

---

[25]We regard ϕ as defined on all of $\mathbb{R}$, and zero off its support. Hence, if $y < \alpha I(x)$, then $p(x,y) = 0$.
[26]The latter is the distribution that prevails without speculation.

a result the no arbitrage and profit maximization conditions (6.22) and (6.23) become

$$\rho \alpha \mathbb{E}_t p_{t+1} - p_t \leq 0 \tag{6.32}$$

$$\rho \alpha \mathbb{E}_t p_{t+1} - p_t < 0 \text{ implies } I_t = 0 \tag{6.33}$$

As in §6.3.1 we seek a pricing function p^* such that the system

$$p_t = p^*(X_t), \quad I_t = X_t - D(p_t), \quad X_{t+1} = \alpha I_t + W_{t+1} \tag{6.34}$$

satisfies (6.32) and (6.33). This can be accomplished directly using the fixed point arguments in §6.3.1, making only minor adjustments to incorporate ρ. However, instead of going down this path we will introduce a fictitious planner who solves the optimal growth model described in §6.2.1. Through a suitable choice of primitives, we show that the resulting optimal policy can be used to obtain such a p^*. Since the optimal policy exists and can be calculated, p^* likewise exists and can be calculated.

Regarding the planner's primitives, the production function f is given by $f(k,z) = \alpha k + z$; the discount factor is $\rho := 1/(1+r)$; the utility function U is defined by $U(c) := \int_0^c P(x)dx$, where P is the inverse demand function for the commodity pricing model; and the distribution ϕ of the shock is the distribution of the harvest.

We assume that P is such that U is bounded on $\mathbb{R}_+$. By the fundamental theorem of calculus we have $U' = P$. The conditions of assumption 6.3.1 on page 145 also hold, and as a result the function U is strictly increasing, strictly concave and satisfies $U'(c) \uparrow \infty$ as $c \downarrow 0$. We remove the assumption in §6.3.1 that the shock is bounded away from zero, as this restriction is not needed.

Using the arguments in §6.2.1, we know that there exists at least one optimal policy. In fact concavity of the primitives implies that there is only one such policy. (For the proof, see §12.1.2.) The policy, denoted simply by σ, is v^*-greedy, which is to say that

$$\sigma(x) = \underset{0 \leq k \leq x}{\operatorname{argmax}} \left\{ U(x-k) + \rho \int v^*(f(k,z))\phi(z)dz \right\} \tag{6.35}$$

for all x. One can also show that v^* is differentiable with $(v^*)'(x) = U'(x - \sigma(x))$, and that the objective function in (6.35) is likewise differentiable. Using these facts and taking into account the possibility of a corner solution, it can be shown that σ satisfies

$$U' \circ c(x) \geq \rho \int U' \circ c[f(\sigma(x),z)]f'(\sigma(x),z)\phi(z)dz \qquad \forall x \in S \tag{6.36}$$

Moreover if the inequality is strict at some $x > 0$, then $\sigma(x) = 0$. Here $c(x) := x - \sigma(x)$ and $f'(k,z)$ is the partial derivative of f with respect to k. This is the famous Euler (in)equality, and full proofs of all these claims can be found via propositions 12.1.23 and 12.1.24 in §12.1.2.

The main result of this section is that by setting p^* equal to marginal utility of consumption we obtain an equilibrium pricing functional for the commodity pricing model.

Proposition 6.3.15 *If p^* is defined by $p^*(x) := U' \circ c(x) :=: U'(c(x))$, then the system defined in (6.34) satisfies (6.32) and (6.33).*

Proof. Substituting the definition of p^* into (6.36) we obtain

$$p^*(x) \geq \rho \int p^*[f(\sigma(x), z)]f'(\sigma(x), z)\phi(z)dz \qquad \forall x \in S$$

with strict inequality at x implying that $\sigma(x) = 0$. Using $f(k, z) = \alpha k + z$ this becomes

$$p^*(x) \geq \rho\alpha \int p^*(\alpha\sigma(x) + z)\phi(z)dz \qquad \forall x \in S$$

Now observe that since $c(x) = x - \sigma(x)$, we must have

$$p^*(x) = U'(x - \sigma(x)) = P(x - \sigma(x)) \qquad \forall x \in S$$

$$\therefore \quad D(p^*(x)) = x - \sigma(x) \qquad \forall x \in S$$

Turning this around, we get $\sigma(x) = x - D(p^*(x))$, and the right-hand side is precisely $I(x)$. Hence $\sigma = I$, and we have

$$\rho\alpha \int p^*(\alpha I(x) + z)\phi(z)dz - p^*(x) \leq 0 \qquad \forall x \in S$$

with strict inequality at x implying that $I(x) = 0$. Since this holds for all $x \in S$, it holds at any realization $X_t \in S$, so

$$\rho\alpha \int p^*(\alpha I(X_t) + z)\phi(z)dz - p^*(X_t) \leq 0$$

with strict inequality implying $I(X_t) = 0$. Substituting I_t and p_t from (6.34), and using the fact that

$$\mathbb{E}_t p_{t+1} = \int p^*(\alpha I(X_t) + z)\phi(z)dz$$

as shown in §6.3.1, we obtain (6.32) and (6.33). $\qquad \square$

6.4   Commentary

Further theory and applications of stochastic recursive sequences in economics and finance can be found in Sargent (1987), Stokey and Lucas (1989), Farmer (1999), Duffie

(2001), Miranda and Fackler (2002), Adda and Cooper (2003), or Ljungqvist and Sargent (2004). The simulation-based approach to computing marginal and stationary densities of stochastic recursive sequences in §6.1.3 and §6.1.4 was proposed by Glynn and Henderson (2001), and a detailed analysis of the technique and its properties can be found in Stachurski and Martin (2008).

The one-sector, infinite horizon, stochastic optimal growth model in §6.2 is essentially that of Brock and Mirman (1972). Related treatments can be found in Mirman and Zilcha (1975), Donaldson and Mehra (1983), Stokey and Lucas (1989), Amir (1996), and Williams (2004). A survey of the field is given in Olsen and Roy (2006). For discussion of the stability properties of the optimal growth model, see §12.1.3. The commentary to that chapter contains additional references.

We saw in §6.3.3 that optimal policies for the growth model coincide with the market equilibria of certain decentralized economies. For more discussion of the links between dynamic programming and competitive equilibria, see Stokey and Lucas (1989, ch. 16), or Bewley (2007). Seminal contributions to this area include Samuelson (1971), Lucas and Prescott (1971), Prescott and Mehra (1980), and Brock (1982).

Under some variations to the standard environment (incomplete markets, production externalities, distortionary taxes, etc.), equilibria and optima no longer conincide, and the problem facing the researcher is to find equilibria rather than optimal policies. For a sample of the literature, see Huggett (1993), Aiyagari (1994), Greenwood and Huffman (1995), Rios-Rull (1996), Krusell and Smith (1998), Kubler and Schmedders (2002), Reffett and Morand (2003), Krebs (2004), Datta et al. (2005), Miao (2006), and Angeletos (2007).

The commodity price model studied in §6.3 is originally due to Samuelson (1971), who connected equilibrium outcomes with solutions to dynamic programming problems. Our treatment in §6.3.1 and §6.3.2 follows Deaton and Laroque (1992), who were the first to derived the equilibrium price directly via Banach's fixed point theorem. The technique of iterating on the pricing functional is essentially equivalent to Coleman's algorithm (Coleman 1990). For more on the commodity pricing model, see, for example, Scheinkman and Schectman (1983) or Williams and Wright (1991).

Part II

Advanced Techniques

Chapter 7

Integration

Measure and integration theory are among the foundation stones of modern mathematics, and particularly those fields of concern to us. Measure theory also has a reputation for being difficult, and indeed it is both abstract and complex. However, with a little bit of effort and attention to the exercises, you will find that measure-theoretic arguments start to seem quite natural, and that the theory has a unique beauty of its own.

Before attempting this chapter you should have a good grounding in basic real analysis. Anyone who has solved most of the exercises in appendix A should be up to the task.

7.1 Measure Theory

In this first section we give a brisk introduction to measure theory. The longer proofs are omitted, although a flavor of the arguments is provided. If you read this section carefully you will have a good feel for what measure theory is about, and for why things are done the way that they are.

7.1.1 Lebesgue Measure

To understand integration, we need to know about Lebesgue measure. The basic problem of Lebesgue measure is how to assign to each subset of $\mathbb{R}^k$ (each element of $\mathfrak{P}(\mathbb{R}^k)$) a real number that will represent its "size" (length, area, volume) in the most natural sense of the word.[1] For a set like $(a, b] \subset \mathbb{R}^1$ there is no debate: The

---

[1] Recall that $\mathfrak{P}(A)$ denotes the set of all subsets of the set A.

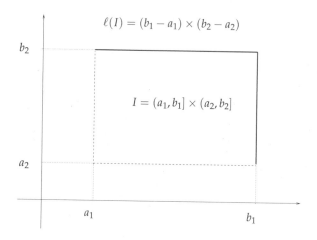

Figure 7.1 The measure of a rectangle $I \subset \mathbb{R}^2$

length is $b - a$. Indeed, for a rectangle such as $\times_{i=1}^{k}(a_i, b_i] = \{x \in \mathbb{R}^k : a_i < x_i \leq b_i,\ i = 1, \ldots, k\}$, the "measure" of this set is the product of the sides $\prod_{i=1}^{k}(b_i - a_i)$. But for an arbitrary set? For example, how large is $\mathbb{Q}$, the set of rational numbers, when taken as a subset of $\mathbb{R}$? And how about the irrational numbers?

A natural approach is to try to extend the notion of size from sets we do know how to measure to sets we don't know how to measure. To begin, let $\mathscr{J}$ be the set of all left-open, right-closed rectangles in $\mathbb{R}^k$:

$$\mathscr{J} := \{\times_{i=1}^{k}(a_i, b_i] \in \mathfrak{P}(\mathbb{R}^k) : a_i, b_i \in \mathbb{R},\ a_i \leq b_i\}$$

Here we admit the possibility that $a_i = b_i$ for some i, in which case the rectangle $\times_{i=1}^{k}(a_i, b_i]$ is the empty set. (Why?) Now let ℓ be the map

$$\ell \colon \mathscr{J} \ni I = \times_{i=1}^{k}(a_i, b_i] \mapsto \ell(I) := \prod_{i=1}^{k}(b_i - a_i) \in \mathbb{R}_+ \tag{7.1}$$

which assigns to each rectangle its natural measure, with $\ell(\varnothing) := 0$ (see figure 7.1). We want to *extend the domain* of ℓ to all of $\mathfrak{P}(\mathbb{R}^k)$. Our extension of ℓ to $\mathfrak{P}(\mathbb{R}^k)$ will be denoted by λ.

The first problem we must address is that there are many possible extensions. For example, $\lambda(A) = 42$ for any $A \notin \mathscr{J}$ is a possible—albeit not very sensible—extension of ℓ. How will we know whether a given extension is the right one?

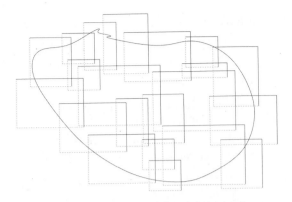

Figure 7.2 A covering of A by elements of $\mathcal{J}$

The solution is to cross-check against our intuition. Our intuition says that size should be nonnegative. Does our extension λ always give nonnegative values? In addition λ should obey the fundamental principle that—at least when it comes to measurement—the whole is the sum of its parts. For example, if $\mathbb{R}^k = \mathbb{R}$ and $A :=$ $(a,b] \cup (c,d]$, where $b \leq c$, then we must have $\lambda(A) = b - a + d - c$. More generally, if A and B are disjoint, then one would expect, from our basic intuition about length or area, that $\lambda(A \cup B) = \lambda(A) + \lambda(B)$. This property is called *additivity*, and we will not be satisfied with our definition of λ unless it holds.

With this in mind, let's go ahead and attempt an extension of ℓ to $\mathfrak{P}(\mathbb{R}^k)$. Given arbitrary $A \in \mathfrak{P}(\mathbb{R}^k)$, let C_A be the set of all *countable covers of A*. That is,

$$C_A := \{(I_n)_{n\geq 1} \subset \mathcal{J} : \cup_n I_n \supset A\}$$

Figure 7.2 shows a (necessarily finite) covering of A by elements of $\mathcal{J}$. Now we define

$$\lambda(A) := \inf \left\{ \sum_{n\geq 1} \ell(I_n) \; : \; (I_n)_{n\geq 1} \in C_A \right\} \qquad (A \in \mathfrak{P}(\mathbb{R}^k)) \qquad (7.2)$$

Thus we are approximating our arbitrary set A by covering it with sets we already know how to measure, and taking the infimum of the value produced by all such covers.[2] The set function λ is called *Lebesgue outer measure*. If $\sum_{n\geq 1} \ell(I_n) = \infty$ for all $(I_n)_{n\geq 1} \in C_A$, then we set $\lambda(A) = \infty$.

Exercise 7.1.1 (Monotonicity) Show that if $A \subset B$, then $\lambda(A) \leq \lambda(B)$.[3]

---

[2]We are using half-open rectangles here, but it turns out that other kinds of rectangles (closed, open, etc.) produce the same number.

[3]Hint: Apply lemma A.2.25 on page 333.

Exercise 7.1.2 (Sub-additivity) Show that if A and B are any two subsets of $\mathbb{R}^k$, then $\lambda(A \cup B) \leq \lambda(A) + \lambda(B)$.[4]

Exercise 7.1.3 Extend sub-additivity to countable sub-additivity if you can: Show that for any $(A_n) \subset \mathfrak{P}(\mathbb{R})$ we have $\lambda(\cup_n A_n) \leq \sum_n \lambda(A_n)$.[5]

Before we go on, we need to consider whether Lebesgue outer measure is actually an extension of ℓ—the function defined on $\mathscr{J}$—to $\mathfrak{P}(\mathbb{R}^k)$. Clearly, λ is a well defined function on $\mathfrak{P}(\mathbb{R}^k)$ (why?), but does it actually agree with the original function ℓ on $\mathscr{J}$? In other words, we need to check that λ assigns "volume" to rectangles.

Lemma 7.1.4 $\lambda: \mathfrak{P}(\mathbb{R}^k) \to [0, \infty]$ *defined in (7.2) agrees with ℓ on $\mathscr{J}$.*

Although the result seems highly likely, the proof is not entirely trivial. It can be found in any text on measure theory.

Now the task is to see whether λ agrees with our intuition in the ways that we discussed above (nonnegativity, additivity, etc.). Nonnegativity is obvious, but when it comes to additivity we run into problems: Additivity *fails*. In 1905 G. Vitali succeeded in constructing sets $A, B \in \mathfrak{P}(\mathbb{R})$ that are so nasty and intertwined that $\lambda(A) + \lambda(B) > \lambda(A \cup B)$.[6]

Well, we said at the start that we were not prepared to accept the validity of our extension unless it preserves additivity. So must (7.2) be abandoned? It might seem so, but no obvious alternatives present themselves.[7] The solution of Henri Lebesgue was to *restrict the domain* of the set function λ to exclude those nasty sets that cause additivity to break down. The method succeeds because the sets remaining in the domain after this exclusion process turn out to be *all those sets we will ever need in day to day analysis.*

The actual restriction most commonly used in modern texts is due to the Greek mathematician Constantin Carathéodory, who considers the class of sets $A \in \mathfrak{P}(\mathbb{R}^k)$ satisfying

$$\lambda(B) = \lambda(B \cap A) + \lambda(B \cap A^c) \quad \text{for all } B \in \mathfrak{P}(\mathbb{R}^k) \tag{7.3}$$

This collection of sets is denoted by $\mathscr{L}$ and called the *Lebesgue measurable sets*. The restriction of λ to $\mathscr{L}$ is called *Lebesgue measure*.

Exercise 7.1.5 Show that $\mathbb{R}^k \in \mathscr{L}$ and $\emptyset \in \mathscr{L}$. Using monotonicity and sub-additivity, show that if $N \subset \mathbb{R}$ and $\lambda(N) = 0$, then $N \in \mathscr{L}$.

---

[4]Hint: Fix $\epsilon > 0$ and choose covers $(I_n^A)_{n \geq 1}$ and $(I_n^B)_{n \geq 1}$ of A and B respectively such that $\sum_n \ell(I_n^A) \leq \lambda(A) + \epsilon/2$ and $\sum_n \ell(I_n^B) \leq \lambda(B) + \epsilon/2$. Now consider $(\cup_n I_n^A) \cup (\cup_n I_n^B)$ as a cover of $A \cup B$.

[5]Hint: Fix $\epsilon > 0$ and choose for each A_n a cover $(I_j^n)_{j \geq 1}$ such that $\sum_j \ell(I_j^n) \leq \lambda(A_n) + \epsilon 2^{-n}$. Now consider $\cup_n \cup_j I_j^n$ as a cover of $\cup_n A_n$.

[6]His construction uses the dreaded Axiom of Choice.

[7]We could try approximating sets from the inside ("inner" measure rather than outer measure), but this is less convenient and it turns out that the same problem reappears.

Our first important observation is that Lebesgue measure is additive on $\mathscr{L}$. In fact one of the central facts of measure theory is that, restricted to this domain, λ is not just additive, but *countably* additive (definition below). This turns out to be crucial when trying to show that λ interacts well with limiting operations. Equally important, $\mathscr{L}$ is *very* large, containing the open sets, closed sets, countable sets, and more.[8]

Let's state without proof the countable additivity property of λ on $\mathscr{L}$.

Theorem 7.1.6 (Countable additivity) *If* $(A_n)_{n \geq 1}$ *is a disjoint sequence in* $\mathscr{L}$, *in that* $A_i \cap A_j = \varnothing$ *whenever* $i \neq j$, *then* $\lambda(\cup_n A_n) = \sum_{n \geq 1} \lambda(A_n)$.[9]

Exercise 7.1.7 Show that countable additivity implies (finite) additivity.[10]

Exercise 7.1.8 Prove that for any $A, B \in \mathscr{L}$ with $A \subset B$ and $\lambda(B) < \infty$ we have $\lambda(B \setminus A) = \lambda(B) - \lambda(A)$.

To learn a bit more about the properties of λ, consider the following exercise: We know that $\{x\} \in \mathscr{L}$ for all $x \in \mathbb{R}^k$ because $\mathscr{L}$ contains all closed sets. Let's show that $\lambda(\{x\}) = 0$ for any point $x \in \mathbb{R}$. It is enough to show that for any $\epsilon > 0$, we can find a sequence $(I_n)_{n \geq 1} \subset \mathscr{J}$ containing x and having $\sum_{n \geq 1} \ell(I_n) \leq \epsilon$. (Why?) So pick such an ϵ, and take a cover $(I_n)_{n \geq 1}$ such that the first rectangle I_1 satisfies $I_1 \ni x$ and $\ell(I_1) \leq \epsilon$, and then $I_n = \varnothing$ for $n \geq 2$. Now $\sum_{n \geq 1} \ell(I_n) \leq \epsilon$ as required.

Exercise 7.1.9 Show that $\lambda(\mathbb{R}^k) = \infty$.[11]

Exercise 7.1.10 Using countable additivity and the fact that singletons $\{x\}$ have zero measure, show that countable sets have zero measure.[12]

You may be wondering if there are uncountable sets of measure zero. The answer is yes. An often cited example is the Cantor set, which can be found in any text on measure theory.

7.1.2 Measurable Spaces

We mentioned above that the set of Lebesgue measurable sets $\mathscr{L}$ contains all of the sets we typically deal with in analysis, In addition it has nice "algebraic" properties. In particular, it is a σ-algebra:

---

[8]For the definition of countable (and uncountable) sets see page 322.

[9]The value $+\infty$ is permitted here. Also, readers might be concerned that for an arbitrary sequence $(A_n)_{n \geq 1} \subset \mathscr{L}$ the union $\cup_n A_n$ may not be in $\mathscr{L}$, in which case $\lambda(\cup_n A_n)$ is not defined. In fact the union is always in $\mathscr{L}$ as we will see below.

[10]Hint: We must show that $\lambda(\cup_{n=1}^N A_n) = \sum_{n=1}^N \lambda(A_n)$ for any finite collection of disjoint sets $(A_n)_{n=1}^N$. Try thinking about the countable sequence $(B_n)_{n \geq 1}$ where $B_n = A_n$ for $n \leq N$ and $B_n = \varnothing$ for $n > N$.

[11]Hint: Using lemma 7.1.4 and monotonicity (exercise 7.1.1), show that $\lambda(\mathbb{R}^k)$ is bigger than any real number.

[12]This implies that in $\mathbb{R}$ we must have $\lambda(\mathbb{Q}) = 0$. By additivity, then, $\lambda(\mathbb{R}) = \lambda(\mathbb{Q}^c)$. In this sense there are "many more" irrational numbers than rational numbers.

Definition 7.1.11 Let S be any nonempty set. A family of sets $\mathscr{S} \subset \mathfrak{P}(S)$ is called a *σ-algebra* if

1. $S \in \mathscr{S}$,

2. $A \in \mathscr{S}$ implies $A^c \in \mathscr{S}$, and

3. if $(A_n)_{n \geq 1}$ is a sequence with A_n in $\mathscr{S}$ for all n, then $\cup_n A_n \in \mathscr{S}$.

The pair $(S, \mathscr{S})$ is called a *measurable space*, and elements of $\mathscr{S}$ are called *measurable sets*. The second and third properties are usually expressed by saying that the collection $\mathscr{S}$ is "closed" or "stable" under complementation and countable unions, in the sense that these operations do not take us outside the collection $\mathscr{S}$. From De Morgan's law $(\cap_n A_n)^c = \cup_n A_n^c$, we see that $\mathscr{S}$ is also stable under countable *intersections*. (Why?) Finally, note that $\varnothing \in \mathscr{S}$.

An example of a *σ-algebra* on S is $\mathfrak{P}(S)$. This is true for every set S. For example, if $A \in \mathfrak{P}(S)$, then $A^c := \{x \in S : x \notin A\}$ is also a subset of S by its very definition. On the other hand, the collection $\mathscr{O}$ of open subsets of $\mathbb{R}$ is not a *σ-algebra* because it is not stable under the taking of complements.

Incidentally, the concept of *σ-algebras* plays a major role in measure theory, and these collections of sets sometimes seem intimidating and abstract to the outsider. But the use of *σ-algebras* is less mysterious than it appears. When working with a *σ-algebra*, we know that if we start with some measurable sets, take unions, then complements, then intersections, and so on, the new sets we create are still measurable sets. By definition, *σ-algebras* are stable under the familiar set operations, so we need not worry that we will leave our safe environment when using these standard operations.

Exercise 7.1.12 Check that $\{\varnothing, S\}$ is a *σ-algebra* on S for any set S. Show that, on the other hand, $\mathscr{J}$ is not a *σ-algebra* on $\mathbb{R}^k$.

Exercise 7.1.13 If $\{\mathscr{S}_\alpha\}_{\alpha \in \Lambda}$ is any collection of *σ-algebras* on S, then their intersection $\cap_\alpha \mathscr{S}_\alpha$ is all $B \subset S$ such that $B \in \mathscr{S}_\alpha$ for every $\alpha \in \Lambda$. Show that $\cap_\alpha \mathscr{S}_\alpha$ is itself a *σ-algebra* on S.

One of the most common ways to define a particular *σ-algebra* is to take a collection $\mathscr{C}$ of subsets of S, and consider the smallest *σ-algebra* that contains this collection.

Definition 7.1.14 If S is any set and $\mathscr{C}$ is any collection of subsets of S, then the *σ-algebra generated by* $\mathscr{C}$ is the smallest *σ-algebra* on S that contains $\mathscr{C}$, and is denoted by $\sigma(\mathscr{C})$. More precisely, $\sigma(\mathscr{C})$ is the intersection of all *σ-algebras* on S that contain $\mathscr{C}$. In general, if $\sigma(\mathscr{C}) = \mathscr{S}$, then $\mathscr{C}$ is called a *generating class* for $\mathscr{S}$.

Exercise 7.1.15 Show that if $\mathscr{C}$ is a σ-algebra, then $\sigma(\mathscr{C}) = \mathscr{C}$, and that if $\mathscr{C}$ and $\mathscr{D}$ are two collections of sets with $\mathscr{C} \subset \mathscr{D}$, then $\sigma(\mathscr{C}) \subset \sigma(\mathscr{D})$.

Now let's return to the Lebesgue measurable sets. As discussed above, it can be shown that $\mathscr{L}$ is a σ-algebra, and that it contains all the sets we need in day-to-day analysis.[13] In fact $\mathscr{L}$ contains more sets than we actually need, and is not easily abstracted to more general spaces. As a result we will almost invariably work with a smaller domain called the Borel sets, and denoted $\mathscr{B}(\mathbb{R}^k)$. The collection $\mathscr{B}(\mathbb{R}^k)$ is just the σ-algebra generated by the open subsets of $\mathbb{R}^k$.[14] More generally,

Definition 7.1.16 Let S be any metric space. The *Borel sets* on S are the sets $\mathscr{B}(S) := \sigma(\mathscr{O})$, where $\mathscr{O}$ is the open subsets of S.

Exercise 7.1.17 Explain why $\mathscr{B}(S)$ must contain the closed subsets of S.

In the case of $S = \mathbb{R}^k$ this collection $\mathscr{B}(\mathbb{R}^k)$ is surprisingly large. In fact it is quite difficult to construct a subset of $\mathbb{R}^k$ that is not in $\mathscr{B}(\mathbb{R}^k)$. Moreover $\mathscr{B}(\mathbb{R}^k)$ is a subset of $\mathscr{L}$, and hence all of the nice properties that λ has on $\mathscr{L}$ it also has on $\mathscr{B}(\mathbb{R}^k)$. In particular, λ is countably additive on $\mathscr{B}(\mathbb{R}^k)$.

Exercise 7.1.18 Show that $\mathbb{Q} \in \mathscr{B}(\mathbb{R})$.[15]

The following theorem gives some indication as to why the Borel sets are so natural and important to analysis.

Theorem 7.1.19 *Let $\mathscr{O}, \mathscr{C}$ and $\mathscr{K}$ be the open, closed and compact subsets of $\mathbb{R}^k$ respectively. We have*

$$\mathscr{B}(\mathbb{R}^k) := \sigma(\mathscr{O}) = \sigma(\mathscr{C}) = \sigma(\mathscr{K}) = \sigma(\mathscr{J})$$

Let's just show that $\mathscr{B}(\mathbb{R}^k) = \sigma(\mathscr{K})$. To see that $\mathscr{B}(\mathbb{R}^k) \supset \sigma(\mathscr{K})$, note that $\mathscr{B}(\mathbb{R}^k)$ is a σ-algebra containing all the closed sets, and hence all the compact sets. (Why?) In which case it also contains $\sigma(\mathscr{K})$. (Why?) To show that $\mathscr{B}(\mathbb{R}^k) \subset \sigma(\mathscr{K})$, it suffices to prove that $\sigma(\mathscr{K})$ contains $\mathscr{C}$. (Why?) To see that $\sigma(\mathscr{K})$ contains $\mathscr{C}$, pick any $C \in \mathscr{C}$ and let $D_n := \{x \in \mathbb{R}^k : \|x\| \leq n\}$. Observe that $C_n := C \cap D_n \in \mathscr{K}$ for every $n \in \mathbb{N}$ (why?), and that $C = \cup_n C_n$. Since $C_n \in \mathscr{K}$ for all n, we have $C = \cup_n C_n \in \sigma(\mathscr{K})$, as was to be shown.

Exercise 7.1.20 Let $\mathscr{A}$ be the set of all open intervals $(a, b) \subset \mathbb{R}$. Show that $\sigma(\mathscr{A}) = \mathscr{B}(\mathbb{R})$.[16]

---

[13]To prove that $\mathscr{L}$ is a σ-algebra, one can easily check from the definition (7.3) that $\mathbb{R} \in \mathscr{L}$ and that $A \in \mathscr{L}$ implies $A^c \in \mathscr{L}$. To show that $\mathscr{L}$ is stable under countable unions is a bit more subtle and the proof is omitted.

[14]The open subsets of $\mathbb{R}^k$ are determined by the euclidean metric d_2. But any metric defined from a norm on $\mathbb{R}^k$ gives us the same open sets (theorem 3.2.30), and hence the same Borel sets.

[15]Hint: Singletons are closed, and $\mathbb{Q}$ = a countable union of singletons.

[16]You should be able to show that $\sigma(\mathscr{A}) \subset \sigma(\mathscr{O})$. To show that $\sigma(\mathscr{O}) \subset \sigma(\mathscr{A})$ it is sufficient to prove that

7.1.3   General Measures and Probabilities

Measure theory forms the foundations of modern probability. Before we study this in earnest, let's think a little bit about why it might be fruitful to generalize the concept of "measures." To start, imagine a space S, which you might like to visualize as a subset of the plane $\mathbb{R}^2$. Like a satellite photograph of the earth at night, the space is sprinkled with lots of tiny glowing particles. If we take a region E of the space S, you might ask how many of these particles are contained in E, or alternatively, what fraction of the total quantity of the particles are in E?

Let μ be a set function on $\mathfrak{P}(S)$ such that $\mu(E)$ is the fraction of particles in E. It seems that μ is going to be nonnegative, monotone ($E \subset F$ implies $\mu(E) \leq \mu(F)$), and additive (E, F disjoint implies $\mu(E \cup F) = \mu(E) \cup \mu(F)$), just as the Lebesgue measure λ was. So perhaps μ is also some kind of "measure," and can be given a neat treatment by using similar ideas.

These considerations motivate us to generalize the notion of Lebesgue measure with an abstract definition. As is always the case in mathematics, abstracting in a clever way will save us saying things over and over again, and lead to new insights.

Definition 7.1.21 Let $(S, \mathscr{S})$ be a measurable space. A *measure* μ on $(S, \mathscr{S})$ is a function from $\mathscr{S}$ to $[0, \infty]$ such that

1. $\mu(\varnothing) = 0$, and

2. μ is countably additive: If $(A_n) \subset \mathscr{S}$ is disjoint, then $\mu(\cup_n A_n) = \sum_n \mu(A_n)$.

The triple $(S, \mathscr{S}, \mu)$ is called a *measure space*.

Exercise 7.1.22 Show that if there exists an $A \in \mathscr{S}$ with $\mu(A) < \infty$, then (2) implies (1).[17]

Exercise 7.1.23 Show that (2) implies monotonicity: If $E, F \in \mathscr{S}$ and $E \subset F$, then $\mu(E) \leq \mu(F)$.

Exercise 7.1.24 Show that a measure μ on S is always sub-additive: If A and B are any elements of $\mathscr{S}$ (disjoint or otherwise), then $\mu(A \cup B) \leq \mu(A) + \mu(B)$.

Let $(A_n)_{n \geq 1} \subset \mathscr{S}$ have the property that $A_{n+1} \subset A_n$ for all $n \in \mathbb{N}$, and let $A := \cap_n A_n$. We say that the sequence $(A_n)_{n \geq 1}$ decreases down to A, and write $A_n \downarrow A$. On the other hand, if $A_{n+1} \supset A_n$ for all $n \in \mathbb{N}$ and $A := \cup_n A_n$, then we say that $(A_n)_{n \geq 1}$ increases up to A, and write $A_n \uparrow A$. For such "monotone" sequences of sets, an arbitrary measure μ on $(S, \mathscr{S})$ has certain continuity properties, as detailed in the next exercise.

$\sigma(\mathscr{A})$ contains the open sets. (Why?) To do so, use the fact that every open subset of $\mathbb{R}$ can be expressed as the countable union of open intervals.

[17]In this sense (1) is just to rule out trivial cases, and (2) is the real condition of interest.

Exercise 7.1.25 Let $(A_n)_{n \geq 1}$ be a sequence in $\mathscr{S}$. Show that

1. if $A_n \uparrow A$, then $\mu(A_n) \uparrow \mu(A)$,[18] and

2. if $\mu(A_1) < \infty$ and $A_n \downarrow A$, then $\mu(A_n) \downarrow \mu(A)$.[19]

Example 7.1.26 Consider the measurable space $(\mathbb{N}, \mathfrak{P}(\mathbb{N}))$ formed by the set of natural numbers $\mathbb{N}$ and the set of all its subsets. On this measurable space define the *counting measure* c, where $c(B)$ is simply the number of elements in B, or $+\infty$ if B is infinite. With some thought you will be able to convince yourself that c is a measure on $(\mathbb{N}, \mathfrak{P}(\mathbb{N}))$.[20]

Exercise 7.1.27 Let $(a_n) \subset \mathbb{R}_+$. Let μ be defined on $(\mathbb{N}, \mathfrak{P}(\mathbb{N}))$ by $\mu(A) = \sum_{n \in A} a_n$. Show that μ is a measure on $(\mathbb{N}, \mathfrak{P}(\mathbb{N}))$.

Let's now specialize this to "probability measures," which are the most important kind of measures for us after Lebesgue measure. In what follows $(S, \mathscr{S})$ is any measurable space.

Definition 7.1.28 A *probability measure* μ is a measure on $(S, \mathscr{S})$ such that $\mu(S) = 1$. The triple $(S, \mathscr{S}, \mu)$ is called a *probability space*. The set of all probability measures on $(S, \mathscr{S})$ is denoted by $\mathscr{P}(S, \mathscr{S})$. When S is a metric space, elements of $\mathscr{P}(S, \mathscr{B}(S))$ are called *Borel probability measures*. For brevity we will write $\mathscr{P}(S, \mathscr{B}(S))$ as $\mathscr{P}(S)$.

In the context of probability theory, a set E in $\mathscr{S}$ is usually called an *event*, and $\mu(E)$ is interpreted as the probability that when uncertainty is realized the event E occurs. Informally, $\mu(E)$ is the probability that $x \in E$ when x is drawn from the set S according to μ. The empty set $\varnothing$ is called the *impossible event*, and S is called the *certain event*.

Defining μ on a σ-algebra works well in terms of probabilistic intuition. For example, the impossible and certain events are always in $\mathscr{S}$. Also, we want E^c to be in $\mathscr{S}$ whenever E is: If the probability of E occurring is defined, then so is the probability of E not occurring. And if E and F are in $\mathscr{S}$, we want $E \cap F \in \mathscr{S}$ so that we can talk about the probability that both E and F occur, and so forth. These properties are assured by the definition of σ-algebras.

Given measurable space $(S, \mathscr{S})$ and point $x \in S$, the *Dirac probability measure* $\delta_x \in \mathscr{P}(S, \mathscr{S})$ is the distribution that puts all mass on x. More formally, $\delta_x(A) = \mathbb{1}_A(x)$ for all $A \in \mathscr{S}$.

Exercise 7.1.29 Confirm that δ_x is a probability measure on $(S, \mathscr{S})$.

---

[18]Hint: Let $B_1 = A_1$ and $B_n = A_n \setminus A_{n-1}$ for $n \geq 2$. Show that (B_n) is a disjoint sequence with $\cup_{n=1}^k B_n = A_k$ and $\cup_n B_n = A$. Now apply countable additivity.

[19]Hint: Use part 1 of the exercise.

[20]When we come to integration, series of real numbers will be presented as integrals of the counting measure. Repackaging summation as integration may not sound very useful, but actually it leads to a handy set of results on passing limits through infinite sums.

7.1.4 Existence of Measures

Suppose that we release at time zero a large number of identical, tiny particles into water at a location defined as zero, and then measure the horizontal distance of the particles from the origin at later point in time t. As Albert Einstein pointed out, the independent action of the water molecules on the particles and the central limit theorem tell us that, at least in this idealized setting, and at least for $E = (a, b] \in \mathscr{J}$, the fraction of the total mass contained in E should now be approximately

$$\mu(E) = \mu((a, b]) = \int_a^b \frac{1}{\sqrt{2\pi t}} \exp \frac{-x^2}{2t} dx \tag{7.4}$$

One can think of $\mu(E)$ as the probability that an individual particle finds itself in E at time t.

In (7.4) we have a way of computing probabilities for intervals $(a, b] \in \mathscr{J}$, but no obvious way of measuring the probability of more complex events. For example, what is $\mu(\mathbb{Q})$, where $\mathbb{Q}$ is the rational numbers? Measuring intervals is all well and good, but there will certainly be times that we want to assign probabilities to more complex sets. In fact *we need to do this to develop a reasonable theory of integration.*

How to extend μ from $\mathscr{J}$ to a larger class of subsets of $\mathbb{R}$? Taking our cue from the process for Lebesgue measure, we could assign probability to an arbitrary set A by setting

$$\mu^*(A) := \inf \sum_{n \geq 1} \mu(I_n) = \inf \sum_{n \geq 1} \int_{a_n}^{b_n} \frac{1}{\sqrt{2\pi t}} \exp \frac{-x^2}{2t} dx \qquad (A \subset \mathbb{R}) \tag{7.5}$$

where the infimum is over all sequences of intervals $(I_n)_{n \geq 1}$, with $I_n := (a_n, b_n] \in \mathscr{J}$ for each n, and with the sequence covering A (i.e., $\cup_n I_n \supset A$). This extension would be suitable if (1) it agrees with μ on $\mathscr{J}$, and (2) it is a measure, at least when restricted to a nice subset of $\mathfrak{P}(\mathbb{R})$ such as $\mathscr{B}(\mathbb{R})$. Being a measure implies attractive properties such as nonnegativity, monotonicity and additivity.[21]

Instead of dealing directly with μ, let's look at this extension process in an abstract setting. As with all abstraction, the advantage is that we can cover a lot of cases in one go. The disadvantage is that the statement of results is quite technical. It is provided mainly for reference purposes, rather than as material to be worked through step by step. The idea is to formulate a system for constructing measures out of set functions that behave like measures (i.e., are countably additive) on small, concrete classes of sets called semi-rings.

---

[21] For probabilities, monotonicity should be interpreted as follows: $A \subset B$ means that whenever A happens, B also happens. In which case B should be at least as likely to occur as A, or $\mu^*(A) \leq \mu^*(B)$. Additivity is familiar from elementary probability. For example, the probability of getting an even number when you roll a dice is the probability of getting a 2 plus that of getting a 4 plus that of getting a 6. (Note that monotonicity is implied by nonnegativity and additivity—see exercise 7.1.23).

Definition 7.1.30 Let S be a nonempty set. A nonempty collection of subsets $\mathscr{R}$ is called a *semi-ring* if, given arbitrary sets I and J in $\mathscr{R}$, we have

1. $\varnothing \in \mathscr{R}$,

2. $I \cap J \in \mathscr{R}$, and

3. $I \setminus J$ can be expressed as a finite union of elements of $\mathscr{R}$.

The definition is not particularly attractive, but all we need to know at this stage is that $\mathscr{J}$, the half-open rectangles $\times_{i=1}^{k}(a_i, b_i]$ in $\mathbb{R}^k$, form a semi-ring. Although the proof is omitted, a little thought will convince you that $\mathscr{J}$ is a semi-ring when $k = 1$. You might like to give a sketch of the proof for the case of $\mathbb{R}^2$ by drawing pictures.

We now give a general result for existence of measures. Let S be any nonempty set, and let $\mathscr{R}$ be a semi-ring on S. A set function $\mu : \mathscr{R} \to [0, \infty]$ is called a *pre-measure* on $\mathscr{R}$ if $\mu(\varnothing) = 0$ and $\mu(\cup_n I_n) = \sum_n \mu(I_n)$ for any disjoint sequence $(I_n)_{n \geq 1} \subset \mathscr{R}$ with $\cup_n I_n \in \mathscr{R}$. For any $A \subset S$, let C_A be the set of all countable covers of A formed from elements of $\mathscr{R}$. That is,

$$C_A := \{(I_n)_{n \geq 1} \subset \mathscr{R} : \cup_n I_n \supset A\}$$

Now define the *outer measure* generated by μ as

$$\mu^*(A) := \inf \left\{ \sum_{n \geq 1} \mu(I_n) \ : \ (I_n)_{n \geq 1} \in C_A \right\} \qquad (A \in \mathfrak{P}(S)) \tag{7.6}$$

The restriction of μ^* to $\sigma(\mathscr{R})$ turns out to be a measure, and is typically denoted simply by μ. Formally,

Theorem 7.1.31 *Let S be any nonempty set and let $\mathscr{R}$ be a semi-ring on S. If μ is a pre-measure on $\mathscr{R}$, then the outer measure (7.6) agrees with μ on $\mathscr{R}$ and is a measure on $(S, \sigma(\mathscr{R}))$. If there exists a sequence $(I_n) \subset \mathscr{R}$ with $\cup_n I_n = S$ and $\mu(I_n) < \infty$ for all n, then the extension is unique in the sense that if ν is any other pre-measure that agrees with μ on $\mathscr{R}$, then its extension agrees with μ on all of $\sigma(\mathscr{R})$.*

The proof is quite similar to the construction of Lebesgue measure sketched in §7.1.1. First one defines the outer measure μ^* by (7.6). Since μ^* is not necessarily additive over all of $\mathfrak{P}(S)$ (think of the case of Lebesgue measure), we then restrict attention to those sets $\mathscr{S}$ that satisfy Carathéodory's condition: All $A \in \mathfrak{P}(S)$ such that

$$\mu^*(B) = \mu^*(B \cap A) + \mu^*(B \cap A^c) \quad \text{for all } B \in \mathfrak{P}(S) \tag{7.7}$$

It can be proved that $\mathscr{S}$ is a σ-algebra containing $\mathscr{R}$, and that μ^* is a measure on $\mathscr{S}$. Evidently $\sigma(\mathscr{R}) \subset \mathscr{S}$ (why?), and the restriction of μ^* to $\sigma(\mathscr{R})$ is simply denoted by μ.

Let's consider applications of theorem 7.1.31. One is Lebesgue measure on $\mathbb{R}^k$. For the semi-ring we take $\mathscr{J}$. It can be shown that ℓ defined on $\mathscr{J}$ by (7.1) on page 160 is a pre-measure. As a result ℓ extends uniquely to a measure on $\sigma(\mathscr{J}) = \mathscr{B}(\mathbb{R}^k)$. This gives us Lebesgue measure on $\mathscr{B}(\mathbb{R}^k)$.

A second application is probabilities on $\mathbb{R}$, such as the Gaussian probability defined in (7.4). Recall that $F \colon \mathbb{R} \to \mathbb{R}$ is called a *cumulative distribution function* on $\mathbb{R}$ if it is nonnegative, increasing, right-continuous, and satisfies $\lim_{x \to -\infty} F(x) = 0$ and $\lim_{x \to \infty} F(x) = 1$. We imagine $F(x)$ represents the probability that random variable X takes values in $(-\infty, x]$. More generally, for interval $(a, b]$ the probability that $X \in (a, b]$ is given by $F(b) - F(a)$.

Fix any distribution function F, and let $\mathscr{J}$ be the semi-ring of all intervals $(a, b]$, where $a \leq b$. Although the proof is not trivial, one can show using the properties of F that $\mu_F \colon \mathscr{J} \to \mathbb{R}_+$ defined by

$$\mu_F((a, b]) = F(b) - F(a)$$

is a pre-measure on $\mathscr{J}$. Clearly, there exists a sequence $(I_n) \subset \mathscr{J}$ with $\cup_n I_n = \mathbb{R}$ and $\mu_F(I_n) < \infty$ for all n. As a result there exists a unique extension of μ_F to a measure on $\sigma(\mathscr{J}) = \mathscr{B}(\mathbb{R})$ with $\mu_F(A) := \inf \sum_{n \geq 1}(F(b_n) - F(a_n))$ for all $A \in \mathscr{B}(\mathbb{R})$. The infimum is over all sequences of intervals (I_n) with $I_n := (a_n, b_n] \in \mathscr{J}$ for each n and $\cup_n I_n \supset A$.

It follows that to each cumulative distribution function on $\mathbb{R}$ there corresponds a unique Borel probability measure. Conversely, suppose that $\mu \in \mathscr{P}(\mathbb{R})$, and let F be defined by $F(x) = \mu((-\infty, x])$ for $x \in \mathbb{R}$.

Exercise 7.1.32 Show that F is a cumulative distribution function on $\mathbb{R}$.

Putting this together and filling in some details, one can show that

Theorem 7.1.33 *There is a one-to-one pairing between the collection of all distribution functions on $\mathbb{R}$ and $\mathscr{P}(\mathbb{R})$, the set of all Borel probability measures on $\mathbb{R}$. If F is a distribution function, then the corresponding probability μ_F satisfies*

$$\mu_F((-\infty, x]) = F(x) \qquad (x \in \mathbb{R})$$

Let's look back and see what we have accomplished. The main result here is theorem 7.1.31, which helps us construct measures. To understand it's importance, suppose that we propose a would-be measure such as (7.4). To check that this is indeed a measure on the Borel sets is a tough ask. After all, what does an arbitrary Borel set look like? But to show that $\mu((a, b]) = \int_a^b (2\pi t)^{-1/2} e^{-x^2/2t} dx$ is a pre-measure on the nice semi-ring $\mathscr{J}$ of intervals $(a, b]$ is much easier. Once this is done theorem 7.1.31 can be applied.

7.2   Definition of the Integral

Elementary calculus courses use the Riemann definition of integrals. As a result of its construction the Riemann integral is inherently limited, in terms of both its domain of definition and its ability to cope with limiting arguments. We want to construct an integral that *extends* the Riemann integral to a wider domain, and has nice analytical properties to boot. With this goal in mind, let's start to develop a different theory of integration (the Lebesgue theory), beginning with the case of functions from $\mathbb{R}$ to $\mathbb{R}$, and working up to more abstract settings.

7.2.1   Integrating Simple Functions

Let's start with the easiest case. A *simple function* is any real-valued function taking only finitely many different values. Consider a simple function $s \colon \mathbb{R} \to \mathbb{R}$ that takes values $\alpha_1, \ldots, \alpha_N$ on a corresponding disjoint intervals $I_1, \ldots, I_N$ in $\mathscr{J}$. Exploiting the assumption that the intervals are disjoint, the function s can be expressed as a linear combination of indicator functions: $s = \sum_{n=1}^N \alpha_n \mathbb{1}_{I_n}$.[22] Take a moment to convince yourself of this.

Since our integral is to be an extension of the Riemann integral, and since the Riemann integral of s is well defined and equal to the sum over n of α_n times the length of I_n, the Lebesgue integral must also be

$$\lambda(s) :=: \int s d\lambda := \sum_{n=1}^N \alpha_n (b_n - a_n) = \sum_{n=1}^N \alpha_n \lambda(I_n) \tag{7.8}$$

where λ on the right-hand side is the Lebesgue measure.

The symbol $\int s d\lambda$ is reminiscent of the traditional notation $\int s(x) dx$. Using $d\lambda$ reminds us that we are integrating with respect to Lebesgue measure. Later, more general integrals are defined. The alternative notation $\lambda(s)$ for the integral of function s with respect to measure λ is also common. It reminds us that we are defining a map that sends functions into numbers.

Little effort is needed to shift our theory up from $\mathbb{R}$ to $\mathbb{R}^k$. For a function $s \colon \mathbb{R}^k \to \mathbb{R}$ defined by $s = \sum_{n=1}^N \alpha_n \mathbb{1}_{I_n}$, where each $I_n = \times_{i=1}^k (a_i, b_i]$ is an element of $\mathscr{J} \subset \mathfrak{P}(\mathbb{R}^k)$ and the rectangles are disjoint, we set

$$\lambda(s) :=: \int s d\lambda := \sum_{n=1}^N \alpha_n \lambda(I_n) \tag{7.9}$$

Here λ on the right-hand side is Lebesgue measure on $\mathbb{R}^k$. Figure 7.3 shows an example of such a function s on $\mathbb{R}^2$.

---

[22]In other words, $s(x) = \sum_{n=1}^N \alpha_n \mathbb{1}_{I_n}(x)$ for every $x \in \mathbb{R}$, where $\mathbb{1}_{I_n}(x) = 1$ if $x \in I_n$ and zero otherwise.

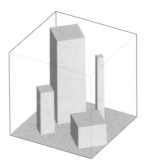

Figure 7.3 A simple function on the plane

Having defined an integral for simple functions that are constant on rectangles, the next step is to extend the definition to the $\mathscr{B}(\mathbb{R}^k)$-*simple* functions $s\mathscr{B}(\mathbb{R}^k)$, each of which takes only finitely many values, but on *Borel* sets rather than just rectangles. More succinctly, $s\mathscr{B}(\mathbb{R}^k)$ is all functions of the form $\sum_{n=1}^N \alpha_n \mathbb{1}_{B_n}$, where the B_n's are disjoint Borel sets. For now let's think about nonnegative simple functions ($\alpha_n \geq 0$ for all n), the set of which we denote $s\mathscr{B}(\mathbb{R}^k)^+$. A natural extension of our integral (7.9) to $s\mathscr{B}(\mathbb{R}^k)^+$ is given by

$$\lambda(s) :=: \int s d\lambda := \sum_{n=1}^N \alpha_n \lambda(B_n) \tag{7.10}$$

This is already a generalization of the Riemann integral. For example, the Riemann integral is not defined for $\mathbb{1}_{\mathbb{Q}}$, which is an element of $s\mathscr{B}(\mathbb{R})^+$. Note also that $\lambda(s) = \infty$ is a possibility, and we do not exclude this case.

Exercise 7.2.1 Explain why the integral of $\mathbb{1}_{\mathbb{Q}}$ is zero.

So far we have defined integrals of (finite-range) functions defined over $\mathbb{R}^k$, where integration was with respect to Lebesgue measure. Next, just as we abstracted from Lebesgue measure to arbitrary measures, let us now introduce integrals of simple functions using general measures.

Suppose that we have a measure μ on an arbitrary measurable space $(S, \mathscr{S})$. We can define the real-valued simple functions $s\mathscr{S}$ on $(S, \mathscr{S})$ in the same way that we defined the Borel simple functions $s\mathscr{B}(\mathbb{R}^k)$ on $\mathbb{R}^k$, replacing $\mathscr{B}(\mathbb{R}^k)$ with $\mathscr{S}$ in the definition. In other words, $s\mathscr{S}$ is those functions of the form $s = \sum_{n=1}^{N} \alpha_n \mathbb{1}_{A_n}$, where the sets $A_1, \ldots, A_N$ are disjoint and $A_n \in \mathscr{S}$ for all n. The set $s\mathscr{S}^+$ is the nonnegative functions in $s\mathscr{S}$.

By direct analogy with (7.10), the integral of $s \in s\mathscr{S}^+$ is defined as

$$\mu(s) :=: \int s\,d\mu := \sum_{n=1}^{N} \alpha_n \mu(A_n) \tag{7.11}$$

To give an illustration of (7.11), consider an experiment where a point ω is selected from some set Ω according to probability measure $\mathbb{P}$. Here $\mathbb{P}$ is defined on some σ-algebra $\mathscr{F}$ of subsets of Ω, and $\mathbb{P}(E)$ is interpreted as the probability that $\omega \in E$ for each $E \in \mathscr{F}$. Suppose that we have a discrete random variable X taking $\omega \in \Omega$ and sending it into one of N values. Specifically, X sends points in $A_n \in \mathscr{F}$ into $\alpha_n \in \mathbb{R}$, where $A_1, \ldots, A_N$ is a partition of Ω. Intuitively, the expectation of X is then

$$\sum_{n=1}^{N} \alpha_n \operatorname{Prob}\{X = \alpha_n\} = \sum_{n=1}^{N} \alpha_n \operatorname{Prob}\{\omega \in A_n\} = \sum_{n=1}^{N} \alpha_n \mathbb{P}(A_n)$$

Comparing the right-hand side of this expression with (7.11), it becomes clear that the expectation of X is precisely the integral $\mathbb{P}(X) :=: \int X\,d\mathbb{P}$. A more traditional notation is $\mathbb{E}X$. We will come back to expectations later on.

Returning to general $(S, \mathscr{S}, \mu)$, integrals of simple functions have some useful properties.

Proposition 7.2.2 *For $s, s' \in s\mathscr{S}^+$ and $\gamma \geq 0$, the following properties hold:*

1. *$\gamma s \in s\mathscr{S}^+$ and $\mu(\gamma s) = \gamma \mu(s)$.*

2. *$s + s' \in s\mathscr{S}^+$ and $\mu(s + s') = \mu(s) + \mu(s')$.*

3. *If $s \leq s'$ pointwise on S, then $\mu(s) \leq \mu(s')$.*

We say that on $s\mathscr{S}^+$ the integral μ is positive homogeneous, additive, and monotone respectively. The full proofs are a bit messy and not terribly exciting so we omit them.

Exercise 7.2.3 Prove part 1 of proposition 7.2.2. Prove parts 2 and 3 in the special case where $s = \alpha \mathbb{1}_A$ and $s' = \beta \mathbb{1}_B$.

7.2.2 Measurable Functions

So far we have extended the integral to $s\mathscr{S}^+$. This is already quite a large class of functions. The next step is to extend it further by a limiting operation (a method of definition so common in analysis!). To do this, we need to define a class of functions that can be approximated well by simple functions. This motivates the definition of a measurable function:

Definition 7.2.4 Let $(S, \mathscr{S})$ and $(R, \mathscr{R})$ be two measurable spaces, and let $f \colon S \to R$. The function f is called $\mathscr{S}, \mathscr{R}$-measurable if $f^{-1}(B) \in \mathscr{S}$ for all $B \in \mathscr{R}$. If $(R, \mathscr{R}) = (\mathbb{R}, \mathscr{B}(\mathbb{R}))$, then f is called $\mathscr{S}$-measurable. If, in addition, S is a metric space and $\mathscr{S} = \mathscr{B}(S)$, then f is called *Borel measurable*.

While this definition is very succinct, it is also rather abstract, and the implications of measurability are not immediately obvious. However, we will see that—for the kinds of functions we want to integrate—measurability of a function f is equivalent to the existence of a sequence of simple functions $(s_n)_{n \geq 1}$ that converges to f in a suitable way (see lemma 7.2.11 below). We will then be able to define the integral of f as the limit of the integrals of the sequence $(s_n)_{n \geq 1}$.

Exercise 7.2.5 Show that if $(S_1, \mathscr{S}_1)$, $(S_2, \mathscr{S}_2)$ and $(S_3, \mathscr{S}_3)$ are any three measurable spaces, $f \colon S_1 \to S_2$ is $\mathscr{S}_1, \mathscr{S}_2$-measurable and $g \colon S_2 \to S_3$ is $\mathscr{S}_2, \mathscr{S}_3$-measurable, then $h := g \circ f \colon S_1 \to S_3$ is $\mathscr{S}_1, \mathscr{S}_3$-measurable.

With measure theory the notation keeps piling up. Here is a summary of the notation we will use for functions from $(S, \mathscr{S})$ into $\mathbb{R}$:

- $m\mathscr{S}$ is defined to be the $\mathscr{S}$-measurable functions on S,

- $m\mathscr{S}^+$ is defined to be the nonnegative functions in $m\mathscr{S}$, and

- $b\mathscr{S}$ is defined to be the bounded functions in $m\mathscr{S}$.

Exercise 7.2.6 Let S be any set. Argue that every $f \colon S \to \mathbb{R}$ is $\mathfrak{P}(S)$-measurable, while only the constant functions are $\{S, \varnothing\}$-measurable.

Exercise 7.2.7 Let $(S, \mathscr{S})$ be any measurable space. Show that $s\mathscr{S} \subset m\mathscr{S}$.

The following lemma is *very useful* when checking measurability. The proof is typical of measure-theoretic arguments.

Lemma 7.2.8 *Let $(E, \mathscr{E})$ and $(F, \mathscr{F})$ be two measurable spaces, and let $f \colon E \to F$. Let $\mathscr{G}$ be a generator of $\mathscr{F}$, in the sense that $\sigma(\mathscr{G}) = \mathscr{F}$. Then f is $\mathscr{E}, \mathscr{F}$-measurable if and only if $f^{-1}(B) \in \mathscr{E}$ for all $B \in \mathscr{G}$.*

Proof. Necessity is obvious. Regarding sufficiency, let

$$\mathscr{M} := \{ B \in \mathscr{F} : f^{-1}(B) \in \mathscr{E} \}$$

It is left to the reader to verify that $\mathscr{M}$ is a σ-algebra containing $\mathscr{G}$.[23] But then $\mathscr{F} = \sigma(\mathscr{G}) \subset \sigma(\mathscr{M}) = \mathscr{M}$. (Why?) Hence $\mathscr{F} \subset \mathscr{M}$, which is precisely what we wish to show.   □

In other words, to check measurability of a function, *we need only check measurability on a generating class*. For example, to verify measurability of a function into $(\mathbb{R}, \mathscr{B}(\mathbb{R}))$, we need only check that the preimages of open sets are measurable. (Why?)

Exercise 7.2.9 Let S be any metric space. Show that if $f: S \to \mathbb{R}$ is continuous, then it is Borel measurable (i.e., in $m\mathscr{B}(S)$).[24]

In fact one can show that families such as

$$[a, b] \text{ with } a \leq b, \quad (a, \infty) \text{ with } a \in \mathbb{R}, \quad (-\infty, b] \text{ with } b \in \mathbb{R}$$

all generate $\mathscr{B}(\mathbb{R})$. So for $f: S \to \mathbb{R}$ to be in $\mathscr{S}$-measurable (given σ-algebra $\mathscr{S}$ on S), it is sufficient that, for example, $\{x \in S : f(x) \leq b\} \in \mathscr{S}$ for all $b \in \mathbb{R}$. To get a feel for why this is useful consider the next example:

Example 7.2.10 Let $(S, \mathscr{S})$ be a measurable space, and let $(f_n) \subset m\mathscr{S}$. If $f: S \to \mathbb{R}$ is a function satisfying $f(x) = \sup_n f_n(x)$ for $x \in S$, then $f \in m\mathscr{S}$ because, given any $b \in \mathbb{R}$, we have $\{x \in S : f(x) \leq b\} = \cap_n \{x \in S : f_n(x) \leq b\} \in \mathscr{S}$.[25]

When defining integrals of measurable functions, our approach will be to approximate them with simple functions, which we saw how to integrate in §7.2.1. While the definition of measurability is rather abstract, it turns out that functions are measurable precisely when they can be well approximated by simple functions. In particular,

Lemma 7.2.11 *A function $f: S \to \mathbb{R}_+$ is $\mathscr{S}$-measurable if and only if there is a sequence $(s_n)_{n \geq 1}$ in $s\mathscr{S}^+$ with $s_n \uparrow f$ pointwise on S.*

That the existence of such an approximating sequence is sufficient for measurability follows from example 7.2.10. (Why?) Let's sketch the proof of necessity in the case of $S = \mathbb{R}$ and $\mathscr{S} = \mathscr{B}(\mathbb{R})$. Figure 7.4 might help with intuition. In this case the function f is bounded above by c. The range space $[0, c]$ is subdivided into the intervals $[0, a)$, $[a, b)$, and $[b, c]$. Using f, this partition also divides the domain (the x-axis) into the sets $f^{-1}([0, a))$, $f^{-1}([a, b))$ and $f^{-1}([b, c])$. We can now define a simple function s by

$$s = 0 \times \mathbb{1}_{f^{-1}([0,a))} + a \times \mathbb{1}_{f^{-1}([a,b))} + b \times \mathbb{1}_{f^{-1}([b,c])}$$

---

[23]Hint: See lemma A.1.2 on page 321.
[24]Hint: Use theorem 3.1.33 on page 43.
[25]Below we often write $\{g \leq b\}$ for the set $\{x \in S : g(x) \leq b\}$.

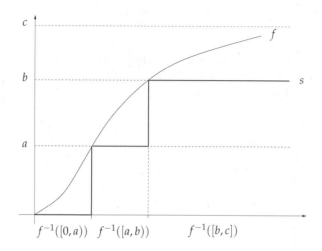

Figure 7.4 A measurable function

Notice that as drawn, $s \in s\mathscr{B}(\mathbb{R})^+$, because s takes only finitely many values on finitely many disjoint sets, and these sets $f^{-1}([0,a))$, $f^{-1}([a,b))$, and $f^{-1}([b,c])$ are all intervals, which qualifies them as members of $\mathscr{B}(\mathbb{R})$. Notice also that s lies below f.

By looking at the figure, you can imagine that if we refine our partition of the range space, we would get another function s' that dominates s but still lies below f, and is again an element of $s\mathscr{B}(\mathbb{R})^+$. Continuing in this way, it seems that we can indeed approximate f from below by an increasing sequence of elements of $s\mathscr{B}(\mathbb{R})^+$.

The function f we chose was a bit special—in particular, it is increasing, which means that the sets $f^{-1}([0,a))$, $f^{-1}([a,b))$ and $f^{-1}([b,c])$ are intervals, and therefore elements of $\mathscr{B}(\mathbb{R})$. Were they not elements of $\mathscr{B}(\mathbb{R})$, we could not say that $s \in s\mathscr{B}(\mathbb{R})^+$. This is where the definition of Borel measurability comes in. Even if f is not increasing, we require in lemma 7.2.11 that it is at least Borel measurable. In which case sets like $f^{-1}([a,b))$ are always elements of $\mathscr{B}(\mathbb{R})$, because $[a,b)$ is a Borel set. As a result the approximating simple functions are always in $s\mathscr{B}(\mathbb{R})^+$.

For arbitrary $(S, \mathscr{S})$, elements of $m\mathscr{S}$ play nicely together, in the sense that when standard algebraic and limiting operations are applied to measurable functions the resulting functions are themselves measurable:

Theorem 7.2.12 *If $f, g \in m\mathscr{S}$, then so is $\alpha f + \beta g$ for any $\alpha, \beta \in \mathbb{R}$. The product fg is also in $m\mathscr{S}$. If $(f_n)_{n\geq 1}$ is a sequence in $m\mathscr{S}$ with $f_n \to f$ pointwise, where $f: S \to \mathbb{R}$, then $f \in m\mathscr{S}$. If $f \in m\mathscr{S}$, then $|f| \in m\mathscr{S}$.*

Exercise 7.2.13 Show that if $f \in m\mathscr{S}$, then $|f| \in m\mathscr{S}$.[26]

7.2.3 Integrating Measurable Functions

Now we are ready to extend our notion of integral from simple functions to measurable functions. Let $(S, \mathscr{S}, \mu)$ be any measure space and consider first integration of a *nonnegative* measurable function $f \colon S \to \mathbb{R}_+$ (i.e., an element of $m\mathscr{S}^+$). We define the integral of f on S with respect to μ by

$$\mu(f) :=: \int f d\mu := \lim_{n \to \infty} \mu(s_n) \text{ where } (s_n)_{n \geq 1} \subset s\mathscr{S}^+ \text{ with } s_n \uparrow f \qquad (7.12)$$

We are appealing to lemma 7.2.11 for the existence of at least one sequence $(s_n)_{n \geq 1} \subset s\mathscr{S}^+$ with $s_n \uparrow f$. Note that $\mu(s_n)$ always converges in $[0, \infty]$ as a result of monotonicity (see proposition 7.2.2).

Regarding notation, all of the following are common alternatives for the integral $\mu(f)$:

$$\mu(f) :=: \int f d\mu :=: \int f(x)\mu(dx)$$

In the case of Lebesgue measure we will also use $\int f(x)dx$:

$$\lambda(f) :=: \int f d\lambda :=: \int f(x)\lambda(dx) :=: \int f(x)dx$$

Isn't it possible that the number we get in (7.12) depends on the particular approximating sequence $(s_n)_{n \geq 1}$ that we choose? The answer is no. If $(s_n)_{n \geq 1}$ and $(s'_n)_{n \geq 1}$ are two sequences in $s\mathscr{S}^+$ with $s_n \uparrow f$ and $s'_n \uparrow f$, then $\mu(s_n)$ and $\mu(s'_n)$ always have the same limit.[27] The number given by taking any of these limits is in fact equal to

$$\sup \left\{ \mu(s) \; : \; s \in s\mathscr{S}^+, \; 0 \leq s \leq f \right\} \in [0, \infty] \qquad (7.13)$$

Example 7.2.14 Recall the counting measure c on $(\mathbb{N}, \mathfrak{P}(\mathbb{N}))$ introduced on page 167. A nonnegative function $f \colon \mathbb{N} \to \mathbb{R}_+$ is just a nonnegative sequence, and we emphasize this by writing f as (f_n). Since $\mathbb{N}$ is paired with its power set (the set of all subsets), there is no issue with measurability—all such functions (sequences) are measurable.

The function (f_n) is simple if it takes only finitely many values. Suppose in particular that $f_n = 0$ for all $n \geq N \in \mathbb{N}$. Then, by the definition of the integral on simple functions,

$$c(f) :=: \int f dc = \sum_{n=1}^{N} f_n \, c(n) = \sum_{n=1}^{N} f_n$$

---

[26]Hint: Use the fact that $g \in m\mathscr{S}$ whenever $\{g \leq b\} \in \mathscr{S}$ for all $b \in \mathbb{R}$.

[27]The reason that the approximating sequence in (7.12) is required to be monotone is to ensure that this independence holds. The proof is not difficult but let's take it as given.

so integration with respect to c is equivalent to summation.

Now consider the case of a general nonnegative sequence $f = (f_n)$. For simple functions converging up to f we can take $f^N := (f_n^N)$, which is defined as equal to f_n if $n \leq N$ and to zero if $n > N$. In light of (7.12) we have

$$c(f) := \int f dc = \lim_{N \to \infty} \int f^N dc = \lim_{N \to \infty} \sum_{n=1}^{N} f_n$$

which is the standard definition of the infinite series $\sum_n f_n$. Again, integration with respect to c is equivalent to summation.

So far we have only defined the integral of nonnegative measurable functions. Integration of general measurable functions is also straightforward: Split the function f into its positive part $f^+ := \max\{0, f\}$ and its negative part $f^- := \max\{0, -f\}$, so $f = f^+ - f^-$. Then set

$$\mu(f) := \mu(f^+) - \mu(f^-) \tag{7.14}$$

The only issue here is that we may end up with the expression $\infty - \infty$, which is *definitely not allowed*—in this case the integral is not defined.

Definition 7.2.15 Let $(S, \mathscr{S}, \mu)$ be a measure space, and let $f \in m\mathscr{S}$. The function f is called *integrable* if both $\mu(f^+)$ and $\mu(f^-)$ are finite. If f is integrable, then its integral $\mu(f)$ is given by (7.14). The set of all integrable functions on S is denoted $\mathscr{L}_1(S, \mathscr{S}, \mu)$, or simply $\mathscr{L}_1(\mu)$.

How do we define integration of a function $f \in \mathscr{L}_1(\mu)$ over a subset E of S, rather than over the whole space? The answer is by setting

$$\int_E f d\mu := \int \mathbb{1}_E f d\mu =: \mu(\mathbb{1}_E f)$$

Here the function $\mathbb{1}_E f$ evaluated at x is the product $\mathbb{1}_E(x) \cdot f(x)$.

Although we omit the proof, if f is a continuous real function on $\mathbb{R}$ with the property that $f = 0$ outside an interval $[a, b]$—so that the Riemann integral is well-defined—then $\lambda(f)$ is precisely the Riemann integral of f on $[a, b]$. We can go ahead and integrate f as our high school calculus instinct tells us to. Hence we have succeeded in extending the elementary integral to a larger class of functions.

7.3   Properties of the Integral

Having defined the abstract Lebesgue integral, let's now look at some of its properties. As we will see, the integral has nice algebraic properties, and also interacts well with limiting operations. Section §7.3.1 focuses on nonnegative functions, while §7.3.2 treats the general case.

7.3.1 Basic Properties

Recall that functions constructed from measurable functions using standard algebraic and limiting operations are typically measurable (theorem 7.2.12). This leads us to consider the relationship between the integrals of the original functions and the integrals of the new functions. For example, is the integral of the sum of two measurable functions equal to the sum of the integrals? And is the integral of the limit of measurable functions equal to the limit of the integrals? Here is a summary of the key results:

Theorem 7.3.1 *Given an arbitrary measure space $(S, \mathscr{S}, \mu)$, the integral has the following properties on $m\mathscr{S}^+$:*

M1. If $A \in \mathscr{S}$ and $f := \mathbb{1}_A$, then $\mu(\mathbb{1}_A) = \mu(A)$.

M2. If $f = \mathbb{1}_\varnothing \equiv 0$, then $\mu(f) = 0$.

M3. If $f, g \in m\mathscr{S}^+$ and $\alpha, \beta \in \mathbb{R}_+$, then $\mu(\alpha f + \beta g) = \alpha\mu(f) + \beta\mu(g)$.

M4. If $f, g \in m\mathscr{S}^+$ and $f \le g$ pointwise on S, then $\mu(f) \le \mu(g)$.

M5. If $(f_n)_{n \ge 1} \subset m\mathscr{S}^+$, $f \in m\mathscr{S}^+$ and $f_n \uparrow f$, then $\mu(f_n) \uparrow \mu(f)$.

Property M1 is immediate from the definition of the integral, and M2 is immediate from M1. Properties M3 and M4 can be derived as follows:

Exercise 7.3.2 Using proposition 7.2.2 (page 173), show that if $\gamma \in \mathbb{R}_+$ and $f \in m\mathscr{S}^+$, then $\mu(\gamma f) = \gamma\mu(f)$. Show further that if $f, g \in m\mathscr{S}^+$, then $\mu(f + g) = \mu(f) + \mu(g)$. Combine these two results to establish M3.

Exercise 7.3.3 Prove M4.[28]

Property M5 is fundamental to the success of Lebesgue's integral, and is usually referred to as the monotone convergence theorem (although we give a more general result with that name below). Note that $\mu(f) = \infty$ is permitted. The proof of M5 is based on countable additivity of μ, and is one of the reasons that countable additivity (as opposed to finite additivity) is so useful. You can consult any book on measure theory and integration to see the proof.

The next result clarifies the relationship between measures and integrals.

Theorem 7.3.4 *Let $(S, \mathscr{S})$ be any measurable space. For each measure $\mu \colon \mathscr{S} \ni B \mapsto \mu(B) \in [0, \infty]$, there exists a function $\mu \colon m\mathscr{S}^+ \ni f \mapsto \mu(f) \in [0, \infty]$ with properties M1–M5. Conversely, each function $\mu \colon m\mathscr{S}^+ \to [0, \infty]$ satisfying properties M2–M5 creates a unique measure on $(S, \mathscr{S})$ via M1.*

---

[28]Hint: Use the expression for the integral in (7.13).

One can think of a measure μ on $(S, \mathscr{S})$ as a kind of "pre-integral," defined on the subset of $m\mathscr{S}^+$ that consists of all indicator functions (the integral of $\mathbb{1}_A$ being $\mu(A)$). The process of creating an integral on $m\mathscr{S}^+$ via simple functions and then monotone limits can be thought of as one that *extends the domain* of μ from indicator functions in $m\mathscr{S}^+$ to all functions in $m\mathscr{S}^+$.

Exercise 7.3.5 Show that if $\mu\colon m\mathscr{S}^+ \to [0,\infty]$ satisfies M2–M5, then the map $\hat{\mu}\colon \mathscr{S} \to [0,\infty]$ defined by $\hat{\mu}(A) = \mu(\mathbb{1}_A)$ is a measure on $\mathscr{S}$.[29]

Exercise 7.3.6 Use M1–M5 to prove the previously stated result (see exercise 7.1.25 on page 167) that if $(E_n) \subset \mathscr{S}$ with $E_n \subset E_{n+1}$ for all n, then $\mu(\cup_n E_n) = \lim_{n\to\infty} \mu(E_n)$.[30]

Lemma 7.3.7 *Let $A, B \in \mathscr{S}$, and let $f \in m\mathscr{S}^+$. If A and B are disjoint, then*

$$\int_{A\cup B} f d\mu = \int_A f d\mu + \int_B f d\mu$$

Proof. We have $\mathbb{1}_{A\cup B}f = (\mathbb{1}_A + \mathbb{1}_B)f = \mathbb{1}_A f + \mathbb{1}_B f$. Now apply M3. $\square$

One of the most important facts about the integral is that integrating over sets of zero measure cannot produce a positive number:

Theorem 7.3.8 *If $f \in m\mathscr{S}^+$, $E \in \mathscr{S}$, and $\mu(E) = 0$, then $\int_E f d\mu = 0$.*

Proof. Since $f \in m\mathscr{S}^+$, there is a sequence $(s_n)_{n\geq 1} \subset s\mathscr{S}^+$ with $s_n \uparrow f$, and hence $\mathbb{1}_E s_n \uparrow \mathbb{1}_E f$.[31] But

$$\mathbb{1}_E s_n = \sum_{k=1}^K \alpha_k(\mathbb{1}_E \cdot \mathbb{1}_{A_k}) = \sum_{k=1}^K \alpha_k \mathbb{1}_{E\cap A_k}$$

$$\therefore \quad \mu(\mathbb{1}_E s_n) = \sum_{k=1}^K \alpha_k \mu(E \cap A_k) = 0 \qquad \forall n \in \mathbb{N}$$

since $\mu(E) = 0$. By property M5, we have $\mu(\mathbb{1}_E f) = \lim_{n\to\infty} \mu(\mathbb{1}_E s_n) = 0$. $\square$

7.3.2 Finishing Touches

Let $(S, \mathscr{S}, \mu)$ be any measure space. By using the five fundamental properties M1–M5, one can derive the classical theorems about integrals on $\mathscr{L}_1(\mu) := \mathscr{L}_1(S, \mathscr{S}, \mu)$. The next few results show that many results from the previous section that hold for non-negative functions also hold for the (not necessarily nonnegative) elements of $\mathscr{L}_1(\mu)$.

---

[29]Hint: Regarding countable additivity, if $(A_n) \subset \mathscr{S}$ are disjoint then $\mathbb{1}_{\cup_n A_n} = \sum_n \mathbb{1}_{A_n}$.

[30]Hint: Use the fact that $\mathbb{1}_{\cup_n E_n} = \lim_{n\to\infty} \mathbb{1}_{E_n}$ pointwise on S.

[31]That is, the convergence $\mathbb{1}_E(x)s_n(x) \to \mathbb{1}_E(x)s_n(x)$ holds at each point $x \in S$, and that $\mathbb{1}_E(x)s_n(x)$ gets progressively larger with n for each $x \in S$.

To state the results we introduce the concept of properties holding "almost everywhere." Informally, if $f, g \in m\mathscr{S}$ and $P(x)$ is a statement about f and g at x (e.g., $f(x) = g(x)$ or $f(x) \leq g(x)$), then we will stay f and g have property P μ-almost everywhere (μ-a.e.) whenever the set of all x such that $P(x)$ fails has μ-measure zero. For example, if the set of $x \in S$ such that $f(x) \neq g(x)$ has measure zero then f and g are said to be equal μ-a.e. In addition we say that $f_n \to f$ μ-a.e. if $\lim f_n = f$ μ-a.e. This sounds a bit complicated, but the basic idea is that null sets don't matter when it comes to integration, so it's enough for properties to hold everywhere except on a null set.

Theorem 7.3.9 *Let $f, g \in \mathscr{L}_1(\mu)$, and let $\alpha, \beta \in \mathbb{R}$. The following results hold:*

1. *$\alpha f + \beta g \in \mathscr{L}_1(\mu)$ and $\mu(\alpha f + \beta g) = \alpha\mu(f) + \beta\mu(g)$.*

2. *If $E \in \mathscr{S}$ with $\mu(E) = 0$, then $\int_E f d\mu = 0$.*

3. *If $f \leq g$ μ-a.e., then $\mu(f) \leq \mu(g)$.*

4. *$|f| \in \mathscr{L}_1(\mu)$, and $|\mu(f)| \leq \mu(|f|)$.*

5. *$\mu(|f|) = 0$ if and only if $f = 0$ μ-a.e.*

This list is not minimal. For example, part 2 follows from parts 4 and 5. Part 1 follows from the definitions and M3 (i.e., linearity of the integral on the space of nonnegative measurable functions). See, for example, Dudley (2002, thm. 4.1.10). Part 2 can also be obtained from the identity $f = f^+ - f^-$, part 1 and theorem 7.3.8:

$$\mu(\mathbb{1}_E f) = \mu(\mathbb{1}_E f^+ - \mathbb{1}_E f^-) = \mu(\mathbb{1}_E f^+) - \mu(\mathbb{1}_E f^-) = 0 - 0$$

Exercise 7.3.10 Prove that if $\mu(E) = 0$ and $f \in \mathscr{L}_1(\mu)$, then $\int_{E^c} f d\mu = \int f d\mu$, and if $f, g \in \mathscr{L}_1(\mu)$ with $f = g$ μ-a.e., then $\mu(f) = \mu(g)$.[32]

Exercise 7.3.11 Prove part 3 using properties M1–M5.[33]

Exercise 7.3.12 Prove part 4 using the identity $|f| = f^+ + f^-$.

Regarding part 5, suppose that $f \neq 0$ on a set E with $\mu(E) > 0$. We will prove that $\mu(|f|) > 0$. To see this, define $E_n := \{x : |f(x)| > 1/n\}$. Observe that $E_n \subset E_{n+1}$ for each n, and that $E = \cup_n E_n$. From exercise 7.1.25 (page 167) there is an N with $\mu(E_N) > 0$. But then $\mu(|f|) \geq \mu(\mathbb{1}_{E_N}|f|) \geq \mu(E_N)/N > 0$. The converse implication follows from exercise 7.3.10.

Now we come to the classical convergence theorems for integrals, which can be derived from M1–M5. They are among the foundation stones of modern real analysis.

---

[32]Hint: Let E be the set on which f and g disagree. Then $\mathbb{1}_{E^c} f = \mathbb{1}_{E^c} g$.

[33]Hint: Convert to a statement about nonnegative functions. Note that if $f \leq g$ μ-a.e., then $\mathbb{1}_{E^c} f^+ + \mathbb{1}_{E^c} g^- \leq \mathbb{1}_{E^c} g^+ + \mathbb{1}_{E^c} f^-$, where E is all x such that $f(x) > g(x)$.

Theorem 7.3.13 (Monotone convergence theorem) *Let $(S, \mathscr{S}, \mu)$ be a measure space, and let $(f_n)_{n \geq 1}$ be a sequence in $m\mathscr{S}$. If $f_n \uparrow f \in m\mathscr{S}$ μ-almost everywhere and $\mu(f_1) > -\infty$, then $\lim_{n \to \infty} \mu(f_n) = \mu(f)$.*[34]

Theorem 7.3.14 (Dominated convergence theorem) *Let $(S, \mathscr{S}, \mu)$ be a measure space, let $g \in \mathscr{L}_1(\mu)$ and let $(f_n)_{n \geq 1} \subset m\mathscr{S}$ with $|f_n| \leq g$ for all n. If $f_n \to f$ μ-almost everywhere, then $f \in \mathscr{L}_1(\mu)$ and $\lim_{n \to \infty} \mu(f_n) = \mu(f)$.*

It's almost impossible to overemphasize what a useful result the dominated convergence theorem is, and the proof can be found in any text on measure theory. For a neat little illustration, consider

Corollary 7.3.15 *Consider the collection of real sequences*

$$a = (a_1, a_2, \ldots), \quad a^k = (a_1^k, a_2^k, \ldots) \qquad (k \in \mathbb{N})$$

Suppose that a^k is dominated pointwise by a sequence $b = (b_n)$ for all k, in the sense that $|a_n^k| \leq b_n$ for all k, n. Suppose further that $\lim_{k \to \infty} a_n^k = a_n$ for each n. If $\sum_n b_n < \infty$, then

$$\lim_{k \to \infty} \sum_{n \geq 1} a_n^k = \sum_{n \geq 1} \lim_{k \to \infty} a_n^k = \sum_{n \geq 1} a_n$$

Proof. Apply the dominated convergence theorem with $(S, \mathscr{S}, \mu) = (\mathbb{N}, \mathfrak{P}(\mathbb{N}), c)$, where c is the counting measure (recall example 7.2.14). $\square$

Let's conclude with the topic of image measures. To define image measures, let $(S, \mathscr{S}, \mu)$ be any measure space, let $(S', \mathscr{S}')$ be a measurable space, and let $T \colon S \to S'$ be $\mathscr{S}, \mathscr{S}'$-measurable. If E is some element of $\mathscr{S}'$, then $T^{-1}(E) \in \mathscr{S}$, so $\mu \circ T^{-1}(E) = \mu(T^{-1}(E))$ is well defined. In fact $E \mapsto \mu \circ T^{-1}(E)$ is a measure on $(S', \mathscr{S}')$, called the *image measure* of μ under T. Figure 7.5 provides a picture. The value of $\mu \circ T^{-1}(E)$ is obtained by pulling E back to S and evaluating with μ.

Exercise 7.3.16 Show that $\mu \circ T^{-1}$ is indeed a measure on $(S', \mathscr{S}')$.[35]

The following result shows how to integrate with image measures:[36]

Theorem 7.3.17 *Let $(S, \mathscr{S}, \mu)$ be any measure space, let $(S', \mathscr{S}')$ be a measurable space, let $T \colon S \to S'$ be a measurable function, and let $\mu \circ T^{-1}$ be the image measure of μ under T. If $w \colon S' \to \mathbb{R}$ is $\mathscr{S}'$-measurable and either w is nonnegative or $\mu(|w \circ T|)$ is finite, then $\mu \circ T^{-1}(w) = \mu(w \circ T)$, where the first integral is over S' and the second is over S.*

One application of this theorem is when μ is Lebesgue measure, and $\mu \circ T^{-1}$ is more complex. If we don't know how to integrate with respect to this new measure, we can use the change of variable to get back to an integral of the form $\int f d\mu$.

---

[34] In fact, for this theorem to hold the limiting function f need not be finite everywhere (or anywhere) on S. See Dudley (2002, thm. 4.3.2).

[35] You might find it useful to refer to lemma A.1.2 on page 321.

[36] A full proof can be found in Dudley (2002, thm. 4.1.11).

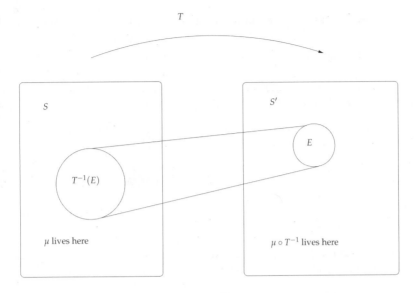

Figure 7.5 Image measure

7.3.3   The Space L_1

In this section we specialize to the case $(S, \mathscr{S}, \mu) = (S, \mathscr{B}(S), \lambda)$, where S is a Borel subset of $\mathbb{R}^k$. Most of the results we discuss hold more generally, but such extra generality is not needed here. Our interest is in viewing the space of integrable functions as a metric space. To this end, we define the "distance" d_1 on $\mathscr{L}_1(\lambda) := \mathscr{L}_1(S, \mathscr{B}(S), \lambda)$ by

$$d_1(f, g) := \int |f - g| d\lambda =: \lambda(|f - g|) \tag{7.15}$$

Alternatively, we can set

$$d_1(f, g) := \|f - g\|_1 \quad \text{where} \quad \|h\|_1 := \lambda(|h|)$$

From the pointwise inequalities $|f - g| \le |f| + |g|$ and $|f + g| \le |f| + |g|$ plus linearity and monotonicity of the integral, we have

$$\|f - g\|_1 \le \|f\|_1 + \|g\|_1 \quad \text{and} \quad \|f + g\|_1 \le \|f\|_1 + \|g\|_1$$

The first inequality tells us that $d_1(f, g)$ is finite for any $f, g \in \mathscr{L}_1(\lambda)$. From the second we can show that d_1 satisfies the triangle inequality on $\mathscr{L}_1(\lambda)$ using add and subtract:

$$\|f - g\|_1 = \|(f - h) + (h - g)\|_1 \le \|f - h\|_1 + \|h - g\|_1$$

Since d_1 satisfies the triangle inequality it seems plausible that d_1 is a metric (see the definition on page 36) on $\mathscr{L}_1(\lambda)$. However, there is a problem: We may have $f \neq g$ and yet $d_1(f, g) = 0$, because functions that are equal almost everywhere satisfy $\int |f - g| d\lambda = 0$. (Why?) For example, when $S = \mathbb{R}$, the functions $\mathbb{1}_\mathbb{Q}$ and $0 := \mathbb{1}_\emptyset$ are at zero distance from one another. Hence $(\mathscr{L}_1(\lambda), d_1)$ fails to be a metric space. Rather it is what's called a pseudometric space:

Definition 7.3.18 A *pseudometric space* is a nonempty set M and a function $\rho \colon M \times M \to \mathbb{R}$ such that, for any $x, y, v \in M$,

1. $\rho(x, y) = 0$ if $x = y$,

2. $\rho(x, y) = \rho(y, x)$, and

3. $\rho(x, y) \leq \rho(x, v) + \rho(v, y)$.

In contrast to a metric space, in a pseudometric space distinct points are permitted to be at zero distance from one another.

Exercise 7.3.19 On the space $\mathbb{R}^2$ consider the function $\rho(x, y) = |x_1 - y_1|$, where x_1 and y_1 are the first components of $x = (x_1, x_2)$ and $y = (y_1, y_2)$ respectively. Show that $(\mathbb{R}^2, \rho)$ is a pseudometric space.

It is not difficult to convert a pseudometric space into a metric space: We simply regard all points at zero distance from each other as the same point. In other words, we partition the original space into *equivalence classes* of points at zero distance from one another, and consider the set of these classes as a new space. Figure 7.6 illustrates for the space in exercise 7.3.19.

The distance between any two equivalence classes is just the distance between arbitrarily chosen members of each class. This value does not depend on the particular members chosen: If x and x' are equivalent, and y and y' are equivalent, then $\rho(x, y) = \rho(x', y')$ because

$$\rho(x, y) \leq \rho(x, x') + \rho(x', y') + \rho(y', y)$$
$$= \rho(x', y') \leq \rho(x', x) + \rho(x, y) + \rho(y, y') = \rho(x, y)$$

The space of equivalence classes and the distance just described form a metric space. In particular, distinct elements of the derived space are at positive distance from one another (otherwise they would not be distinct).

The metric space derived from the pseudometric space $(\mathscr{L}_1(\lambda), d_1)$ is traditionally denoted $(L_1(\lambda), d_1)$, and has a major role to play in the rest of this book.[37] Since two functions in $\mathscr{L}_1(\lambda)$ are at zero distance if and only if they are equal almost everywhere,

---

[37] To simplify notation, we are using the symbol d_1 to represent distance on both spaces.

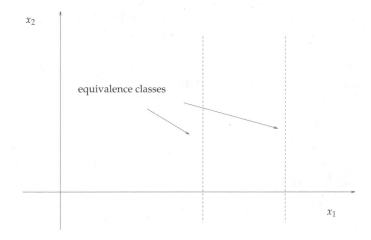

Figure 7.6 Equivalence classes for $(\mathbb{R}^2, \rho)$

the new space $(L_1(\lambda), d_1)$ consists of equivalences classes of functions that are equal almost everywhere.

A *density* on S is a nonnegative measurable function that integrates to one. We are interested in describing the set of densities as a metric space, with the intention of studying Markov chains such that their marginal distributions evolve in the space of densities. The densities are embedded into $L_1(\lambda)$ as follows:

Definition 7.3.20 The space of *densities* on S is written as $D(S)$ and defined by

$$D(S) := \{f \in L_1(\lambda) : f \geq 0 \text{ and } \|f\|_1 = 1\}$$

In the definition, f is actually an equivalence class of functions f', f'', etc., that are all equal almost everywhere. The statement $f \geq 0$ means that all of these functions are nonnegative almost everywhere, while $\|f\|_1 = 1$ means that all integrate to one. (More generally, if $f \in L_1(\lambda)$, then $\|f\|_1$ is the integral of the absolute value of any element in the equivalence class.)

Theorem 7.3.21 *The spaces $(L_1(\lambda), d_1)$ and $(D(S), d_1)$ are both complete.*

A proof of completeness of $(L_1(\lambda), d_1)$ can be found in any good text on measure theory. Completeness of $(D(S), d_1)$ follows from the fact that $D(S)$ is closed as a subset of $(L_1(\lambda), d_1)$ and theorem 3.2.3 (page 45). The proof that $D(S)$ is closed is left as an exercise for enthusiastic readers.

Densities are used to represent distributions of random variables. Informally, the statement that X has density $f \in D(S)$ means that X is in $B \subset S$ with probability

$\int_B f'(x)dx :=: \int_B f' d\lambda$, where f' is some member of the equivalence class. Note that it does not matter which member we pick, as all give the same value here. In this sense it is equivalence classes that represent distributions rather than individual densities.

Finally, *Scheffés identity* provides a nice quantitative interpretation of d_1 distance between densities: For any f and g in $D(S)$,

$$\|f - g\|_1 = 2 \times \sup_{B \in \mathscr{B}(S)} \left| \int_B f(x)dx - \int_B g(x)dy \right| \tag{7.16}$$

It follows that if $\|f - g\|_1 \leq \epsilon$, then for any event B of interest, the deviation in the probability assigned to B by f and g is less than $\epsilon/2$.[38]

7.4   Commentary

The treatment of measure theory and integration in this chapter is fairly standard among modern expositions. There are many excellent references with which to round out the material. Good starting points are Williams (1991) and Taylor (1997). Aliprantis and Burkinshaw (1998) is more advanced, and contains many exercises. The books by Pollard (2002), Dudley (2002), and Schilling (2005) are also highly recommended.

---

[38] A proof of this identity is given later in a more general context (see lemma 11.1.29).

Chapter 8

Density Markov Chains

In this chapter we take an in-depth look at Markov chains on state space $S \subset \mathbb{R}^n$ with the property that conditional (and hence marginal) distributions can be represented by densities. These kinds of processes were previously studied in chapter 6. Now that we have measure theory under our belts, we will be able to cover a number of deeper results.

Not all Markov chains fit into the density framework (see §9.2 for the general case). For those that do, however, the extra structure provided by densities aids us in analyzing dynamics and computing distributions. In addition densities have a concreteness that abstract probability measures lack, in the sense that they are easy to represent visually. This concreteness makes them a good starting point when building intuition.

8.1 Outline

We start with the basic theory of Markov chains with density representations. After defining density Markov chains, we will illustrate the connection to stochastic recursive sequences (stochastic difference equations). In §8.1.3 we introduce the Markov operator for the density case, and show how the marginal distributions of a density Markov chain can be generated by iterating on the Markov operator. This theory closely parallels the finite case, as discussed in §4.2.2.

8.1.1 Stochastic Density Kernels

We met some examples of density kernels in chapter 6. Let's now give the formal definition of a density kernel.

Definition 8.1.1 Let S be a Borel subset of $\mathbb{R}^n$. A *stochastic density kernel* on S is a Borel measurable function $p\colon S \times S \to \mathbb{R}_+$ such that

$$\int p(x,y)dy :=: \int p(x,y)\lambda(dy) :=: \lambda(p(x,\cdot)) = 1 \quad \text{for all } x \in S \qquad (8.1)$$

In particular, the function $y \mapsto p(x,y)$ is a density for each $x \in S$. We can think of p as a family of density functions, one for each point in the state space. In what follows, *we will use the notation $p(x,y)dy$ to represent the density function $y \mapsto p(x,y)$.* A second point is that $S \times S$ is a subset of $\mathbb{R}^{2n}$, and Borel measurability of p refers to Borel subsets of this space. In practice, one rarely encounters stochastic kernels where Borel measurability is problematic.

To illustrate the definition, consider the kernel p defined by

$$p(x,y) = \frac{1}{\sqrt{2\pi}} \exp\left(-\frac{(y - ax - b)^2}{2}\right) \qquad ((x,y) \in S \times S = \mathbb{R} \times \mathbb{R})$$

In other words, $p(x,y)dy = N(ax+b, 1)$. The kernel is presented visually in figure 8.1. Each point on the x-axis picks out a distribution $N(ax+b, 1)$, which is represented as a density running along the y-axis. In this case a is positive, so an increase in x leads to an increase in the mean of the corresponding density $p(x,y)dy$, and the density puts probability mass on larger y.

From an initial condition $\psi \in D(S)$ and a density kernel p, we can generate a Markov chain $(X_t)_{t \geq 0}$. Here is a definition paralleling the finite case (page 71):

Definition 8.1.2 Let $\psi \in D(S)$. A random sequence $(X_t)_{t \geq 0}$ on S is called *Markov-(p, ψ)* if

1. at time zero, X_0 is drawn from ψ, and

2. at time $t+1$, X_{t+1} is drawn from $p(X_t, y)dy$.

In the case of the kernel $p(x,y)dy = N(ax+b, 1)$, we draw X_0 from some given ψ and then, at each time t, draw $X_{t+1} \sim N(aX_t + b, 1)$. Listing 8.1 generates one observation (of length 100) for the process when $\psi = N(0, 1)$. An observation (sample path) is shown in figure 8.2.

There is another way to visualize the dynamics associated with our stochastic kernel. Recall the 45 degree diagram technique for studying univariate deterministic dynamic systems we introduced in figure 4.3 (page 58). Now consider figure 8.3, each panel of which shows a series generated by the kernel $p(x,y)dy = N(ax+b, 1)$. The kernel itself is represented in the graphs by shading. You should understand this shading as a "contour" representation of the 3D graph in figure 8.1, with lighter areas corresponding to higher probability. For each graph the sequence of arrows traces out

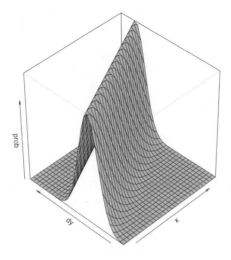

Figure 8.1 The stochastic kernel $p(x, y)dy = N(ax + b, 1)$

an individual time series. The initial condition is $X_0 = -4$, and X_1 is then drawn from $N(aX_0 + b, 1)$. We trace this value back to the 45 degree line to obtain the distribution for $X_2 \sim N(aX_1 + b, 1)$, and so on.

When most of the probability mass lies above the 45 degree line, the value of the state tends to increase. When most is below the 45 degree line, the value tends to decrease. The actual outcome, however, is random, depending on the sequence of draws that generate the time series.

8.1.2 Connection with SRSs

Suppose that we wish to investigate a stochastic recursive sequence (SRS) where the state space S is a Borel subset of $\mathbb{R}^n$, Z is a Borel subset of $\mathbb{R}^k$, $F: S \times Z \to S$ is a given function, and

$$X_{t+1} = F(X_t, W_{t+1}), \quad X_0 \sim \psi \in D(S), \quad (W_t)_{t \geq 1} \overset{\text{IID}}{\sim} \phi \in D(Z) \tag{8.2}$$

We would like to know when there exists a stochastic density kernel p on S that represents (8.2), in the sense that $(X_t)_{t \geq 0}$ defined in (8.2) is Markov-(p, ψ). Put differently, we wish to know when there exists a density kernel p such that, for all $x \in S$,

$$p(x, y)dy = \text{ the density of } F(x, W) \text{ when } W \sim \phi \tag{8.3}$$

Listing 8.1 (ar1.py) Simulation of $(X_t)_{t \geq 0}$ for $p(x, y)dy = N(ax + b, 1)$

```
from random import normalvariate as N

a, b = 0.5, 1          # Parameters
X = {}                 # An empty dictionary to store path
X[0] = N(0, 1)         # X_0 has distribution N(0, 1)

for t in range(100):
    X[t+1] = N(a * X[t] + b, 1)
```

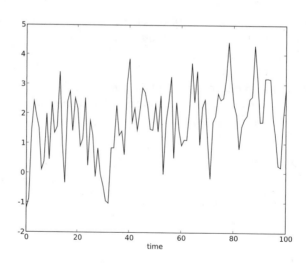

Figure 8.2 Time series

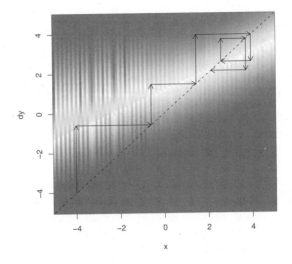

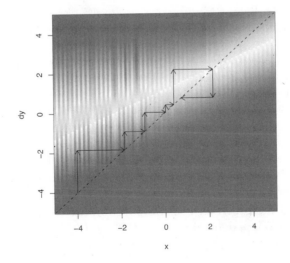

Figure 8.3 Two time series

Such a p will do the job for us because, if the state X_t arrives at any $x \in S$, then drawing X_{t+1} from $p(x,y)dy$ is—by definition of p—probabilistically equivalent to drawing W_{t+1} from ϕ and setting $X_{t+1} = F(x, W_{t+1})$.

The reason for our interest in this question is that the theory of Markovian dynamics is more easily developed in the framework of stochastic kernels than in that of SRSs such as (8.2). This is largely because the stochastic density kernel captures the stochastic law of motion for the system in one single object. To apply the theory developed below, it is necessary to be able to take a given model of the form (8.2) and obtain its density kernel representation.

The existence of a stochastic density kernel p satisfying (8.3) is not guaranteed, as not all random variables have distributions that can be represented by densities. Let us try to pin down simple sufficient conditions implying that $F(x, W)$ can be represented by a density.

Here is a more general question: If Y is a random variable on S, when does there exist a density $\phi \in D(S)$ such that ϕ represents Y in the sense that, for every $B \in \mathscr{B}(S)$, the number $\int_B \phi d\lambda :=: \int_B \phi(x)dx$ gives the probability that $Y \in B$? In essence the answer is that Y *must not take values in any Lebesgue null set with positive probability*. To see this, suppose that $N \in \mathscr{B}(S)$ with $\lambda(N) = 0$ and $Y \in N$ with positive probability. Then regardless of which $\phi \in D(S)$ we choose, theorem 7.3.8 implies that $\int_N \phi(x)dx = 0$, and hence ϕ does not represent Y.

Now let us consider the distribution of $Y = F(x, W)$, $W \sim \phi \in D(Z)$. The following theorem, though not as general as some, will be sufficient for our purposes:

Theorem 8.1.3 *Let W be a random variable on $\mathbb{R}^n$ with density ϕ, let $\gamma \in \mathbb{R}^n$, and let Γ be an $n \times n$ matrix. If $\det \Gamma \neq 0$, then $Y = \gamma + \Gamma W$ has density ϕ_Y on $\mathbb{R}^n$, where*

$$\phi_Y(y) := \phi(\Gamma^{-1}(y - \gamma))|\det \Gamma^{-1}| \qquad (y \in \mathbb{R}^n)$$

Why is $\det \Gamma \neq 0$ required? If $\det \Gamma = 0$ then $Y = \gamma + \Gamma W$ takes values in a subspace of dimension less than n. In $\mathbb{R}^n$, such subspaces have Lebesgue measure zero. Hence Y takes values in a Lebesgue null set with positive probability, in which case it cannot be represented by a density.

Before looking at the proof, let's see how we can use the theorem. In §8.1.1 we looked at a process $(X_t)_{t \geq 0}$ defined by kernel $p(x,y)dy = N(ax + b, 1)$. This corresponds to the SRS

$$X_{t+1} = aX_t + b + W_{t+1}, \quad (W_t)_{t \geq 1} \overset{\text{IID}}{\sim} \phi = N(0,1)$$

In other words, $p(x,y)dy = N(ax + b, 1)$ is the density of $Y = ax + b + W$ when $W \sim \phi$. This claim is easily verified. From theorem 8.1.3 the density ϕ_Y of Y is $\phi_Y(y) =$

$\phi(y - ax - b)$. Since $\phi = N(0,1)$, this becomes

$$\phi_Y(y) = \frac{1}{\sqrt{2\pi}} \exp\left(-\frac{(y - ax - b)^2}{2}\right) = N(ax + b, 1)$$

Exercise 8.1.4 Consider the $\mathbb{R}^n$-valued SRS

$$X_{t+1} = AX_t + b + W_{t+1}, \quad (W_t)_{t \geq 1} \overset{\text{IID}}{\sim} \phi \in D(\mathbb{R}^n) \tag{8.4}$$

where A is an $n \times n$ matrix and b is an $n \times 1$ vector. Show that the stochastic density kernel corresponding to this model is $p(x, y) = \phi(y - Ax - b)$.

Exercise 8.1.5 Consider the well-known threshold autoregression model (Chan and Tong 1986). The model is a nonlinear AR(1) process

$$X_{t+1} = \sum_{k=1}^{K} (A_k X_t + b_k) \mathbb{1}_{B_k}(X_t) + W_{t+1}, \quad (W_t)_{t \geq 1} \overset{\text{IID}}{\sim} \phi \in D(\mathbb{R}^n) \tag{8.5}$$

where X_t takes values in $\mathbb{R}^n$, the family of sets $(B_k)_{k=1}^{K}$ is a (measurable) partition of $\mathbb{R}^n$, and $(A_k)_{k=1}^{K}$ and $(b_k)_{k=1}^{K}$ are $n \times n$-dimensional matrices and $n \times 1$-dimensional vectors respectively. The idea is that when X_t is in the region of the state space B_k, the state variable follows the law of motion $A_k X_t + b_k$. Show that the corresponding density kernel is

$$p(x, y) = \phi\left[y - \sum_{k=1}^{K} (A_k x + b_k) \mathbb{1}_{B_k}(x)\right] \tag{8.6}$$

Example 8.1.6 Consider again the Solow–Swan model. Set $\delta = 1$ and $f(k, W) = f(k)W$. In other words, the state evolves according to

$$k_{t+1} = sf(k_t)W_{t+1}, \quad (W_t)_{t \geq 1} \overset{\text{IID}}{\sim} \phi \tag{8.7}$$

Suppose that $s > 0$ and $f(k) > 0$ whenever $k > 0$, and take $S = Z = (0, \infty)$. We wish to determine the stochastic density kernel $p(x, y)dy$; equivalently, we wish to find the density ϕ_Y of the random variable $Y = sf(x)W$ when $x \in S$ is fixed and $W \sim \phi$.

The only obstacle to applying theorem 8.1.3 in this case is that Z is a proper subset of $\mathbb{R}$, not $\mathbb{R}$ itself. Hence ϕ is not necessarily defined on all of $\mathbb{R}$. However, we can get around this easily enough by setting $\phi = 0$ on the complement $(-\infty, 0]$ of Z.

When $x \in S$ is fixed, $sf(x)$ is a strictly positive constant, and from theorem 8.1.3 the density of $Y = sf(x)W$ is

$$p(x, y) = \phi\left(\frac{y}{sf(x)}\right) \frac{1}{sf(x)} \tag{8.8}$$

Exercise 8.1.7 Consider again example 8.1.6, but this time with $\delta < 1$. In other words, $(k_t)_{t\geq 0}$ evolves according to

$$k_{t+1} = sf(k_t)W_{t+1} + (1-\delta)k_t, \quad (W_t)_{t\geq 1} \overset{\text{IID}}{\sim} \phi$$

Show that the stochastic density kernel is now given by

$$p(x,y) = \phi\left(\frac{y-(1-\delta)x}{sf(x)}\right)\frac{1}{sf(x)} \tag{8.9}$$

Notice that if $y < (1-\delta)x$, then ϕ is evaluated at a negative number. This is why we need to extend ϕ to all of $\mathbb{R}$, with $\phi(z) = 0$ for all $z \leq 0$.

Exercise 8.1.8 In §6.1.2 we considered a stochastic threshold externality model with multiple equilibria, with law of motion for capital given by

$$k_{t+1} = sA(k_t)k_t^\alpha W_{t+1}, \quad (W_t)_{t\geq 1} \overset{\text{IID}}{\sim} \phi \tag{8.10}$$

Here $k \mapsto A(k)$ is any function with $A(k) > 0$ when $k > 0$. Set $Z = S = (0,\infty)$. Derive the stochastic density kernel p corresponding to this model.

Now let's sketch the proof of theorem 8.1.3. We will use the following standard change-of-variable result:

Theorem 8.1.9 *Let A and B be open subsets of $\mathbb{R}^k$, and let $T\colon B \to A$ be a C^1 bijection. If $f \in \mathscr{L}_1(A, \mathscr{B}(A), \lambda)$, then*

$$\int_A f(x)dx = \int_B f \circ T(y) \cdot |\det J_T(y)|dy \tag{8.11}$$

Here $J_T(y)$ is the Jacobian of T evaluated at y, and C^1 means that T is continuously differentiable on B.[1]

Example 8.1.10 Let $A = \mathbb{R}$, let $B = (0,\infty)$, and let $Tx = \ln(x)$. Then $|\det J_T(y)| = 1/y$, and for any measurable f on $\mathbb{R}$ with finite integral we have

$$\int_{\mathbb{R}} f(x)dx = \int_{(0,\infty)} f(\ln y)\frac{1}{y}dy$$

Now suppose we have a random variable W with density $\phi \in D(\mathbb{R}^n)$, and we transform W with some function h to create a new random variable $Y = h(W)$. In the case of theorem 8.1.3 the transformation h is the linear function $z \mapsto \gamma + \Gamma z$. The following generalization of theorem 8.1.3 holds:

---

[1] theorem 8.1.9 is actually a special case of theorem 7.3.17 on page 182.

Theorem 8.1.11 *Let S and T be open subsets of $\mathbb{R}^n$, and let W be a random vector on S, distributed according to density ϕ on S. Let $Y = h(W)$, where $h\colon S \to T$ is a bijection and the inverse h^{-1} is a C^1 function. In that case Y is a random vector on T with density ϕ_Y, where*

$$\phi_Y(y) := \phi(h^{-1}(y)) \cdot |\det J_{h^{-1}}(y)| \qquad (y \in T) \tag{8.12}$$

The proof can be constructed along the following lines: The statement that ϕ_Y is the density of Y means that $\mathbb{P}\{Y \in B\} = \int_B \phi_Y(y)dy$ holds for every $B \in \mathcal{B}(T)$. By an application of theorem 8.1.9 we get

$$\mathbb{P}\{Y \in B\} = \mathbb{P}\{W \in h^{-1}(B)\}$$
$$= \int_{h^{-1}(B)} \phi(x)dx = \int_B \phi(h^{-1}(y)) |\det J_{h^{-1}}(y)|dy = \int_B \phi_Y(y)dy$$

8.1.3   The Markov Operator

Let p be a density kernel on $S \in \mathcal{B}(\mathbb{R}^n)$, let $\psi \in D(S)$ be an initial condition, and let $(X_t)_{t \geq 0}$ be Markov-(p, ψ). As usual, let ψ_t be the (marginal) distribution of X_t for each $t \geq 0$. In §6.1.3 we saw that if X and Y are random variables on S with marginals p_X and p_Y, and if $p_{Y|X}$ is the conditional density of Y given X, then $p_Y(y) = \int p_{Y|X}(x, y)p_X(x)dx$ for all $y \in S$. Letting $X_{t+1} = Y$ and $X_t = X$, we obtain

$$\psi_{t+1}(y) = \int p(x, y)\psi_t(x)dx \qquad (y \in S) \tag{8.13}$$

Equation (8.13) is just (6.7) on page 126, translated to a more general setting. It is the continuous state version of (4.10) on page 75 and the intuition is roughly the same: The probability of being at y tomorrow is the probability of moving from x today to y tomorrow, summed over all x and weighted by the probability $\psi_t(x)dx$ of observing x today.

Now define an operator $\mathbf{M}$ sending $\psi \in D(S)$ into $\psi\mathbf{M} \in D(S)$ by

$$\psi\mathbf{M}(y) = \int p(x, y)\psi(x)dx \qquad (y \in S) \tag{8.14}$$

This operator is called the *Markov operator* corresponding to stochastic density kernel p, and parallels the definition of the Markov operator for the finite state case given on page 75. As in that case, $\mathbf{M}$ acts on distributions (densities) to the left rather than the right, and is understood as updating the distribution of the state: If the state is currently distributed according to ψ then next period its distribution is $\psi\mathbf{M}$. In particular, (8.13) can now be written as

$$\psi_{t+1} = \psi_t\mathbf{M} \tag{8.15}$$

This a density version of (4.12) on page 76. Iterating backward we get the representation $\psi_t = \psi M^t$ for the distribution of X_t given $X_0 \sim \psi$.

Technical note: An element of $D(S)$ such as ψ is actually an equivalence class of functions on S that are equal almost everywhere—see §7.3.3. This does not cause problems for the definition of the Markov operator, since applying M to any element of the equivalence class yields the same function: If ψ' and ψ'' are equal off a null set E, then the integrands in (8.14) are equal off E for every y, and hence both integrate to the same number. Thus ψM in (8.14) is a well-defined function on S, and we embed it in $D(S)$ by identifying it with the equivalence class of functions to which it is equal almost everywhere.

That M does in fact map $D(S)$ into itself can be verified by showing that ψM is nonnegative and integrates to 1. That ψM integrates to 1 can be seen by changing the order of integration:

$$\int \psi M(y)dy = \int \int p(x,y)\psi(x)dxdy = \int \left[\int p(x,y)dy \right] \psi(x)dx$$

Since $\int p(x,y)dy = 1$ and $\int \psi(x)dx = 1$, the proof is done. Regarding nonnegativity, fix any $y \in S$. Since $p \geq 0$ and since $\psi \geq 0$ almost everywhere, the integrand in (8.14) is nonnegative almost everywhere, and $\psi M(y) \geq 0$.

Before moving on let us briefly investigate the iterates of the Markov operator M. Recall that when we studied finite Markov chains, the t-th order kernel $p^t(x,y)$ was defined by

$$p^1 := p, \quad p^t(x,y) := \sum_{z \in S} p^{t-1}(x,z)p(z,y)$$

Analogously, let p be a stochastic *density* kernel p, and define a sequence $(p^t)_{t \geq 1}$ of kernels by

$$p^1 := p, \quad p^t(x,y) := \int p^{t-1}(x,z)p(z,y)dz \qquad (8.16)$$

Below p^t is called the t-th order density kernel corresponding to p. As in the finite state case, $p^t(x,y)dy$ can be interpreted as the distribution (density) of X_t when $X_0 = x$.

Exercise 8.1.12 Using induction, verify that p^t is density kernel on S for each $t \in \mathbb{N}$.

Lemma 8.1.13 *If M is the Markov operator associated with stochastic density kernel p on S, then M^t is the Markov operator associated with p^t. In other words, for any $\psi \in D(S)$, we have*

$$\psi M^t(y) = \int p^t(x,y)\psi(x)dx \qquad (y \in S)$$

This lemma is the continuous state version of lemma 4.2.8 on page 76. Essentially it is telling us what we claimed above: that $p^t(x,y)dy$ is the distribution of X_t given $X_0 = x$. The proof is an exercise.

Given any kernel p on S, the Markov operator $\mathbf{M}$ is always continuous on $D(S)$ (with respect to d_1). In fact it is nonexpansive. To see this, observe that for any $\phi, \psi \in D(S)$, we have

$$\|\phi\mathbf{M} - \psi\mathbf{M}\|_1 = \int \left| \int p(x,y)(\phi(x) - \psi(x))dx \right| dy$$

$$\leq \int \int p(x,y)|\phi(x) - \psi(x)|dxdy$$

$$= \int \int p(x,y)dy|\phi(x) - \psi(x)|dx = \|\phi - \psi\|_1$$

8.2 Stability

Let's now turn to the topic of stability. The bad news is that for density Markov chains on infinite state spaces, the theory is considerably more complex than for the finite case (see §4.3.3). The good news is that we can build on the intuition gained from studying the finite case, and show how the concepts can be extended to cope with infinite state spaces. After reviewing the density analogue of the Dobrushin coefficient, we look at drift conditions that keep probability mass within a "bounded" region of the state space as the Markov chain evolves. Combining drift conditions with a concept related to positivity of the Dobrushin coefficient, we obtain a rather general sufficient condition for stability of density Markov chains, and apply it to several applications.

8.2.1 The Big Picture

Before plunging into the formal theory of stability, we are going spend some time building intuition. In particular, we would like to know under what circumstances stability will *fail*, with the aim of developing conditions that rule out these kinds of circumstances. This section considers these issues in a relatively heuristic way. We begin with the definitions of stationary densities and global stability.

Let p be a stochastic density kernel on S, and let $\mathbf{M}$ be the corresponding Markov operator. As before, S is a Borel subset of $\mathbb{R}^n$ endowed with the standard euclidean metric d_2. Since $\mathbf{M}$ sends $D(S)$ into $D(S)$, and since $D(S)$ is a well-defined metric space with the distance d_1 (see §7.3.3), the pair $(D(S), \mathbf{M})$ is a dynamical system in the sense of chapter 4. The trajectory $(\psi\mathbf{M}^t)_{t \geq 0}$ of a point $\psi \in D(S)$ corresponds to the sequence of marginal distributions for a Markov-(p, ψ) process $(X_t)_{t \geq 0}$.

Now consider the stability properties of the dynamical system $(D(S), \mathbf{M})$. We are interested in existence of fixed points and global stability. A fixed point ψ^* of $\mathbf{M}$ is also called a *stationary density,* and, by definition, satisfies

$$\psi^*(y) = \int p(x,y)\psi^*(x)dx \qquad (y \in S)$$

Exercise 8.2.1 Consider the linear AR(1) model

$$X_{t+1} = aX_t + W_{t+1}, \quad (W_t)_{t \geq 1} \overset{\text{IID}}{\sim} \phi = N(0,1) \tag{8.17}$$

with $|a| < 1$. The corresponding density kernel is $p(x,y)dy = N(ax,1)$. Using pencil and paper, show that the normal density $N(0, 1/(1-a^2))$ is stationary for this kernel.

In the finite state case every Markov chain has a stationary distribution (theorem 4.3.5, page 86). When S is not finite, however, stationary distributions can easily fail to exist. Consider, for example, the model (8.17) with $\alpha = 1$, which is called a random walk. With a bit of thought, you will be able to convince yourself that $X_t \sim N(X_0, t)$, and hence

$$p^t(x,y) = \frac{1}{\sqrt{2\pi t}} \exp\left(\frac{-(y-x)^2}{2t}\right) \qquad ((x,y) \in \mathbb{R} \times \mathbb{R})$$

Exercise 8.2.2 Show that p has no stationary distribution by arguing that if ψ^* is stationary, then

$$\psi^*(y) = \int p^t(x,y)\psi^*(x)dx \qquad (t \in \mathbb{N}, \ y \in \mathbb{R})$$

in which case the dominated convergence theorem implies that $\psi^*(y) = 0$ for all $y \in \mathbb{R}$. (Contradicting what?)

Translated to the present context, global stability of $(D(S), \mathbf{M})$ is equivalent to the existence of a unique stationary density ψ^* such that

$$\psi\mathbf{M}^t \to \psi^* \text{ in } d_1 \text{ as } t \to \infty \text{ for every } \psi \in D(S)$$

Let's try to work out when such stability can be expected. First, we have to rule out the kind of behavior exhibited by the random walk above. In this case the density of X_t becomes more and more spread out over $\mathbb{R}$. In fact ψ_t converges to zero everywhere because

$$\psi_t(y) = (\psi\mathbf{M}^t)(y) = \int p^t(x,y)\psi(x)dx \to 0 \qquad (t \to \infty)$$

for all $y \in \mathbb{R}$ by the dominated convergence theorem.

A similar problem arises when densities are diverging off to the right or the left, as illustrated in figure 8.4. In either case probability mass is escaping from the "center" of the state space. In other words, it is not concentrating in any one place.

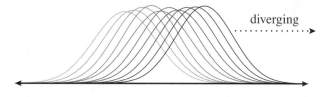

Figure 8.4 Divergence to $+\infty$

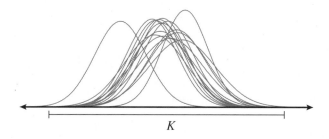

Figure 8.5 Nondiverging sequence

What we need then is to ensure that densities concentrate in one place, or that "most" of the probability mass stays in the "center" for all densities in the trajectory. Another way of putting this is to require that, for each density, most of the mass is on a bounded set K, as in figure 8.5. More mathematically, we require the existence of a bounded set K such that $\int_K \psi_t(x)dx \cong 1$ for all t. (Requiring that *all* mass stays on K is too strict because there will always be tails of the densities poking out.)

Now, having said that K must be bounded, the truth of the matter is that boundedness is not enough. Suppose that we are studying a system not on $S = \mathbb{R}$, but rather on $S = (-1, 1)$. And suppose, for example, that the densities are shifting all their mass toward 1.[2] Now this is also an example of instability (after all, there is no limiting density for such a sequence to converge to), and it has a similar feel to the divergence in figure 8.4 (in the sense that probability mass is leaving the center of the state space). But we cannot rule this problem out by requiring that probability remains on some bounded set $K \subset (-1, 1)$; it already stays on the bounded set $(-1, 1)$. To keep probability mass in the center of the state space, what we really need is an interval $[a, b]$ with $[a, b] \subset (-1, 1)$ and most of the mass remaining on $[a, b]$.

A suitable condition, then, is to require that K be not only bounded but also *com-*

---

[2]We can imagine this might be the case if we took the system in figure 8.4 and transformed it via a change of variables such as $y = \arctan(x)$ into a system on $(-1, 1)$.

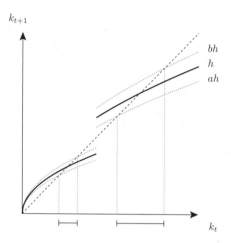

Figure 8.6 Multiple invariant sets

pact. That is, we require that most of the probability mass stays on a *compact* subset of S. For example, we might require that, given any $\epsilon > 0$, there is a compact set $K \subset S$ such that $\int_K \psi_t(x)dx \geq 1 - \epsilon$ for all t. Indeed this is precisely the definition of *tightness,* and we will see it plays a crucial role in what follows.

Tightness is necessary for stability, but it is not sufficient. For example, consider the model represented in figure 8.6, which is a stochastic version of the model in example 4.1.8 (page 57). The deterministic model is $k_{t+1} = h(k_t) := sA(k_t)k_t^\alpha$, and the function h is the bold curve in figure 8.6. The stochastic version is given by $k_{t+1} = W_{t+1}h(k_t)$, where the shock sequence $(W_t)_{t \geq 1}$ is supported on a bounded interval $[a, b]$. The functions $k \mapsto ah(k)$ and $k \mapsto bh(k)$ are represented by dashed lines.

A little thought will convince you that the two intervals marked in the figure are *invariant sets,* which is to say that if the state enters either of these sets, then it cannot escape—it remains there with probability one. As a result global stability fails, even though tightness appears likely to hold. (It does, as we'll see later on.)

The problem here is that there is insufficient *mixing* for global stability to obtain. What we need, then, is a condition to ensure sufficient mixing, as well as a condition for tightness. Finally, we need a minor technical condition called uniform integrability that ensures trajectories do not pile up on a set of measure zero—as does, for example, the sequence of densities $N(0, 1/n)$—in which case the limiting distribution cannot be represented by a density, and hence does not exist in $D(S)$.

We now turn to a formal treatment of these conditions for stability.

8.2.2   Dobrushin Revisited

On page 89 we introduced the Dobrushin coefficient for the finite case. Replacing sums with integrals, we get

$$\alpha(p) := \inf \left\{ \int p(x,y) \wedge p(x',y)dy \, : \, (x,x') \in S \times S \right\}$$

For densities f and g the pointwise minimum $f \wedge g$ is sometimes called the *affinity* between f and g, with a maximum of 1 when $f = g$ and a minimum of zero when f and g have disjoint supports. The Dobrushin coefficient reports the infimum of the affinities for all density pairs in the stochastic kernel.

The proof of theorem 4.3.17 (page 89) carries over to the density case almost unchanged, replacing sums with integrals at each step. That is to say,

$$\|\phi\mathbf{M} - \psi\mathbf{M}\|_1 \leq (1 - \alpha(p))\|\phi - \psi\|_1 \qquad \forall \, \phi, \psi \in D(S)$$

and, moreover, this bound is the best available, in the sense that

$$\forall \lambda < 1 - \alpha(p), \; \exists \, \phi, \psi \in D(S) \text{ such that } \|\phi\mathbf{M} - \psi\mathbf{M}\|_1 > \lambda\|\phi - \psi\|_1 \qquad (8.18)$$

From completeness of $D(S)$, Banach's fixed point theorem, and lemma 4.1.21 (see page 61), it now follows (supply the details) that $(D(S), \mathbf{M})$ is globally stable whenever there exists a $t \in \mathbb{N}$ such that $\alpha(p^t) > 0$.

On the other hand, positivity of $\alpha(p^t)$ for some t is no longer *necessary* for global stability.[3] This is fortunate because in many applications we find that $\alpha(p^t) = 0$ for all $t \in \mathbb{N}$. To give an example, consider the stochastic density kernel p given by $p(x,y)dy = N(ax,1)$, which corresponds to the AR(1) process (8.17). If $|a| < 1$ then this process is globally stable, as was shown in chapter 1 using elementary arguments. However, it turns out that $\alpha(p^t) = 0$ for every $t \in \mathbb{N}$. Indeed, for fixed t, $p^t(x,y)dy = N(cx,d)$ for some constants c and d. Choosing x, x' so that $cx = n \in \mathbb{N}$ and $cx' = -n$, the integral of $p^t(x,y) \wedge p^t(x',y)$ is the area of the two tails shown in figure 8.7. This integral can be made arbitrarily small by choosing n sufficiently large.

Thus, in view of (8.18), the Markov operator $\mathbf{M}$ associated with the AR(1) process is not uniformly contracting, and neither is any iterate $\mathbf{M}^t$. Hence Banach's fixed point theorem does not apply.

Fortunately we can get around this negative result and produce a highly serviceable sufficient condition for stability. However, a bit of fancy footwork is required. The rest of this section explains the details.

---

[3]If you did the proof of necessity in exercise 4.3.22 (page 90) you will understand that finiteness of S is critical.

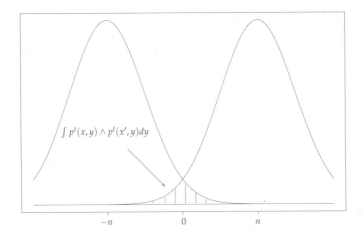

$$\int p^t(x,y) \wedge p^t(x',y)dy$$

$-n$ 0 n

Figure 8.7 Dobrushin coefficient is zero

First, even when $\mathbf{M}^t$ fails to be a uniform contraction on $D(S)$, it may still be a *contraction*, in the sense that

$$\|\phi\mathbf{M}^t - \psi\mathbf{M}^t\|_1 < \|\phi - \psi\|_1 \text{ whenever } \phi \neq \psi \qquad (8.19)$$

In fact the following result holds (see the appendix to this chapter for a proof.)

Lemma 8.2.3 *Let $t \in \mathbb{N}$, let p be a stochastic density kernel on S, and let $\mathbf{M}$ be the Markov operator corresponding to p. If*

$$\int p^t(x,y) \wedge p^t(x',y)dy > 0 \text{ for all } x, x' \in S \qquad (8.20)$$

then $\mathbf{M}^t$ is a contraction on $D(S)$; that is, (8.19) holds.

The existence of a t such that (8.20) holds is a mixing condition. (The need for mixing was discussed in §8.2.1.) A simple but important special case is when p is strictly positive on $S \times S$ (an example is the kernel p associated with the AR(1) process in exercise 8.2.1), in which case $p(x,y) \wedge p(x',y) > 0$ for each y. Integrating a positive function over a set of positive measure produces a positive number (theorem 7.3.9, page 181), and condition (8.20) is satisfied.

Contractiveness can be used to prove global stability when paired with compactness of the state space. In particular, if $h\colon U \to U$ is contracting and U is compact, then the dynamical system (U, h) is globally stable (see theorem 3.2.38 on page 54). In our case this result is potentially helpful, but does not immediately apply. The reasons is that when S is infinite, $D(S) = (D(S), d_1)$ is not compact. The next two exercises help to illustrate why this is so.

Exercise 8.2.4 Let $S = \mathbb{R}$, and let $(\phi_n)_{n \geq 1} \subset D(S)$ be given by $\phi_n := \mathbb{1}_{[n,n+1)}$. Show that $d_1(\phi_n, \phi_m) = 2$ whenever $n \neq m$. Conclude that this sequence has no subsequence converging to a point in $D(S)$.

Exercise 8.2.5 Let $S = (0,1)$, and let $(\phi_n)_{n \geq 1} \subset D(S)$ be given by $\phi_n := n \cdot \mathbb{1}_{(0,1/n)}$. Suppose that $d_1(\phi_n, \phi) \to 0$ for some $\phi \in D(S)$. Using your measure-theoretic bag of tricks, show that $\lambda(\phi) = 0$, contradicting $\phi \in D(S)$. You have now shown that the sequence itself cannot converge to any point in $D(S)$. Argue that the same is true for any subsequence.

Fortunately there is a way around this problem created by lack of compactness of $D(S)$. Recall from exercise 4.1.9 (page 58) that a dynamical system (U, h) is called Lagrange stable if every trajectory is precompact in U. Lagrange stability is weaker than compactness of U,[4] but it turns out that *if (U, h) is contracting and Lagrange stable then it is globally stable.*[5]

So suppose that (8.20) holds for some $t \in \mathbb{N}$, and hence $\mathbf{M}^t$ is contracting. If we can prove that all trajectories of $(D(S), \mathbf{M})$ are precompact, then all trajectories of $(D(S), \mathbf{M}^t)$ are also precompact (subsets of precompact sets are precompact), and $(D(S), \mathbf{M}^t)$ is globally stable. Finally, lemma 4.1.21 (page 61) implies that if $(D(S), \mathbf{M}^t)$ is globally stable, then so is $(D(S), \mathbf{M})$. Let's record this as a theorem.

Theorem 8.2.6 *Let $(D(S), \mathbf{M})$ be Lagrange stable. If $\mathbf{M}^t$ is a contraction for some $t \in \mathbb{N}$, then $(D(S), \mathbf{M})$ is globally stable.*

But how to prove precompactness of trajectories? We need the following two definitions:

Definition 8.2.7 Let $\mathcal{M}$ be a subset of $D(S)$. The collection of densities $\mathcal{M}$ is called *tight* if

$$\forall \epsilon > 0, \ \exists \text{ a compact set } K \subset S \text{ such that } \sup_{\psi \in \mathcal{M}} \int_{K^c} \psi(x)dx \leq \epsilon$$

It is called *uniformly integrable* if

$$\forall \epsilon > 0, \ \exists \delta > 0 \text{ such that } \lambda(A) < \delta \text{ implies } \sup_{\psi \in \mathcal{M}} \int_A \psi(x)dx \leq \epsilon$$

---

[4]If U is compact then (U, h) is always Lagrange stable (why?), but the converse is not true (example?).

[5]Proof: Fix $x \in U$, and define $\Gamma(x)$ to be the closure of $\{h^n(x) : n \in \mathbb{N}\}$. The set $\Gamma(x)$ is a compact subset of U. (Why?) Moreover h sends $\Gamma(x)$ into itself (see exercise 4.1.3, page 56). Hence $(\Gamma(x), h)$ is a dynamical system where h is contracting and $\Gamma(x)$ is compact, implying the existence of a unique fixed point $x^* \in \Gamma(x)$ with $h^n(x) \to x^*$ (theorem 3.2.38, page 54). Finally, h has at most one fixed point in U by contractiveness over U. Therefore x^* does not depend on x, and (U, h) is globally stable.

Here λ is Lebesgue measure. Essentially, tightness rules out the violation of compactness seen in exercise 8.2.4, while uniform integrability rules out that seen in exercise 8.2.5. Tightness and uniform integrability are important to us because of the following result:

Theorem 8.2.8 *Let p be a stochastic density kernel on S, and let $\mathbf{M}$ be the corresponding Markov operator. Let $\psi \in D(S)$. If the sequence $(\psi \mathbf{M}^t)_{t \geq 0}$ is both tight and uniformly integrable, then it is also precompact in $D(S)$.*

While the proof is omitted, those readers familiar with functional analysis will understand that tightness and uniform integrability together imply weak precompactness in L_1. Further, the fact that $\mathbf{M}$ is an integral operator means it is sufficiently smoothing that it sends weakly precompact sets into (strongly) precompact sets. The proof of the latter result can be found in Lasota (1994, thm. 4.1).

8.2.3 Drift Conditions

How might one verify the tightness and uniform integrability of a given trajectory $(\psi \mathbf{M}^t)_{t \geq 0}$? Tightness is usually established using drift inequalities. We will use a variety that rely on the concept of norm-like functions (Meyn and Tweedie 1993):

Definition 8.2.9 A measurable function $w \colon S \to \mathbb{R}_+$ is called *norm-like* if all of its sublevel sets (i.e., sets of the form $C_a := \{x \in S : w(x) \leq a\}$, $a \in \mathbb{R}_+$) are precompact in (S, d_2).

Example 8.2.10 Let $S = \mathbb{R}^n$, and let $w(x) := \|x\|$, where $\|\cdot\|$ is any norm on $\mathbb{R}^n$ (definition 3.1.2, page 37). This function w is norm-like on S because the sublevel sets of w are bounded with respect to the metric induced by $\|\cdot\|$, and hence bounded for d_2 (theorem 3.2.30, page 50). For subsets of $(\mathbb{R}^n, d_2)$, boundedness implies precompactness (theorem 3.2.19, page 48).

We can now state our drift condition:

Definition 8.2.11 Let p be a stochastic density kernel on S. We say that p satisfies *geometric drift to the center* if there exists a norm-like function w on S and positive constants $\alpha < 1$ and $\beta < \infty$ such that

$$\int w(y) p(x, y) dy \leq \alpha w(x) + \beta \qquad (x \in S)$$

As we will see, this condition is often easy to check in applications. Moreover

Proposition 8.2.12 *If p satisfies geometric drift to the center, then $(\psi \mathbf{M}^t)$ is tight for every $\psi \in D(S)$.*

The proof is given in the appendix to the chapter, but the intuition is not difficult: Under geometric drift, the ratio $\int w(y)p(x,y)dy/w(x)$ is dominated by $\alpha + \beta/w(x)$. Norm-like functions tend to get large as x moves away from the center of the state space, so if x is sufficiently far from the center, then $\alpha + \beta/w(x) < 1$. In which case $\int w(y)p(x,y)dy$, the expectation of $w(X_{t+1})$ given $X_t = x$, is less than $w(x)$. This in turn means that probability mass is moving back toward the center, where w is smaller. Keeping probability mass in the center of the state space is the essence of tightness.

Let's conclude this section by showing that the AR(1) model

$$X_{t+1} = aX_t + b + W_{t+1}, \quad (W_t)_{t \geq 1} \overset{\text{IID}}{\sim} \phi = N(0,1), \quad |a| < 1$$

is Lagrange stable. Since its stochastic kernel

$$p(x,y) = \phi(y - ax - b), \quad \phi(z) := \frac{1}{\sqrt{2\pi}} \exp(-z^2/2)$$

is strictly positive on $\mathbb{R} \times \mathbb{R}$ (and hence its Markov operator is contracting), this implies global stability.

Regarding tightness, let w be the norm-like function $|\cdot|$. The change of variable $z = y - ax - b$ gives

$$\int |y|p(x,y)dy = \int |y|\phi(y - ax - b)dy$$

$$= \int |ax + b + z|\phi(z)dz \leq \alpha|x| + \beta, \quad \alpha := |a|, \ \beta := |b| + \int |z|\phi(z)dz$$

Since $|a| < 1$ the geometric drift condition is satisfied, and, in view of proposition 8.2.12, every trajectory is tight.

Only uniform integrability of trajectories remains to be checked. To do this, pick any $\psi \in D(S)$. Observe there is a constant K such that $p(x,y) \leq K$ for every x,y. Hence, for any $A \in \mathcal{B}(\mathbb{R}), t \in \mathbb{N}$,

$$\int_A \psi\mathbf{M}^t(y)dy = \int_A \left[\int p(x,y)\psi\mathbf{M}^{t-1}(x)dx \right] dy$$

$$= \int \left[\int_A p(x,y)dy \right] \psi\mathbf{M}^{t-1}(x)dx \leq \int K\lambda(A)\psi\mathbf{M}^{t-1}(x)dx = K\lambda(A)$$

Now fix $\epsilon > 0$. If $\lambda(A) < \epsilon/K$, then $\int_A \psi\mathbf{M}^t(y)dy < \epsilon$, independent of t. Hence uniform integrability of $(\psi\mathbf{M}^t)_{t \geq 0}$ is established. Theorem 8.2.8 now tells us that $(D(S), \mathbf{M})$ is Lagrange stable, and hence globally stable.

We will see in the next section that these ideas can be used to prove the stability of much more complex models. Before discussing applications in earnest, let's try to package our results in a simple format. On the Lagrange stability side, we can make our life easier with the following result:

Proposition 8.2.13 *Let $\psi \in D(S)$, let p be a stochastic density kernel on S, and let $\mathbf{M}$ be the corresponding Markov operator. If the sequence $(\psi \mathbf{M}^t)_{t \geq 0}$ is tight, and in addition there exists a continuous function $m \colon S \to \mathbb{R}$ such that $p(x,y) \leq m(y)$ for all $x, y \in S$, then $(\psi \mathbf{M}^t)_{t \geq 0}$ is also uniformly integrable.*

The proof is an extension of the proof of uniform integrability of the AR(1) system above, where we used the fact that p is bounded above by a constant (which is certainly a continuous function). It is given in the appendix to the chapter and should be skipped on first pass.

Now let's put this all together:

Theorem 8.2.14 *Let p be a stochastic density kernel on S, and let $\mathbf{M}$ be the corresponding Markov operator. If*

1. *$\exists\, t \in \mathbb{N}$ such that $\int p^t(x,y) \wedge p^t(x',y) dy > 0$ for all $(x, x') \in S \times S$,*

2. *p satisfies geometric drift to the center, and*

3. *$\exists$ a continuous $m \colon S \to \mathbb{R}$ such that $p(x,y) \leq m(y)$ for all $x, y \in S$,*

then the dynamical system $(D(S), \mathbf{M})$ is globally stable.

Proof. In view of theorem 8.2.6, we need only show that $(\psi \mathbf{M}^t)_{t \geq 0}$ is precompact for every $\psi \in D(S)$. So pick any $\psi \in D(S)$. Since p satisfies geometric drift, $(\psi \mathbf{M}^t)_{t \geq 0}$ is tight. By proposition 8.2.13 it is also uniformly integrable, and therefore precompact (theorem 8.2.8). □

Just as for the finite state case, stability is connected with the law of large numbers (recall theorem 4.3.33 on page 94). For example, take the stochastic recursive sequence

$$X_{t+1} = F(X_t, W_{t+1}), \quad X_0 \sim \psi, \quad (W_t)_{t \geq 1} \stackrel{\text{IID}}{\sim} \phi \tag{8.21}$$

where the state space S is a Borel subset of $\mathbb{R}^n$, Z is a Borel subset of $\mathbb{R}^k$, $\phi \in D(Z)$ and $\psi \in D(S)$. Let kernel p represent this SRS on S in the sense of (8.3) on page 189. Let $\mathbf{M}$ be the corresponding Markov operator. We have the following result:

Theorem 8.2.15 *Let $h \colon S \to \mathbb{R}$ be a Borel measurable function, and let $(X_t)_{t \geq 0}$, the kernel p and the Markov operator $\mathbf{M}$ be as above. If $(D(S), \mathbf{M})$ is globally stable with stationary distribution ψ^*, then*

$$\frac{1}{n} \sum_{t=1}^{n} h(X_t) \to \int h(x)\psi^*(x) dx \quad \text{as } n \to \infty$$

with probability one whenever $\int |h(x)| \psi^*(x) dx$ is finite.

The meaning of probability one convergence will be discussed later, but for now you can understand it as it sounds: The probability of generating a path $(W_t)_{t\geq 1}$ such that this convergence fails is zero. Notice that convergence holds independent of the initial condition ψ.

The proof of the theorem is beyond the scope of this book. See Nummelin (1984, prop. 6.3) and Meyn and Tweedie (1993, thm. 17.1.7). Note that theorem 8.2.15 justifies the stationary density look-ahead estimator introduced in §6.1.4, at least when stability holds. Also, as we saw in the finite state case, the LLN leads to a new *interpretation* of the stationary density that is valid in the globally stable case:

$$\int_B \psi^*(x)dx \cong \text{the fraction of time that } (X_t)_{t\geq 0} \text{ spends in } B$$

for any $B \in \mathscr{B}(S)$. To see this, take $h = \mathbb{1}_B$. Then theorem 8.2.15 gives

$$\frac{1}{n}\sum_{t=1}^{n} h(X_t) = \frac{1}{n}\sum_{t=1}^{n} \mathbb{1}_B(X_t) \to \int_B \psi^*(x)dx \quad \text{as } n \to \infty \qquad (8.22)$$

8.2.4   Applications

Let's turn to applications. To start, recall the model of commodity dynamics with speculation discussed in §6.3.2. The state evolves according to

$$X_{t+1} = \alpha I(X_t) + W_{t+1}, \quad (W_t)_{t\geq 1} \overset{\text{IID}}{\sim} \phi$$

where I is the equilibrium investment function defined in (6.27), page 146. Suppose for now that ϕ is a lognormal density. The stochastic kernel is

$$p(x,y) = \phi(y - \alpha I(x)) \qquad ((x,y) \in S \times S)$$

where $\phi(z) = 0$ when $z < 0$. The state space is $S = \mathbb{R}_+$.

This model is easily seen to be globally stable. Regarding condition 1 of theorem 8.2.14, pick any $x, x' \in S$. Let E be all $y \in S$ such that $y > \alpha I(x)$ and $y > \alpha I(x')$. On E the function $y \mapsto p(x,y) \wedge p(x',y)$ is strictly positive, and integrals of strictly positive functions on sets of positive measure are positive (theorem 7.3.9, page 181). Hence condition 1 holds for $t = 1$.

Regarding condition 2, let $w(x) = x$. This function is norm-like on S. (Proof?) Moreover geometric drift to the center holds because, using the change of variable $z = y - \alpha I(x)$,

$$\int yp(x,y)dy = \int y\phi(y - \alpha I(x))dy = \alpha I(x) + \int z\phi(z)dz \leq \alpha x + \int z\phi(z)dz$$

Condition 3 is trivial because $p(x, y) \leq K$ for some constant K, and constant functions are continuous. Hence global stability holds.

Next, recall the STAR model of example 6.1.3, where $Z = S = \mathbb{R}$, and the state evolves according to

$$X_{t+1} = g(X_t) + W_{t+1}, \quad (W_t)_{t \geq 1} \overset{\text{IID}}{\sim} \phi \in D(\mathbb{R}) \tag{8.23}$$

with $g(x) := (\alpha_0 + \alpha_1 x)(1 - G(x)) + (\beta_0 + \beta_1 x)G(x)$. Here $G \colon S \to [0, 1]$ is a smooth transition function satisfying $G' > 0$, $\lim_{x \to -\infty} G(x) = 0$, and $\lim_{x \to \infty} G(x) = 1$. Suppose that ϕ is bounded, everywhere positive on $\mathbb{R}$,

$$\gamma := |\alpha_1| \vee |\beta_1| < 1, \quad \text{and} \quad \int |z| \phi(z) dz < \infty$$

Exercise 8.2.16 Show that under these assumptions there exists a constant c such that $|g(x)| \leq \gamma |x| + c$ for all $x \in S = \mathbb{R}$.

Since the stochastic density kernel $p(x, y) = \phi(y - g(x))$ is strictly positive on $S \times S$, condition 1 of theorem 8.2.14 holds. Regarding condition 2, set $w(x) = |x|$. The change of variable $z = y - g(x)$ gives

$$\int |y| p(x, y) dy = \int |y| \phi(y - g(x)) dy = \int |g(x) + z| \phi(z) dz \leq \gamma |x| + c + \int |z| \phi(z) dz$$

Since $\gamma < 1$, condition 2 is satisfied.

Condition 3 is trivial because ϕ and hence p are bounded by some constant K. Setting $m(y) = K$ for all y gives a continuous upper bound.

As another application, consider again the threshold autoregression model, where $S = Z = \mathbb{R}^n$, and

$$X_{t+1} = \sum_{k=1}^{K} (A_k X_t + b_k) \mathbb{1}_{B_k}(X_t) + W_{t+1}, \quad (W_t)_{t \geq 1} \overset{\text{IID}}{\sim} \phi \in D(\mathbb{R}^n)$$

In exercise 8.1.5 you showed that the stochastic kernel is given by

$$p(x, y) = \phi \left[y - \sum_{k=1}^{K} (A_k x + b_k) \mathbb{1}_{B_k}(x) \right] \quad ((x, y) \in S \times S) \tag{8.24}$$

Assume that ϕ is strictly positive on $\mathbb{R}^n$, bounded, and that $\int \|z\| \phi(z) dz < \infty$ for some norm $\| \cdot \|$ on $\mathbb{R}^n$. Conditions 1 and 3 of theorem 8.2.14 can be verified in much the same way as the previous example. Regarding condition 2, let γ_k be a real number such that $\|A_k x\| \leq \gamma_k \|x\|$ for all x. Assume that $\gamma := \max_k \gamma_k < 1$. Then, using a change of variable again,

$$\int \|y\| p(x,y) dy = \int \left\| \sum_{k=1}^{K} (A_k x + b_k) \mathbb{1}_{B_k}(x) + z \right\| \phi(z) dz$$

$$\leq \sum_{k=1}^{K} \|A_k x + b_k\| \mathbb{1}_{B_k}(x) + \int \|z\| \phi(z) dz$$

$$\leq \sum_{k=1}^{K} \gamma_k \|x\| \mathbb{1}_{B_k}(x) + \sum_{k=1}^{K} \|b_k\| + \int \|z\| \phi(z) dz$$

$$\leq \gamma \|x\| + \beta, \qquad \beta := \sum_{k=1}^{K} \|b_k\| + \int \|z\| \phi(z) dz$$

Since $\gamma < 1$ and $\| \cdot \|$ is norm-like, condition 2 is also satisfied, and the model is globally stable.

Before starting the next application, let's think a little more about norm-like functions (i.e., nonnegative functions with precompact sublevel sets). For the metric space $(\mathbb{R}^n, d_2)$ we can equate precompactness with boundedness (theorem 3.2.19, page 48). For $S \subset \mathbb{R}^n$, bounded subsets of (S, d_2) are *not* necessarily precompact (e.g., see exercise 3.2.14 on page 47), making precompactness and hence the norm-like property harder to check. The next result gives some guidance when S is an open interval in $\mathbb{R}$. The proof is an exercise.

Lemma 8.2.17 *If $S = (u,v)$, where $u \in \{-\infty\} \cup \mathbb{R}$ and $v \in \{+\infty\} \cup \mathbb{R}$, then $w \colon S \to \mathbb{R}_+$ is norm-like if and only if $\lim_{x \to u} w(x) = \lim_{x \to v} w(x) = \infty$.*[6]

As a consequence $w(x) := |\ln x|$ is a norm-like function on $S = (0, \infty)$. We exploit this fact below.

Now for the last application. In exercise 8.1.8 you derived the stochastic kernel for the nonconvex growth model $k_{t+1} = sA(k_t)k_t^\alpha W_{t+1}$, where $S = Z = (0, \infty)$ and $(W_t)_{t \geq 1}$ is IID with density $\phi \in D(Z)$. It has the form

$$p(x,y) = \phi \left(\frac{y}{sA(x)x^\alpha} \right) \frac{1}{sA(x)x^\alpha} \qquad ((x,y) \in S \times S) \qquad (8.25)$$

Suppose that A takes values in $[a_1, a_2] \subset S$ and that $\alpha < 1$. Regarding the density ϕ, assume that ϕ is strictly positive on $(0, \infty)$, that $\int |\ln z| \phi(z) dz$ is finite, and that $\phi(z)z \leq M$ for some $M < \infty$ and all $z \in (0, \infty)$. For example, the lognormal density satisfies all of these conditions.

---

[6] Here $\lim_{x \to a} f(x) = \infty$ means that for any $x_n \to a$ and any $M \in \mathbb{N}$, there exists an $N \in \mathbb{N}$ such that $n \geq N$ implies $f(x_n) \geq M$. Hint: Show that $K \subset S$ is precompact in S if and only if no sequence in K converges to either u or v.

Now let's check the conditions of theorem 8.2.14. Condition 1 holds because p is strictly positive on $S \times S$. Regarding condition 2, set $w(x) = |\ln x|$, so

$$\int w(sA(x)x^\alpha z)\phi(z)dz = \int |\ln s + \ln A(x) + \alpha \ln x + \ln z|\phi(z)dz$$

$$\leq |\ln s| + |\ln A(x)| + \alpha |\ln x| + \int |\ln z|\phi(z)dz$$

Setting $\beta := |\ln s| + \max\{|\ln a_1|, |\ln a_2|\} + \int |\ln z|\phi(z)dz$ we obtain

$$\int w(sA(x)x^\alpha z)\phi(z)dz \leq \alpha |\ln x| + \beta = \alpha w(x) + \beta$$

Since w is norm-like on $(0, \infty)$, condition 2 is proved.

Finally, consider condition 3. Given any x and y in S we have

$$p(x,y) = p(x,y)\frac{y}{y} = \phi\left(\frac{y}{sA(x)x^\alpha}\right)\frac{y}{sA(x)x^\alpha}\frac{1}{y} \leq \frac{M}{y}$$

Since $m(y) := M/y$ is continuous on S, condition 3 is satisfied.

8.3   Commentary

The material in this chapter draws heavily on Lasota and Mackey (1994), and also on a slightly obscure but fascinating paper of Lasota (1994). Theorem 8.2.14 is from Mirman, Reffett, and Stachurski (2005). More details on the theory and an application to optimal growth can be found in Stachurski (2002). See also Stachurski (2003).

Chapter 9

Measure-Theoretic Probability

In the first few decades of the twentieth century, mathematicians realized that through measure theory it was possible to place the slippery subject of probability in a completely sound and rigorous framework, where manipulations are straightforward and powerful theorems can be proved. This integration of probability and measure yielded a standard language for research in probability and statistics shared by mathematicians and other scientists around the world. A careful read of this chapter will provide sufficient fluency to understand and participate in their conversation.

9.1   Random Variables

The language of probability begins with random variables and their distributions. Let's start with a detailed treatment of these topics, beginning with basic definitions and moving on to key concepts such as expectations and independence.

9.1.1   Basic Definitions

In probability theory, the term random variable is just another way of saying $\mathscr{F}$-measurable real-valued function on some measure space $(\Omega, \mathscr{F})$. In other words, a random variable on $(\Omega, \mathscr{F})$ is a map $X \colon \Omega \to \mathbb{R}$ with the property that $X^{-1}(B) \in \mathscr{F}$ for all $B \in \mathscr{B}(\mathbb{R})$. For historical reasons random variables are typically written with upper-case symbols such as X and Y, rather than lower-case symbols such as f and g. The measurable space $(\Omega, \mathscr{F})$ is usually paired with a probability measure $\mathbb{P}$ (see page 167), which assigns probabilities to events $E \in \mathscr{F}$.

Why restrict attention to $\mathscr{F}$-measurable functions? Well, suppose that $\mathbb{P}$ is a probability on $(\Omega, \mathscr{F})$. We can think of a draw from $\mathbb{P}$ as an experiment that results in a nonnumerical outcome $\omega \in \Omega$, such as "three heads and then two tails." In order to make the outcome of this experiment more amenable to analysis, we specify a function $X \colon \Omega \to \mathbb{R}$ that maps outcomes into numbers. Suppose further that we wish to evaluate the probability that $X \geq a$, or

$$\mathbb{P}\{\omega \in \Omega : X(\omega) \geq a\} :=: \mathbb{P}\{X \geq a\} :=: \mathbb{P}X^{-1}([a, \infty))$$

Since $\mathbb{P}$ is only defined on the sets in $\mathscr{F}$, this requires $X^{-1}([a, \infty)) \in \mathscr{F}$. The latter is guaranteed by $\mathscr{F}$-measurability of X.

Actually, the definition of random variables as real-valued functions is not general enough for our purposes. We need to consider random "objects," which are like (real-valued) random variables except that they take values in other spaces (such as $\mathbb{R}^n$, or abstract metric space). Some authors use the term "random object," but we will call them all random variables:

Definition 9.1.1 Let $(\Omega, \mathscr{F}, \mathbb{P})$ be a probability space, and let $(S, \mathscr{S})$ be any measurable space. An *S-valued random variable* is a function $X \colon \Omega \to S$ that is $\mathscr{F}, \mathscr{S}$-measurable: $X^{-1}(B) \in \mathscr{F}$ whenever $B \in \mathscr{S}$. The *distribution* of X is the unique measure $\mu_X \in \mathscr{P}(S, \mathscr{S})$ defined by

$$\mu_X(B) := \mathbb{P}(X^{-1}(B)) = \mathbb{P}\{\omega \in \Omega : X(\omega) \in B\} \qquad (B \in \mathscr{S})$$

Note that μ_X, which gives the probability that $X \in B$ for each $B \in \mathscr{S}$, is just the image measure $\mathbb{P} \circ X^{-1}$ of $\mathbb{P}$ under X (see page 183 for a discussion of image measures). Distributions play a central role in probability theory.

A quick but important point on notation: In probability theory *it is standard to use the abbreviation* $\{X$ has property $P\}$ *for the set* $\{\omega \in \Omega : X(\omega)$ has property $P\}$. We will follow this convention. Similarly, $\mathbb{1}\{X \in A\}$ is the indicator function for the set $\{\omega \in \Omega : X(\omega) \in A\}$.

Exercise 9.1.2 Let S be a metric space, and let $(\Omega, \mathscr{F})$ be any measurable space. Let $f \colon \Omega \to S$ be $\mathscr{F}, \mathscr{B}(S)$-measurable, and let $g \colon S \to \mathbb{R}$ be a continuous function. Show that $X := g \circ f$ is a random variable on $(\Omega, \mathscr{F})$.

The integral of a real-valued random variable X on $(\Omega, \mathscr{F}, \mathbb{P})$ is called its *expectation*, and written $\mathbb{E}(X)$ or just $\mathbb{E}X$. That is,

$$\mathbb{E}X := \int X \, d\mathbb{P} :=: \int X(\omega)\mathbb{P}(d\omega) :=: \mathbb{P}(X)$$

The definition on the far right is the linear functional notation for the integral, which I personally prefer, as it eliminates the need for the new symbol $\mathbb{E}$, and forces us to specify the underlying probability whenever we want to take expectations. However, the $\mathbb{E}$ notation is more traditional, and is used in all of what follows—albeit with some resentment on my part.

Our definition of $\mathbb{E}$ has a sound probabilistic interpretation. If X is simple, taking values $\alpha_1, \ldots, \alpha_N$ on $A_1, \ldots, A_N$ respectively, then the expectation $\mathbb{E}X = \sum_n \alpha_n \mathbb{P}(A_n)$, which is the sum of all possible values taken by the random variable X multiplied by the probability that each such value occurs. If X is not simple, then to calculate expectation we approximate X by simple functions (see page 177), in which case similar intuition applies.

Exercise 9.1.3 Consider the probability space $(S, \mathscr{S}, \delta_z)$, where δ_z is the degenerate probability measure introduced in §7.1.3. The expectation of $f \in m\mathscr{S}^+$ is $\mathbb{E}f := \int f d\delta_z$. Intuitively, $\mathbb{E}f = f(z)$, since we are sure that δ_z will pick out the point z. Confirm this intuition.

There is a close connection between distributions and expectations:

Theorem 9.1.4 *If X is an S-valued random variable on $(\Omega, \mathscr{F}, \mathbb{P})$ with distribution $\mu_X \in \mathscr{P}(S, \mathscr{S})$, and if $w \in m\mathscr{S}$ is nonnegative or $\mathbb{E}|w(X)| < \infty$, then*

$$\mathbb{E}w(X) := \int w \circ X \, d\mathbb{P} = \int w \, d\mu_X =: \mu_X(w) \tag{9.1}$$

This is just a special case of theorem 7.3.17 on page 182, which gives

$$\int w \, d\mu_X := \int w \, d(\mathbb{P} \circ X^{-1}) = \int w \circ X \, d\mathbb{P}$$

Let $(\Omega, \mathscr{F}, \mathbb{P})$ be a probability space, let X be a real-valued random variable on this space, and let $k \in \mathbb{N}$. The k-th *moment* of X is $\mathbb{E}(X^k)$, which may or may not exist as an expectation. By definition, existence of the expectation requires that $\mathbb{E}|X|^k < \infty$. The elementary inequality $a^j \leq a^k + 1$ for all $j \leq k$ and $a \geq 0$ can be used to prove that if $j \leq k$ and the k-th moment is finite, then so is the j-th: By the previous bound we have $|X|^j \leq |X|^k + \mathbb{1}_\Omega$, where the inequality is understood to hold pointwise on Ω (i.e., for all $\omega \in \Omega$). Referring to properties M1–M5 of the integral (page 179), we conclude that $\mathbb{E}|X|^j \leq \mathbb{E}|X|^k + 1 < \infty$.

Let X and Y be real-valued random variables with finite second moment. The *variance* of X is the real number $\mathrm{Var}(X) := \mathbb{E}[(X - \mathbb{E}X)^2]$, while the *covariance* of X and Y is

$$\mathrm{Cov}(X, Y) := \mathbb{E}[\, (X - \mathbb{E}X)(Y - \mathbb{E}Y)\,]$$

Exercise 9.1.5 Show that if X has finite second moment and if a and b are constants, then $\mathrm{Var}(aX + b) = a^2 \mathrm{Var}(X)$. Show in addition that if $\mathrm{Cov}(X, Y) = 0$, then $\mathrm{Var}(X + Y) = \mathrm{Var}(X) + \mathrm{Var}(Y)$.

In the results that follow, we will often need to say something along the lines of "let X be a random variable on $(\Omega, \mathscr{F}, \mathbb{P})$ taking values in $(S, \mathscr{S})$ and having distribution μ." It is fortunate then that:

Theorem 9.1.6 *Given any measurable space $(S, \mathscr{S})$ and $\mu \in \mathscr{P}(S, \mathscr{S})$, there exists a probability space $(\Omega, \mathscr{F}, \mathbb{P})$ and a random variable $X \colon \Omega \to S$ such that X has distribution μ.*

Exercise 9.1.7 Prove theorem 9.1.6 by setting $(\Omega, \mathscr{F}, \mathbb{P}) = (S, \mathscr{S}, \mu)$ and $X =$ the identity map on S. Show, in particular, that X is measurable and has distribution μ.

Sometimes it's useful to construct a supporting probability space a little more explicitly. Consider an arbitrary distribution in $\mathscr{P}(\mathbb{R})$, which can be represented by a cumulative distribution function H.[1] For simplicity, we assume that H is strictly increasing (a discussion of the general case can be found in Williams 1991, ch. 3). Let $X \colon (0, 1) \to \mathbb{R}$ be the inverse of $H \colon X = H^{-1}$. Then X is a random variable on $(\Omega, \mathscr{F}, \mathbb{P}) = ((0, 1), \mathscr{B}(0, 1), \lambda)$ with distribution H. (Here λ is Lebesgue measure.)

Leaving measurability as an exercise, let's confirm that X has distribution H. Fix $z \in \mathbb{R}$. Note that for each $u \in (0, 1)$ we have $X(u) \leq z$ if and only if $u \leq H(z)$, and hence

$$
\begin{aligned}
\mathbb{P}\{X \leq z\} &:= \mathbb{P}\{u \in (0, 1) : X(u) \leq z\} \\
&= \mathbb{P}\{u \in (0, 1) : u \leq H(z)\} = \lambda((0, H(z)]) = H(z)
\end{aligned}
$$

Thus X has distribution H, as was to be shown.

Next we introduce *Chebychev's inequality*, which allows us to bound tail probabilities in terms of expectations—the latter being easier to calculate in many applications.

Theorem 9.1.8 *Let X be an S-valued random variable on $(\Omega, \mathscr{F}, \mathbb{P})$. If $h \colon S \to \mathbb{R}_+$ is a measurable function and $\delta \in \mathbb{R}$, then $\delta \, \mathbb{P}\{h(X) \geq \delta\} \leq \mathbb{E}h(X)$.*

Proof. Observe that $h(X) \geq h(X) \, \mathbb{1}\{h(X) \geq \delta\} \geq \delta \, \mathbb{1}\{h(X) \geq \delta\}$ pointwise on Ω. Now integrate, using properties M1–M5 as required. □

Two special cases are used repeatedly in what follows: First, if X is a nonnegative random variable then setting $h(x) = x$ gives

$$
\mathbb{P}\{X \geq \delta\} \leq \frac{\mathbb{E}X}{\delta} \qquad (\delta > 0) \tag{9.2}
$$

Second, application of the bound to $Y := X - \mathbb{E}X$ with $h(x) = x^2$ gives

$$
\mathbb{P}\{|X - \mathbb{E}X| \geq \delta\} = \mathbb{P}\{(X - \mathbb{E}X)^2 \geq \delta^2\} \leq \frac{\mathrm{Var}(X)}{\delta^2} \qquad (\delta > 0) \tag{9.3}
$$

As we will see, (9.3) can be used to prove a law of large numbers.

---

[1] Recall theorem 7.1.33 on page 170.

9.1.2   Independence

We have already used the concept of independence repeatedly in the text. It's now time for a formal definition.

Definition 9.1.9 Random variables X and Y taking values in $(S, \mathscr{S})$ and $(T, \mathscr{T})$ respectively are said to be *independent* whenever

$$\mathbb{P}\{X \in A\} \cap \{Y \in B\} = \mathbb{P}\{X \in A\} \cdot \mathbb{P}\{Y \in B\} \text{ for all } A \in \mathscr{S} \text{ and } B \in \mathscr{T}$$

More generally, a finite collection of random variables $X_1, \ldots, X_n$ with X_i taking values in $(S_i, \mathscr{S}_i)$ is called independent if

$$\mathbb{P} \cap_{i=1}^n \{X_i \in A_i\} = \prod_{i=1}^n \mathbb{P}\{X_i \in A_i\} = \prod_{i=1}^n \mu_{X_i}(A_i) \tag{9.4}$$

for any sets with $A_i \in \mathscr{S}_i$. An infinite collection of random variables is called independent if any finite subset of the collection is independent.

In (9.4), the right-hand side is the product of the marginal distributions. Hence the joint distributions of independent random variables are just the product of their marginals. The next result extends this "independence means multiply" rule from probabilities to expectations.

Theorem 9.1.10 *If X and Y are independent real-valued random variables with $\mathbb{E}|X| < \infty$ and $\mathbb{E}|Y| < \infty$, then $\mathbb{E}|XY| < \infty$ and $\mathbb{E}(XY) = \mathbb{E}X\mathbb{E}Y$.*

Exercise 9.1.11 Show that if X and Y are independent, then $\mathrm{Cov}(X, Y) = 0$.

One important consequence of independence is the following result

Theorem 9.1.12 (Fubini) *Let X and Y be as above, and let S and T be subsets of $\mathbb{R}^n$ and $\mathbb{R}^k$ respectively. Let $h \in b\mathscr{B}(S \times T)$ or $m\mathscr{B}(S \times T)^+$. In other words, h is a either a bounded or a nonnegative Borel measurable function on the space where the pair (X, Y) takes values. If X and Y are independent, then*

$$\mathbb{E}h(X, Y) = \int \int h(x, y) \mu_X(dx) \mu_Y(dy) = \int \int h(x, y) \mu_Y(dy) \mu_X(dx)$$

Things start to get interesting when we have an infinite number of random variables on the one probability space indexed by a value that often represents time. These collections of random variables are called stochastic processes. A formal definition follows.

Definition 9.1.13 Let $(S, \mathscr{S})$ be a measurable space. An S-valued *stochastic process* is a tuple

$$(\Omega, \mathscr{F}, \mathbb{P}, (X_t)_{t \in \mathbb{T}})$$

where $(\Omega, \mathscr{F}, \mathbb{P})$ is a probability space, $\mathbb{T}$ is an index set such as $\mathbb{N}$ or $\mathbb{Z}$, and X_t is an S-valued random variable on $(\Omega, \mathscr{F}, \mathbb{P})$ for all $t \in \mathbb{T}$.

With stochastic processes, the idea is that, at the "start of time," a point ω is selected by "nature" from the set Ω according to the probability law $\mathbb{P}$ (i.e., $\mathbb{P}(E)$ is the probability that $\omega \in E$). This is a once-off realization of all uncertainty, and $X_t(\omega)$ simply reports the time t outcome for the variable of interest as a function of that realization.

The simplest kinds of stochastic processes are the IID processes:

Definition 9.1.14 An S-valued stochastic process $(\Omega, \mathscr{F}, \mathbb{P}, (X_t)_{t \in \mathbb{T}})$ is called *independent and identically distributed* (IID) if the sequence $(X_t)_{t \in \mathbb{T}}$ is independent and each X_t has the same distribution, in the sense that

$$\mathbb{P}\{X_t \in B\} = \mathbb{P}\{X_s \in B\} \quad \text{for any } s, t \in \mathbb{T} \text{ and any } B \in \mathscr{S}$$

For an IID process, any event that occurs at each t with nonzero probability occurs eventually with probability one. To see this, suppose that $(\Omega, \mathscr{F}, \mathbb{P}, (X_t)_{t \in \mathbb{N}})$ is a IID, and that the common distribution of each X_t is $\mu \in \mathscr{P}(S, \mathscr{S})$. Consider a set $A \in \mathscr{S}$ with $\mu(A) > 0$.

Exercise 9.1.15 Show that $\{X_t \notin A, \ \forall t \in \mathbb{N}\} \subset \cap_{t \leq T}\{X_t \notin A\}$ for all $T \in \mathbb{N}$.[2] Using this relation, show that $\mathbb{P}\{X_t \notin A, \ \forall t \in \mathbb{N}\} \leq (1 - \mu(A))^T$ for all $T \in \mathbb{N}$. Conclude that this probability is zero, and hence that $X_t \in A$ for at least one $t \in \mathbb{N}$ with probability one.

9.1.3 Back to Densities

We have mentioned a few times that some but not all distributions can be represented by densities. Let's now clarify exactly when distributions do have density representations, as well as collecting some miscellaneous facts about densities.

Let S be a Borel subset of $\mathbb{R}^n$. Recall that a density on S is a function $\phi \in m\mathscr{B}(S)^+$ with the property that $\int \phi(x)dx :=: \int \phi d\lambda :=: \lambda(\phi) = 1$. The set of all densities on S is denoted by $D(S)$. Each density $\phi \in D(S)$ creates a distribution $\mu_\phi \in \mathscr{P}(S)$ via $\mu_\phi(B) = \int_B \phi(x)dx$.

Exercise 9.1.16 Confirm that this μ_ϕ is countably additive.[3]

---

[2] Remember that these are subsets of Ω.

[3] Hint: If (B_n) is a disjoint sequence of sets, show that $\mathbb{1}_{\cup B_n} = \sum_n^\infty \mathbb{1}_{B_n}$. Complete the proof using properties M1–M5 of the integral.

Sometimes we can go the other way, from distributions to associated density. In particular, suppose that $S \in \mathscr{B}(\mathbb{R}^n)$, and let $\mu \in \mathscr{P}(S)$. The distribution μ is said to have a density representation ϕ if $\phi \in D(S)$ and

$$\mu(B) = \int_B \phi(x)\,dx \qquad (B \in \mathscr{B}(S)) \tag{9.5}$$

However, the pairing of $D(S)$ and $\mathscr{P}(S)$ in (9.5) is not a one-to-one correspondence. Every density creates a distribution, but there are distributions in $\mathscr{P}(S)$ without such an "integral" representation by an element of $D(S)$. Here is an example:

Exercise 9.1.17 Let $a \in \mathbb{R}$, and let δ_a be the element of $\mathscr{P}(\mathbb{R})$ that puts unit mass on a. That is, $\delta_a(B) = 1$ if $a \in B$ and zero otherwise. Argue that there is no $\phi \in D(\mathbb{R})$ such that $\delta_a(B) = \int_B \phi(x)dx$ for all $B \in \mathscr{B}(\mathbb{R})$.[4]

So when do density representations exist? The following rather fundamental theorem answers that question. The proof is omitted, but you can find it in any text on measure theory.

Theorem 9.1.18 (Radon–Nikodym) *Let $\mu \in \mathscr{P}(S)$, where $S \in \mathscr{B}(\mathbb{R}^n)$, and let λ be the Lebesgue measure. The distribution μ has a density representation if and only if $\mu(B) = 0$ whenever $B \in \mathscr{B}(S)$ and $\lambda(B) = 0$.*

When densities exist, they can make our life much easier. The next theorem indicates how density representations can be used to compute expectations by changing the measure used to integrate from a given distribution to Lebesgue measure. Often the transformation results in a standard Riemann integral, which can be solved using calculus.

Theorem 9.1.19 *Let $S \in \mathscr{B}(\mathbb{R}^n)$. If distribution $\mu \in \mathscr{P}(S)$ has density representation $\phi \in D(S)$, and if $h \in b\mathscr{B}(S)$ or $h \in m\mathscr{B}(S)^+$, then*

$$\mu(h) = \int h(x)\phi(x)dx \tag{9.6}$$

Proof. The proof follows a very standard argument, and is probably worth reading through at least once. Let's focus on the case of $h \in b\mathscr{B}(S)$. Suppose first that $h = \mathbb{1}_B$, where $B \in \mathscr{B}(S)$. For such an h the equality (9.6) holds by (9.5). Now suppose that h is a simple function: $h \in s\mathscr{B}(S)$, $h = \sum_{n=1}^N \alpha_n \mathbb{1}_{B_n}$, $B_n \in \mathscr{B}(S)$. Since the integral is linear, we have

$$\mu\left(\sum_{n=1}^N \alpha_n \mathbb{1}_{B_n}\right) = \sum_{n=1}^N \alpha_n \mu(\mathbb{1}_{B_n}) = \sum_{n=1}^N \alpha_n \int \mathbb{1}_{B_n}(x)\phi(x)dx = \int \sum_{n=1}^N \alpha_n \mathbb{1}_{B_n}(x)\phi(x)dx$$

---

[4]Hint: Recall from theorem 7.3.9 that if $\lambda(B) = 0$, then $\int_B \phi(x)dx = 0$.

In other words, (9.6) holds for $h \in s\mathscr{B}(S)$. Now let $h \in b\mathscr{B}(S)$ with $h \geq 0$. By lemma 7.2.11 (page 175) there is a sequence $(s_k) \subset s\mathscr{B}(S)^+$ with $s_k \uparrow h$. Since (9.6) holds for each s_k, we have

$$\mu(s_k) = \int s_k(x)\phi(x)dx \qquad (k \in \mathbb{N})$$

Taking limits with respect to k and using the monotone convergence theorem gives (9.6). Finally, for general $h \in b\mathscr{B}(S)$, we have $h = h^+ - h^-$, and another application of linearity completes the proof. □

9.2 General State Markov Chains

It's time to develop a general theory of Markov chains on uncountably infinite state spaces.[5] In chapter 8 we covered uncountable state spaces when the stochastic (density) kernel was a family of densities $p(x, y)dy$, one for each x in the state space. We now drop the assumption that these distributions can be represented as densities and permit them to be arbitrary probability measures.

9.2.1 Stochastic Kernels

For discrete time Markov chains of all shapes and forms, the most important primitive is the stochastic kernel.[6] You have already met some stochastic kernels: The first was the finite kernel p, living on a finite set S, with the property that $p(x, y) \geq 0$ and $\sum_{y \in S} p(x, y) = 1$. The second was the density kernel $p(x, y)dy$ on a Borel subset S of $\mathbb{R}^k$. Here is the general (i.e., probability measure) case:

Definition 9.2.1 Let S be a Borel subset of $\mathbb{R}^n$. A *stochastic kernel* on S is a family of probability measures

$$P(x, dy) \in \mathscr{P}(S) \qquad (x \in S)$$

where $x \mapsto P(x, B)$ is Borel measurable for each $B \in \mathscr{B}(S)$.[7]

Each finite kernel p on finite S defines a general kernel P on S by

$$P(x, B) = \sum_{y \in B} p(x, y) \qquad (x \in S, \ B \subset S)$$

---

[5]In order to be consistent with earlier theory and what lies ahead, we stick to state spaces that are subsets of $\mathbb{R}^k$. The theory for abstract measure spaces differs little.

[6]Stochastic kernels are also called Markov kernels, or transition probability functions.

[7]This last property is just a regularity condition to make sure that various integrals we want to use will make sense.

Each density kernel p on Borel set $S \subset \mathbb{R}^n$ defines a general kernel P on S by

$$P(x, B) = \int_B p(x, y) dy \qquad (x \in S, \ B \in \mathscr{B}(S))$$

The next definition provides a link between Markov chains and kernels.

Definition 9.2.2 Let $\psi \in \mathscr{P}(S)$. A stochastic process $(X_t)_{t \geq 0}$ on S is called Markov-(P, ψ) if

1. at time zero, X_0 is drawn from ψ, and

2. at time $t + 1$, X_{t+1} is drawn from $P(X_t, dy)$.

If $\psi = \delta_x$ for some $x \in S$, then we say $(X_t)_{t \geq 0}$ is Markov-(P, x).

While this definition is intended to parallel the finite and density case definitions given on pages 71 and 188 respectively, it is time to address an issue that needs clarification: If the Markov-(P, ψ) process $(X_t)_{t \geq 0}$ is to be regarded a *stochastic process*, then, by definition of stochastic processes (see page 216) it must be a sequence of S-valued random variables, all defined on a common probability space $(\Omega, \mathscr{F}, \mathbb{P})$. In the definition above no probability space is mentioned, and it is not clear how $(X_t)_{t \geq 0}$ is defined as a sequence of *functions* from Ω to S.

While construction of the underlying probability space can be undertaken without any additional assumptions—interested readers are referred to Pollard (2002, §4.8) or Shiryaev (1996, p. 249)—the construction is usually redundant in economic applications because Markov chains typically present themselves in the form of stochastic recursive sequences (SRSs). Such representations simultaneously determine the stochastic kernel P, provide the probability space $(\Omega, \mathscr{F}, \mathbb{P})$, and furnish us with the random variables $(X_t)_{t \geq 0}$ living on that space. Let's see how this works, starting with the following definition.

Definition 9.2.3 Let $S \in \mathscr{B}(\mathbb{R}^n)$, $Z \in \mathscr{B}(\mathbb{R}^k)$, $\phi \in \mathscr{P}(Z)$, and $\psi \in \mathscr{P}(S)$. Let $F \colon S \times Z \to S$ be Borel measurable. The canonical stochastic recursive sequence $(X_t)_{t \geq 0}$ is defined by

$$X_{t+1} = F(X_t, W_{t+1}), \quad (W_t)_{t \geq 1} \overset{\text{IID}}{\sim} \phi, \quad X_0 \sim \psi \tag{9.7}$$

The random variables $(W_t)_{t \geq 1}$ and X_0 are defined on a common probability space $(\Omega, \mathscr{F}, \mathbb{P})$ and are jointly independent.

In the definition, each X_t is defined as a random variable on $(\Omega, \mathscr{F}, \mathbb{P})$ as follows: Given $\omega \in \Omega$, we have $(W_t(\omega))_{t \geq 1}$ and $X_0(\omega)$. From these, $(X_t(\omega))_{t \geq 0}$ is recursively determined by

$$X_{t+1}(\omega) = F(X_t(\omega), W_{t+1}(\omega))$$

Note that according to the definition, X_t is a function only of X_0 and $W_1, \ldots, W_t$. Hence X_t and the current shock W_{t+1} are independent.

There is a unique stochastic kernel P on S that represents the dynamics implied by F and ϕ. To define P, we need to specify $P(x, B)$ for arbitrary $x \in S$ and $B \in \mathscr{B}(S)$, corresponding to the probability that $X_{t+1} \in B$ given $X_t = x$. Since $X_{t+1} = F(X_t, W_{t+1})$, we get

$$P(x, B) = \mathbb{P}\{F(x, W_{t+1}) \in B\} = \mathbb{E}\mathbb{1}_B[F(x, W_{t+1})]$$

(Recall that the expectation of an indicator function is equal to the probability of the event it refers to.) Since W_{t+1} is distributed according to $\phi \in \mathscr{P}(Z)$, this becomes

$$P(x, B) = \int \mathbb{1}_B[F(x, z)]\phi(dz) \qquad (x \in S, \ B \in \mathscr{B}(S)) \tag{9.8}$$

The integral is over the space Z on which the shock is defined.

In what follows, whenever we introduce a Markov-(P, ψ) process $(X_t)_{t \geq 0}$, it will be implicitly assumed that P is derived from the canonical SRS via (9.8), and that $(X_t)_{t \geq 0}$ is the sequence of random variables defined recursively in (9.7). This way, $(X_t)_{t \geq 0}$ is always a well-defined stochastic process living on the probability space $(\Omega, \mathscr{F}, \mathbb{P})$ that supports the shocks $(W_t)_{t \geq 1}$ and initial condition X_0.[8]

Example 9.2.4 In §6.1 we introduced a stochastic Solow–Swan growth model where output is a function f of capital k and a real-valued shock W. The sequence of productivity shocks $(W_t)_{t \geq 1}$ is IID with distribution $\phi \in \mathscr{P}(\mathbb{R})$. Capital at time $t+1$ is equal to that fraction s of output saved last period, plus undepreciated capital. As a result k_t follows the law

$$k_{t+1} = sf(k_t, W_{t+1}) + (1 - \delta)k_t \tag{9.9}$$

Let $f : \mathbb{R}_+ \times \mathbb{R} \to \mathbb{R}_+$. A suitable state space is $S = \mathbb{R}_+$, and the shock space is $Z = \mathbb{R}$. We set

$$F(x, z) = sf(x, z) + (1 - \delta)x$$

which clearly maps $S \times Z$ into S. Using (9.8), the "Solow-Swan stochastic kernel" is given by

$$P(x, B) = \int \mathbb{1}_B(sf(x, z) + (1 - \delta)x)\phi(dz)$$

---

[8]Two technical notes: Given a distribution ϕ on Z, there always exists a probability space $(\Omega, \mathscr{F}, \mathbb{P})$ and an independent sequence of random variables $(W_t)_{t \geq 1}$ on $(\Omega, \mathscr{F}, \mathbb{P})$ such that the distribution of W_t is ϕ (i.e., $\mathbb{P} \circ W_t^{-1} = \phi$) for each t. See, for example, Pollard (2002, §4.8). Second, there is no loss of generality in assuming the existence of an SRS representation for a given kernel P on S. In fact every kernel on S can be shown to have such a representation. See Bhattacharya and Majumdar (2007, §3.8) for details.

Example 9.2.5 Consider the deterministic model $X_{t+1} = h(X_t)$. Since $P(x, B)$ is the probability that $X_{t+1} \in B$ given $X_t = x$, we can set $P(x, B) = \mathbb{1}_B(h(x)) = \mathbb{1}_{h^{-1}(B)}(x)$.[9]

Example 9.2.6 Consider the linear model with correlated shocks given by

$$Y_{t+1} = \alpha Y_t + \tilde{\zeta}_{t+1}$$
$$\tilde{\zeta}_{t+1} = \rho \tilde{\zeta}_t + W_{t+1}$$

where all variables take values in $\mathbb{R}$ and $(W_t)_{t \geq 1}$ is IID according to $\phi \in \mathscr{P}(\mathbb{R})$. Although $(Y_t)_{t \geq 0}$ is not itself a Markov chain, the bivariate process given by $X_t := (Y_t, \tilde{\zeta}_t)$ is Markov on $\mathbb{R}^2$. It is a special case of the canonical SRS defined in (9.7), with $S = \mathbb{R}^2, Z = \mathbb{R}$ and

$$F(x, z) = F[(y, \tilde{\zeta}), z] = \left(\begin{array}{c} \alpha y + \rho \tilde{\zeta} + z \\ \rho \tilde{\zeta} + z \end{array} \right)$$

If $\max\{|\alpha|, |\rho|\} < 1$, then the model has certain stability properties elaborated on below.

Figures 9.1–9.3 give some idea of the dynamics for $(Y_t)_{t \geq 0}$ that can arise in the linear correlated shock model. In figure 9.1 the shocks W_t are identically zero and the parameters α and ρ are negative, causing oscillation. In figures 9.2 and 9.3 the shock is $N(0, 0.25)$ and the parameters α and ρ are nonnegative. In figure 9.2 the coefficient ρ is relatively large, leading to strong autocorrelation, while in 9.3 we set $\rho = 0$. In this case the shocks are IID, $(Y_t)_{t \geq 0}$ is Markovian, and the autocorrelation is weaker.

Example 9.2.7 This next example is the so-called AR(p) model. It demonstrates that Markov models are more general than they first appear. Suppose that the state variable X takes values in $\mathbb{R}$, that (W_t) is an independent and identically distributed sequence in $\mathbb{R}$, and that

$$X_{t+1} = a_0 X_t + a_1 X_{t-1} + \cdots + a_{p-1} X_{t-p+1} + W_{t+1} \tag{9.10}$$

Define $Y_t := (X_t, X_{t-1}, \ldots, X_{t-p+1})$, and consider the system

$$Y_{t+1} = \begin{pmatrix} a_0 & a_1 & \cdots & a_{p-2} & a_{p-1} \\ 1 & 0 & \cdots & 0 & 0 \\ \vdots & & & & \\ 0 & 0 & \cdots & 1 & 0 \end{pmatrix} Y_t + \begin{pmatrix} 1 \\ 0 \\ \vdots \\ 0 \end{pmatrix} W_{t+1} \tag{9.11}$$

The process (9.11) is an SRS with a well-defined stochastic kernel. At the same time, the first element of Y_t follows the process in (9.10).

---

[9] Another path to the same conclusion is by considering $X_{t+1} = h(X_t) + W_{t+1}$ where $W_t = 0$ with probability one and then appealing to (9.8).

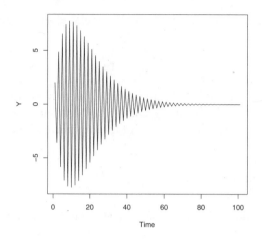

Figure 9.1 Correlated shocks, $\alpha = -0.9$, $\rho = -0.9$, $W_t \equiv 0$

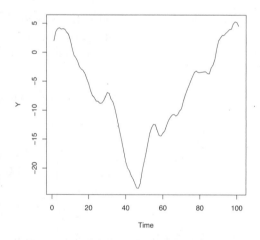

Figure 9.2 Correlated shocks, $\alpha = 0.9$, $\rho = 0.9$, $W_t \sim N(0, 0.25)$

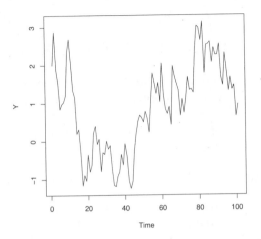

Figure 9.3 Correlated shocks, $\alpha = 0.9$, $\rho = 0.0$, $W_t \sim N(0, 0.25)$

9.2.2 The Fundamental Recursion, Again

On page 68 we showed that for a finite state Markov chain $(X_t)_{t\geq0}$ with Markov operator $\mathbf{M}$, the sequence of marginal distributions $(\psi_t)_{t\geq0}$ satisfies $\psi_{t+1} = \psi_t\mathbf{M}$. An analogous recursion was obtained for the density case by defining the density Markov operator $\psi\mathbf{M}(y) = \int p(x,y)\psi(x)dx$. Since a measure kernel P generalizes the finite and density kernels, perhaps we can identify the general rule.

To begin, let S be a Borel subset of $\mathbb{R}^n$, let P be a stochastic kernel on S, and let $(X_t)_{t\geq0}$ be Markov-(P, ψ) for some $\psi \in \mathscr{P}(S)$. Writing $\psi_t \in \mathscr{P}(S)$ for the distribution of X_t, we claim that the sequence $(\psi_t)_{t\geq0} \subset \mathscr{P}(S)$ satisfies

$$\psi_{t+1}(B) = \int P(x, B)\psi_t(dx) \qquad (B \in \mathscr{B}(S)) \tag{9.12}$$

The intuition is the same as for the finite case: The probability that $X_{t+1} \in B$ is the probability that X_t goes from x into B, summed across all $x \in S$, weighted by the probability $\psi_t(dx)$ that X_t takes the value x.

In verifying (9.12), let us assume that P is defined by the canonical SRS (9.7). Picking any $h \in b\mathscr{B}(S)$, independence of X_t and W_{t+1} plus theorem 9.1.12 (page 215) give

$$\mathbb{E}\,h(X_{t+1}) = \mathbb{E}\,h[F(X_t, W_{t+1})] = \int\int h[F(x, z)]\phi(dz)\psi_t(dx)$$

Specializing to the case $h = \mathbb{1}_B \in b\mathscr{B}(S)$ gives

$$\mathbb{E}\mathbb{1}_B(X_{t+1}) = \int \int \mathbb{1}_B[F(x,z)]\phi(dz)\psi_t(dx) = \int P(x,B)\psi_t(dx)$$

where the second inequality is due to (9.8). But $\mathbb{E}\mathbb{1}_B(X_{t+1}) = \mathbb{P}\{X_{t+1} \in B\} = \psi_{t+1}(B)$, confirming (9.12).

When we studied finite and density Markov chains, we made extensive use of the Markov operator. In both cases this operator was defined in terms of the stochastic kernel and satisfied $\psi_{t+1} = \psi_t M$ for all t. Now let's consider the general measure setting: Given stochastic kernel P, define the Markov operator M by as the map $\mathscr{P}(S) \ni \phi \mapsto \phi M \in \mathscr{P}(S)$, where

$$\phi M(B) := \int P(x,B)\phi(dx) \qquad (B \in \mathscr{B}(S)) \tag{9.13}$$

The next lemma verifies that ϕM is a probability measure, and shows how to compute integrals of the form $(\phi M)(h)$.

Lemma 9.2.8 *If Q is any stochastic kernel on S and $\mu \in \mathscr{P}(S)$, then the set function ν defined by*

$$\nu(B) = \int Q(x,B)\mu(dx) \qquad (B \in \mathscr{B}(S))$$

is an element of $\mathscr{P}(S)$, and for any $h \in b\mathscr{B}(S)$, we have

$$\nu(h) :=: \int h\,d\nu = \int \left[\int h(y)Q(x,dy) \right] \mu(dx) \tag{9.14}$$

Proof. Clearly, $\nu(S) = 1$. Countable additivity of ν can be checked using either the dominated or the monotone convergence theorem. The proof of (9.14) can be obtained along the same lines as that of theorem 9.1.19 (page 217) and is left as an exercise. $\square$

As before, M acts on distributions to the left rather than the right. Using the Markov operator, we can write the recursion (9.12) as $\psi_{t+1} = \psi_t M$, which exactly parallels our expression for the finite case (see (4.12) on page 76) and the density case (see (8.15) on page 195). An inductive argument now confirms that if $X_0 \sim \psi$, then $X_t \sim \psi M^t$.

Example 9.2.9 In the case of the deterministic dynamical system $X_{t+1} = h(X_t)$, recall that $P(x,B) = \mathbb{1}_B(h(x))$. Now suppose that $X_0 = \bar{x}$ (i.e., $X_0 \sim \delta_{\bar{x}}$). Our intuition tells us that the distribution of X_1 must then be $\delta_{h(\bar{x})}$, and indeed

$$\delta_{\bar{x}}M(B) = \int \mathbb{1}_B(h(x))\delta_{\bar{x}}(dx) = \mathbb{1}_B(h(\bar{x})) = \delta_{h(\bar{x})}(B)$$

$$\therefore \quad \delta_{\bar{x}}M = \delta_{h(\bar{x})}$$

Iterating forward, we find that $X_t \sim \delta_{\bar{x}}M^t = \delta_{h^t(\bar{x})}$, as expected.

Given a kernel P, the higher order kernels $(P^t)_{t \geq 1}$ are defined by

$$P^1 := P, \quad P^t(x, B) := \int P(z, B) P^{t-1}(x, dz) \quad (x \in S, \ B \in \mathscr{B}(S))$$

These kernels are defined so that $P^t(x, B)$ gives the probability of moving from x into B in t steps, and $P^t(x, dy)$ is the distribution of X_t given $X_0 = x$. To see this, observe that the distribution of X_t given $X_0 = x$ is precisely $\delta_x \mathbf{M}^t$, so we are claiming that

$$\delta_x \mathbf{M}^t(B) = P^t(x, B) \quad (x \in S, \ B \in \mathscr{B}(S), \ t \in \mathbb{N})$$

This claim is a special case of the more general statement

$$\phi \mathbf{M}^t(B) = \int P^t(x, B) \phi(dx) \quad (\phi \in \mathscr{P}(S), \ B \in \mathscr{B}(S), \ t \in \mathbb{N})$$

which says that $\mathbf{M}^t$, the t-th iterate of $\mathbf{M}$, is the Markov operator corresponding to the t-th-order kernel P^t.[10] For the proof we refer to theorem 9.2.15 below.

9.2.3 Expectations

As before, let $b\mathscr{B}(S)$ be all the bounded measurable functions on S. We now introduce a second operator, also called the Markov operator and also denoted by $\mathbf{M}$, which sends $h \in b\mathscr{B}(S)$ into $\mathbf{M}h \in b\mathscr{B}(S)$, where

$$\mathbf{M}h(x) := \int h(y) P(x, dy) \quad (x \in S) \tag{9.15}$$

Intuitively, $\mathbf{M}h(x)$ represents the expectation of $h(X_{t+1})$ given $X_t = x$. This is the same interpretation as the finite case, where we defined $\mathbf{M}h(x) = \sum_{y \in S} h(y) p(x, y)$.

We now have $\mathbf{M}$ acting on *measures* to the *left* and *functions* to the *right*. This is analogous to the finite state notation, where $\psi \mathbf{M}$ is the row vector ψ postmultiplied by $\mathbf{M}$, and $\mathbf{M}h$ is the column vector h premultiplied by $\mathbf{M}$. Stochastic kernels and the operators $\psi \mapsto \psi \mathbf{M}$ and $h \mapsto \mathbf{M}h$ are in one-to-one correspondence via the identity

$$P(x, B) = \delta_x \mathbf{M}(B) = \mathbf{M} \mathbb{1}_B(x) \quad (x \in S, \ B \in \mathscr{B}(S)) \tag{9.16}$$

In (9.15), h is restricted to be bounded so that the integral and hence $\mathbf{M}h$ are well defined. On occasion it will be convenient to use the same operator notation when h is unbounded. For example, if h is nonnegative and measurable, then the integral is well defined (although possibly infinite) and the notation $\mathbf{M}h$ is useful. In the remainder of this section, however, $\mathbf{M}$ always acts on functions in $b\mathscr{B}(S)$.

The next few exercises show that the operator (9.15) has certain well-defined properties. Knowing these properties makes manipulating expectations of functions of $(X_t)_{t \geq 0}$ straightforward.

---

[10]This (unsurprising) result is the measure analogue of the density case stated in lemma 8.1.13 (page 196).

Exercise 9.2.10 Verify that if $h\colon S \to \mathbb{R}$ is bounded (resp., nonnegative), then so is $\mathbf{M}h$.

Exercise 9.2.11 Show that $\mathbf{M}\mathbb{1}_S = \mathbb{1}_S$ pointwise on S (i.e., that $\mathbb{1}_S$ is a fixed point of $\mathbf{M}$).

Exercise 9.2.12 Show that $\mathbf{M}$ is monotone, in the sense that if $h, g \in b\mathscr{B}(S)$ and $h \leq g$, then $\mathbf{M}h \leq \mathbf{M}g$ (inequalities are pointwise on S).[11]

Exercise 9.2.13 Show that $\mathbf{M}$ is linear, in the sense that if $h, g \in b\mathscr{B}(S)$ and $\alpha, \beta \in \mathbb{R}$, then $\mathbf{M}(\alpha h + \beta g) = \alpha \mathbf{M}h + \beta \mathbf{M}g$.[12]

Often we will be dealing with the kernel $P(x, B) = \int \mathbb{1}_B[F(x, z)]\phi(dz)$ generated by the canonical SRS defined in (9.7) on page 219. In this case $\mathbf{M}h$ takes the form

$$\mathbf{M}h(x) := \int h(y)P(x, dy) = \int h[F(x, z)]\phi(dz) \qquad (9.17)$$

This expression is intuitive because $\mathbf{M}h(x)$ represents the expectation of $h(X_{t+1})$ given $X_t = x$, and $h(X_{t+1}) = h[F(X_t, W_{t+1})]$. Hence

$$\mathbf{M}h(x) = \mathbb{E}h[F(x, W_{t+1})] = \int h[F(x, z)]\phi(dz) \qquad (9.18)$$

Here's another way to get the same answer:

Exercise 9.2.14 Verify (9.17) using theorem 7.3.17 on page 182.[13]

The t-th iterate $\mathbf{M}^t h$ of h under $\mathbf{M}$ can be represented in terms of P^t:

$$\mathbf{M}^t h(x) = \int h(y)P^t(x, dy) \qquad (x \in S) \qquad (9.19)$$

We state a more general result immediately below (theorem 9.2.15). Before doing so, note that since $P^t(x, dy)$ is the distribution of X_t given $X_0 = x$, it follows from (9.19) that $\mathbf{M}^t h$ can be interpreted as the conditional expectation

$$\mathbf{M}^t h(x) = \mathbb{E}[h(X_t) \mid X_0 = x] \qquad (9.20)$$

Theorem 9.2.15 *Let P be a stochastic kernel on S. If $\mathbf{M}$ is the corresponding Markov operator, then for every $\phi \in \mathscr{P}(S)$, $h \in b\mathscr{B}(S)$ and $t \in \mathbb{N}$, we have*

$$(\phi \mathbf{M}^t)(h) = \phi(\mathbf{M}^t h) = \int \left[\int h(y)P^t(x, dy)\right]\phi(dx)$$

---

[11]Hint: Use theorem 7.3.9 on page 181.

[12]Hint: In other words, show that the function on the left-hand side equals the function on the right-hand side at every point $x \in S$. Use theorem 7.3.9 on page 181.

[13]Hint: $P(x, B) = \int \mathbb{1}_B[F(x, z)]\phi(dz)$ is the image measure of ϕ under $z \mapsto F(x, z)$.

We are using the linear functional notation for the first two integrals. In traditional notation,

$$(\phi \mathbf{M}^t)(h) = \int h(y)(\phi \mathbf{M}^t)(dy) \quad \text{and} \quad \phi(\mathbf{M}^t h) = \int (\mathbf{M}^t h)(x)\phi(dx)$$

Theorem 9.2.15 connects the iterates of $\phi \mapsto \phi \mathbf{M}$ and those of $h \mapsto \mathbf{M}h$.[14]

Proof of theorem 9.2.15. Consider the case $t = 1$. We have

$$\phi(\mathbf{M}h) = \int \mathbf{M}h(x)\phi(dx) = \int \left[\int h(y)P(x,dy) \right] \phi(dx) = \phi \mathbf{M}(h)$$

where the final equality is due to (9.14). It is an exercise for the reader to extend this to general t using induction. □

9.3 Commentary

The first monograph to exposit the measure-theoretic foundations of probability is Kolmogorov (1956)—originally published in 1933—which provides excellent historical perspective, and is still well worth reading. For general references on measure-theoretic probability, see Williams (1991), Breiman (1992), Shiryaev (1996), Durrett (1996), Taylor (1997), Pollard (2002), Dudley (2002), and Schilling (2005).

For further background on general state Markov chains see Breiman (1992, ch. 7), Meyn and Tweedie (1993, ch. 3), Durrett (1996, ch. 5), and Taylor (1997, ch. 3). For a reference with economic applications see Stokey and Lucas (1989, ch. 8). Stability of general state Markov chains is discussed in chapter 11, and the commentary at the end of the chapter contains more pointers to the literature.

---

[14]There is an obvious parallel with the finite case. See (4.16) on page 79.

Chapter 10

Stochastic Dynamic Programming

In this chapter we continue our study of intertemporal decision problems begun in §5.1 and §6.2, working our way through a rigorous treatment of stochastic dynamic programming. Intertemporal problems are challenging because they involve optimization in high dimensions; in fact the objective function is often defined over a space of infinite dimension.[1] We will see that studying the theory behind dynamic programming is valuable not only for the understanding it provides, but also for developing numerical solution methods. Value iteration and policy iteration are the most common techniques, and convergence of the algorithms is covered in some detail.

10.1 Theory

Our first step is to give a careful definition of the problem. Once the definition is in place we will go on to state and prove the basic principle of optimality for (infinite horizon, stationary) stochastic dynamic programs.

10.1.1 Statement of the Problem

For an infinite horizon stochastic dynamic program (SDP), the scenario is one where actions taken by an agent affect the future path of a state variable. Actions are spec-

---

[1] We have not defined infinite-dimensional space, which is an algebraic concept, but spaces of sequences and spaces of functions typically have this property.

ified in terms of a policy function, which maps the current state of the system into a given action. Each policy induces a Markov process on the state space, and different processes give different levels of expected reward.

Each SDP has a "state" space $S \in \mathscr{B}(\mathbb{R}^n)$, an "action" space $A \in \mathscr{B}(\mathbb{R}^m)$ and a nonempty correspondence Γ mapping $x \in S$ into $\mathscr{B}(A)$. The set $\Gamma(x)$ will be interpreted as the collection of all feasible actions for the agent when the state is x. We set

$$\operatorname{gr} \Gamma := \{(x, u) \in S \times A : u \in \Gamma(x)\}$$

Below $\operatorname{gr} \Gamma$ ("graph" of Γ) is called the set of feasible state/action pairs.

Next we introduce a measurable "reward" function $r \colon \operatorname{gr} \Gamma \to \mathbb{R}$ and a discount factor $\rho \in (0, 1)$. Finally, let $Z \in \mathscr{B}(\mathbb{R}^k)$ be a shock space, let $(W_t)_{t \geq 1}$ be a sequence of IID shocks with distribution $\phi \in \mathscr{P}(Z)$, and let

$$F \colon \operatorname{gr} \Gamma \times Z \ni (x, u, z) \mapsto F(x, u, z) \in S$$

be a measurable "transition function," which captures the dynamics. At the start of time t the agent observes the state $X_t \in S$ and responds with action $U_t \in \Gamma(X_t) \subset A$. After choosing U_t, the agent receives a reward $r(X_t, U_t)$, and the state is updated according to $X_{t+1} = F(X_t, U_t, W_{t+1})$. The whole process then repeats, with the agent choosing U_{t+1}, receiving reward $r(X_{t+1}, U_{t+1})$ and so on.

If our agent cared only about present rewards, the best action would be to choose $U_t = \operatorname{argmax}_{u \in \Gamma(X_t)} r(X_t, u)$ at each date t. However, the agent cares about the future too, and must therefore trade off maximizing current rewards against positioning the state optimally in order to reap good rewards in future periods. The optimal decision depends on how much he or she cares about the future, which is in turn parameterized by the discount factor ρ. The role of ρ is clarified below.

Example 10.1.1 Consider again the accumulation problem treated in §6.2. At the start of time t, an agent has income y_t, which is divided between consumption c_t and savings k_t. From consumption c the agent receives utility $U(c)$, where $U \colon \mathbb{R}_+ \to \mathbb{R}$. Savings is added to the existing capital stock. For simplicity, assume that depreciation is total: current savings and the stock of capital are equal. After the time t investment decision is made, shock W_{t+1} is observed. Production then takes place, yielding

$$y_{t+1} = f(k_t, W_{t+1}), \quad (W_t)_{t \geq 1} \stackrel{\text{IID}}{\sim} \phi \in \mathscr{P}(Z), \ Z \in \mathscr{B}(\mathbb{R}) \tag{10.1}$$

(See also figure 1.1 on page 2 for a depiction of the timing.) This fits our SDP framework, with $y \in S := \mathbb{R}_+$ the state variable and $k \in A := \mathbb{R}_+$ the control. Γ is the map $S \ni y \mapsto [0, y] \subset A$ that defines feasible savings given income y. The reward function $r(y, k)$ on $\operatorname{gr} \Gamma$ is $U(y - k)$. The transition function is $F(y, k, z) = f(k, z)$. For the present model it is independent of the state.

Clearly, some states in S are more attractive than others. High income positions us well in terms of future consumption. Hence a trade off exists between consuming now, which gives current reward, and saving, which places us at a more attractive point in the state space tomorrow. Such trade-offs are the essence of dynamic programming.

Example 10.1.2 Now consider the same model but with correlated shocks. That is, $y_{t+1} = f(k_t, \eta_{t+1})$, where $\eta_{t+1} = g(\eta_t, W_{t+1})$, $g \colon \mathbb{R}_+ \times \mathbb{R}_+ \to \mathbb{R}_+$. This also fits the SDP framework, with the only modification being that the state space has two elements $(y, \eta) \in S := \mathbb{R}_+ \times \mathbb{R}_+$, and the transition function F is

$$F \colon (y, \eta, k, z) \mapsto \begin{pmatrix} f(k, g(\eta, z)) \\ g(\eta, z) \end{pmatrix}$$

The feasible correspondence Γ is the map sending (y, η) into $[0, y]$.

Returning to the general case, at minimum we need some regularity assumptions on the primitives that will ensure that at least one solution to the SDP exists:

Assumption 10.1.3 The map $r \colon \operatorname{gr} \Gamma \to \mathbb{R}$ is continuous and bounded.

Assumption 10.1.4 $\Gamma \colon S \to \mathscr{B}(A)$ is continuous and compact valued.[2]

Assumption 10.1.5 $\operatorname{gr} \Gamma \ni (x, u) \mapsto F(x, u, z) \in S$ is continuous, $\forall z \in Z$.

These continuity and compactness assumptions are all about guaranteeing existence of maximizers. For us the important implication of assumption 10.1.5 is that for any $w \in bcS$, the function

$$\operatorname{gr} \Gamma \ni (x, u) \mapsto \int w[F(x, u, z)] \phi(dz) \in \mathbb{R}$$

is continuous. To see this, take any $(x_n, u_n) \subset \operatorname{gr} \Gamma$ converging to some arbitrary $(x, u) \in \operatorname{gr} \Gamma$, and any $w \in bcS$. We need to show that

$$\int w[F(x_n, u_n, z)] \phi(dz) \to \int w[F(x, u, z)] \phi(dz) \qquad (n \to \infty)$$

You can verify this using the dominated convergence theorem (page 182).

10.1.2 Optimality

In order to construct a sensible optimization problem, we will restrict the agent to policies in the set of *stationary Markov policies*.[3] For such a policy, the agent makes exactly the same decision after observing $X_t = x$ as after observing $X_{t'} = x$ at some later

---

[2]For the definition of continuity for correspondences and a simple sufficient condition see page 339.

[3]In fact it can be shown that every reasonable optimal policy is of this type.

date t'. This is intuitive because when looking toward the infinite future, the agent faces exactly the same trade-off (i.e., maximizing current rewards versus positioning the state attractively next period), independent of whether the time is t or t'.

Under a stationary Markov policy, the agent's behavior is described by a Borel measurable function σ mapping each possible $x \in S$ into a feasible action $u \in \Gamma(x)$. The interpretation is that if the current state is $x \in S$, then the agent responds with action $\sigma(x) \in \Gamma(x)$. We let Σ denote the set of all Borel measurable $\sigma \colon S \to A$ with $\sigma(x) \in \Gamma(x)$ for all $x \in S$. In what follows we refer to Σ simply as the set of *feasible policies*.

For each $\sigma \in \Sigma$, we obtain a stochastic recursive sequence

$$X_{t+1} = F(X_t, \sigma(X_t), W_{t+1}), \quad (W_t)_{t \geq 1} \overset{\text{IID}}{\sim} \phi \tag{10.2}$$

for the state $(X_t)_{t \geq 0}$, and hence a stochastic kernel $P_\sigma(x, dy)$ on S given by

$$P_\sigma(x, B) := \int \mathbb{1}_B[F(x, \sigma(x), z)]\phi(dz) \qquad (x \in S, \ B \in \mathcal{B}(S))$$

We denote by $\mathbf{M}_\sigma$ the corresponding Markov operator. It is also convenient to define the function

$$r_\sigma \colon S \ni x \mapsto r(x, \sigma(x)) \in \mathbb{R}$$

so that $r_\sigma(x)$ is the reward at x when the agent follows policy σ. Using operator notation, expected rewards next period under policy σ can be expressed as

$$\mathbf{M}_\sigma r_\sigma(x) = \int r_\sigma(y) P_\sigma(x, dy) = \int r_\sigma[F(x, \sigma(x), z)]\phi(dz) \qquad (x \in S)$$

where the last equality follows from (9.17) on page 226.

The shocks $(W_t)_{t \geq 1}$ are defined on a fixed probability space $(\Omega, \mathcal{F}, \mathbb{P})$. Each $\omega \in \Omega$ picks out a sequence $(W_t(\omega))_{t \geq 1}$. Combining this sequence with an initial condition $X_0 = x \in S$ and a policy σ yields a path $(X_t(\omega))_{t \geq 0}$ for the state:

$$X_{t+1}(\omega) = F(X_t(\omega), \sigma(X_t(\omega)), W_{t+1}(\omega)), \quad X_0(\omega) = x$$

The reward corresponding to this path is random variable $Y_\sigma \colon \Omega \to \mathbb{R}$,

$$Y_\sigma(\omega) := \sum_{t=0}^{\infty} \rho^t r_\sigma(X_t(\omega)) \qquad (\omega \in \Omega)$$

Exercise 10.1.6 Using boundedness of r, prove that this random variable is well defined, in the sense that the sum converges for each $\omega \in \Omega$.[4]

---

[4]Hint: if (x_n) is any sequence in $\mathbb{R}$ and $\sum_n |x_n|$ converges, then so does $\sum_n x_n$.

The optimization problem for the agent is $\max_{\sigma \in \Sigma} \mathbb{E}Y_\sigma$. More precisely, if we set

$$v_\sigma(x) := \mathbb{E}Y_\sigma :=: \mathbb{E}\left[\sum_{t=0}^{\infty} \rho^t r_\sigma(X_t)\right] :=: \int \left[\sum_{t=0}^{\infty} \rho^t r_\sigma(X_t(\omega))\right] \mathbb{P}(d\omega)$$

and define the *value function* $v^* : S \to \mathbb{R}$ as

$$v^*(x) = \sup_{\sigma \in \Sigma} v_\sigma(x) \qquad (x \in S) \tag{10.3}$$

then a policy $\sigma^* \in \Sigma$ is called *optimal* if it attains the supremum in (10.3) for every $x \in S$. In other words, $\sigma^* \in \Sigma$ is optimal if and only if $v_{\sigma^*} = v^*$.

Exercise 10.1.7 Using the dominated convergence theorem, show that we can exchange limit and integral to obtain

$$v_\sigma(x) := \mathbb{E}\left[\sum_{t=0}^{\infty} \rho^t r_\sigma(X_t)\right] = \sum_{t=0}^{\infty} \rho^t \mathbb{E}r_\sigma(X_t)$$

Now if $h \in b\mathscr{B}(S)$, then we can express $\mathbb{E}h(X_t)$ as $\mathbf{M}_\sigma^t h(x)$, where $\mathbf{M}_\sigma^t$ is the t-th iterate of the Markov operator $\mathbf{M}_\sigma$ and $X_0 = x$ (see (9.20) on page 226). As a result v_σ can be written as

$$v_\sigma(x) = \sum_{t=0}^{\infty} \rho^t \mathbf{M}_\sigma^t r_\sigma(x) \qquad (x \in S) \tag{10.4}$$

As an aside, note that by theorem 9.2.15 (page 226), we have

$$\mathbf{M}_\sigma^t r_\sigma(x) = (\delta_x \mathbf{M}_\sigma^t)(r_\sigma) = \int r_\sigma(y) P_\sigma^t(x, dy) \qquad (x \in S) \tag{10.5}$$

Thus each policy $\sigma \in \Sigma$ creates a Markov chain $(X_t)_{t \geq 0}$ starting at x, with corresponding marginal distributions $(\delta_x \mathbf{M}_\sigma^t)_{t \geq 0}$. By integrating each distribution with r_σ, the reward corresponding to σ, and computing the discounted sum, we obtain a value for the policy. This is our objective function, to be maximized over $\sigma \in \Sigma$.

Exercise 10.1.8 Show that the sup in (10.3) is well-defined for each $x \in S$.

Definition 10.1.9 Given $w \in b\mathscr{B}(S)$, we define $\sigma \in \Sigma$ to be *w-greedy* if

$$\sigma(x) \in \operatorname*{argmax}_{u \in \Gamma(x)} \left\{ r(x, u) + \rho \int w[F(x, u, z)]\phi(dz) \right\} \qquad (x \in S) \tag{10.6}$$

Lemma 10.1.10 *Let assumptions 10.1.3–10.1.5 hold. If $w \in bcS$, then the objective function on the right-hand side of (10.6) is continuous in u for each $x \in S$, and Σ contains at least one w-greedy policy.*

The proof of this lemma is harder than it looks. On one hand, because $w \in bcS$, assumptions 10.1.3 and 10.1.5 imply that the objective function on the right-hand side of (10.6) is continuous with respect to u for each x. Since the constraint set $\Gamma(x)$ is compact, a solution to the maximization problem exists. Thus for every x we can find at least one u_x^* that attains the maximum, and the map $x \mapsto u_x^*$ certainly defines a function σ from $S \to A$ satisfying (10.6). On the other hand, for this "policy" to be in Σ *it must be Borel measurable*. Measurability is not immediately clear.

Fortunately there are "measurable selection" theorems stating that under the current assumptions we can find at least one such $x \mapsto u_x^*$ that is measurable. We omit the details, referring interested readers to Aliprantis and Border (1999, §17.3).[5]

We are now ready to state our main result on dynamic programming:

Theorem 10.1.11 *Under assumptions 10.1.3–10.1.5, the value function v^* is bounded and continuous. It is the unique function in bcS that satisfies*

$$v^*(x) = \max_{u \in \Gamma(x)} \left\{ r(x, u) + \rho \int v^* [F(x, u, z)] \phi(dz) \right\} \qquad (x \in S) \qquad (10.7)$$

A feasible policy is optimal if and only if it is v^*-greedy. At least one such policy exists.

Before turning to the proof let's make some brief comments and discuss an easy application. As a preliminary observation, note that in view of (10.7), a policy $\sigma \in \Sigma$ is v^*-greedy if and only if

$$v^*(x) = r(x, \sigma(x)) + \rho \int v^* [F(x, \sigma(x), z)] \phi(dz) \qquad (x \in S) \qquad (10.8)$$

In operator notation, this translates to $v^* = r_\sigma + \rho \mathbf{M}_\sigma v^*$.

Next let's discuss how theorem 10.1.11 can be applied. One use is as a sufficient condition: We will see below that v^* can be computed using value iteration. With v^* in hand, one can then compute a v^*-greedy policy. Assuming Borel measurability, we have found an optimal policy. The second way that we can use the theorem is as a necessary condition for optimality. For example, suppose that we want to know about the properties of optimal policies. We know that if σ^* is optimal, then it satisfies (10.8). We can (and do) use this to deduce facts about σ^*.

As an application of theorem 10.1.11, consider again the optimal growth example discussed on page 230. Recall that the state variable is income $y \in S := \mathbb{R}_+$, the control is savings $k \in A := \mathbb{R}_+$, the feasible correspondence is $\Gamma(y) = [0, y]$, the reward function is $r(y, k) = U(y - k)$, and the transition function is $F(y, k, z) = f(k, z)$. The shocks $(W_t)_{t \geq 1}$ are independent and take values in $Z \subset \mathbb{R}$ according to $\phi \in \mathscr{P}(Z)$.

---

[5]In many of the applications considered here, solutions of the form $x \mapsto u_x^*$ described above are easily seen to be measurable (being either continuous or monotone).

Assumption 10.1.12 The map $U\colon \mathbb{R}_+ \to \mathbb{R}_+$ is bounded and continuous. The function f is measurable and maps $\mathbb{R}_+ \times Z$ into $\mathbb{R}_+$. For each fixed $z \in Z$, the map $k \mapsto f(k, z)$ is continuous.

A feasible savings policy $\sigma \in \Sigma$ is a Borel function from S to itself such that $\sigma(y) \in [0, y]$ for all y. Every $\sigma \in \Sigma$ defines a process for income via

$$y_{t+1} = f(\sigma(y_t), W_{t+1}) \tag{10.9}$$

The corresponding stochastic kernel P_σ on S is given by

$$P_\sigma(y, B) = \int \mathbb{1}_B[f(\sigma(y), z)]\phi(dz) \qquad (y \in S,\ B \in \mathscr{B}(S))$$

Proposition 10.1.13 *Under assumption 10.1.12, the value function v^* is bounded and continuous. It is the unique function in bcS that satisfies*

$$v^*(y) = \max_{0 \le k \le y} \left\{ U(y - k) + \rho \int v^*[f(k, z)]\phi(dz) \right\} \qquad (y \in S)$$

At least one optimal policy exists. Moreover a policy σ^* is optimal if and only if

$$v^*(y) = U(y - \sigma^*(y)) + \rho \int v^*[f(\sigma^*(y), z)]\phi(dz) \qquad (y \in S)$$

Exercise 10.1.14 Verify proposition 10.1.13 using assumption 10.1.12. In particular, show that assumptions 10.1.3–10.1.5 on page 231 all hold.

10.1.3 Proofs

Let's turn to the proof of theorem 10.1.11. As a preliminary step, we introduce two important operators and investigate their properties. Many proofs in dynamic programming can be reduced to simple manipulations of these maps.

Definition 10.1.15 The operator $T_\sigma\colon b\mathscr{B}(S) \to b\mathscr{B}(S)$ is defined for all $\sigma \in \Sigma$ by

$$T_\sigma w(x) = r(x, \sigma(x)) + \rho \int w[F(x, \sigma(x), z)]\phi(dz) \qquad (x \in S)$$

The *Bellman operator* $T\colon bcS \to bcS$ is defined by

$$Tw(x) = \max_{u \in \Gamma(x)} \left\{ r(x, u) + \rho \int w[F(x, u, z)]\phi(dz) \right\} \qquad (x \in S)$$

Using the Bellman operator, we can restate the first part of theorem 10.1.11 as: v^* is the unique fixed point of T in bcS.

Exercise 10.1.16 Confirm that T does in fact send bcS into itself. Regarding boundedness, refer to lemma A.2.31 on page 334. Regarding continuity, refer to assumptions 10.1.3–10.1.5 and Berge's theorem on page 340.

Recalling the definition $v_\sigma := \sum_{t=0}^\infty \rho^t \mathbf{M}_\sigma^t r_\sigma$, our first result is as follows:

Lemma 10.1.17 *For every $\sigma \in \Sigma$, the operator T_σ is uniformly contracting on $(b\mathscr{B}(S), d_\infty)$, with*

$$\|T_\sigma w - T_\sigma w'\|_\infty \le \rho \|w - w'\|_\infty \qquad \forall\, w, w' \in b\mathscr{B}(S) \tag{10.10}$$

and the unique fixed point of T_σ in $b\mathscr{B}(S)$ is v_σ. In addition T_σ is monotone on $b\mathscr{B}(S)$, in the sense that if $w, w' \in b\mathscr{B}(S)$ and $w \le w'$, then $T_\sigma w \le T_\sigma w'$.

Here inequalities such as $w \le w'$ are pointwise inequalities on S.

Proof. The proof that T_σ is monotone is not difficult, and is left to the reader (you might want to use monotonicity of $\mathbf{M}_\sigma$, as in exercise 9.2.12, page 226). Regarding the claim that $T_\sigma v_\sigma = v_\sigma$, pointwise on S we have

$$v_\sigma = \sum_{t=0}^\infty \rho^t \mathbf{M}_\sigma^t r_\sigma = r_\sigma + \sum_{t=1}^\infty \rho^t \mathbf{M}_\sigma^t r_\sigma = r_\sigma + \rho \mathbf{M}_\sigma \sum_{t=0}^\infty \rho^t \mathbf{M}_\sigma^t r_\sigma = r_\sigma + \rho \mathbf{M}_\sigma v_\sigma = T_\sigma v_\sigma$$

The only tricky part of this argument is passing $\mathbf{M}_\sigma$ through the limit in the infinite sum. Justifying this is a good exercise for the reader who wants to improve his or her familiarity with the dominated convergence theorem.

The proof that T_σ is uniformly contracting is easy. Pick any $w, w' \in b\mathscr{B}(S)$. Making use of the linearity and monotonicity of $\mathbf{M}_\sigma$, we have

$$|T_\sigma w - T_\sigma w'| = |\rho \mathbf{M}_\sigma w - \rho \mathbf{M}_\sigma w'| = \rho |\mathbf{M}_\sigma (w - w')|$$
$$\le \rho \mathbf{M}_\sigma |w - w'| \le \rho \mathbf{M}_\sigma \|w - w'\|_\infty \mathbb{1}_S = \rho \|w - w'\|_\infty$$

pointwise on S. The inequality (10.10) now follows.   □

Next we turn to the Bellman operator.

Lemma 10.1.18 *The operator T is uniformly contracting on (bcS, d_∞), with*

$$\|Tw - Tw'\|_\infty \le \rho \|w - w'\|_\infty \qquad \forall\, w, w' \in bcS \tag{10.11}$$

In addition T is monotone on bcS, in the sense that if $w, w' \in bcS$ and $w \le w'$, then $Tw \le Tw'$.

Proof. The proof of the second claim (monotonicity) is easy and is left to the reader. Before starting the proof of (10.11), we make the following observation: If w and w' are bounded functions on some arbitrary set, then

$$|\sup w - \sup w'| \le \sup |w - w'| =: \|w - w'\|_\infty \tag{10.12}$$

To see this, pick any such w, w'. We have

$$\sup w = \sup(w - w' + w') \le \sup(w - w') + \sup w' \le \sup |w - w'| + \sup w'$$

$$\therefore \quad \sup w - \sup w' \le \sup |w - w'|$$

The same argument reversing the roles of w and w' finishes the job.

Now consider (10.11). For any $w, w' \in bcS$ and any $x \in S$, the deviation $|Tw(x) - Tw'(x)|$ is equal to

$$\left| \sup_u \left\{ r(x, u) + \rho \int w[F(x, u, z)]\phi(dz) \right\} - \sup_u \left\{ r(x, u) + \rho \int w'[F(x, u, z)]\phi(dz) \right\} \right|$$

Using (10.12), we obtain

$$|Tw(x) - Tw'(x)| \le \rho \sup_u \left| \int \{w[F(x, u, z)] - w'[F(x, u, z)]\}\phi(dz) \right|$$

$$\le \rho \sup_u \int |w[F(x, u, z)] - w'[F(x, u, z)]|\phi(dz)$$

$$\le \rho \sup_u \int \|w - w'\|_\infty \phi(dz) = \rho\|w - w'\|_\infty$$

Taking the supremum over $x \in S$ gives the desired inequality.[6] □

Exercise 10.1.19 Give an alternative proof that T is a uniform contraction of modulus ρ by applying theorem 6.3.8 (page 149).

Now we turn to the first claim in theorem 10.1.11. In operator notation, this translates to the following assertion:

Lemma 10.1.20 *The value function v^* is the unique fixed point of T in bcS.*

Proof. Since T is uniformly contracting on the complete space (bcS, d_∞), it follows from Banach's fixed point theorem (theorem 3.2.36) that T has one and only one fixed point w^* in this set.[7] It remains to show that $w^* = v^*$.

---

[6] This proof is due to Hernández-Lerma and Lasserre (1996, lmm. 2.5).
[7] Completeness of (bcS, d_∞) is proved in theorem 3.2.9, page 46.

To begin, note that by lemma 10.1.10 there exists a policy $\sigma \in \Sigma$ satisfying $Tw^* = T_\sigma w^*$. (Why?) For this policy σ we have $w^* = Tw^* = T_\sigma w^*$. But v_σ is the only fixed point of T_σ, so $w^* = v_\sigma$. In which case $w^* \leq v^*$, since, by definition, $v_\sigma \leq v^*$ for any $\sigma \in \Sigma$.

To check the reverse inequality, pick an arbitrary $\sigma \in \Sigma$, and note that $w^* = Tw^* \geq T_\sigma w^*$. Iterating on this inequality and using the monotonicity of T_σ, we obtain $w^* \geq T_\sigma^k w^*$ for all $k \in \mathbb{N}$. Taking limits and using the fact that $T_\sigma^k w^* \to v_\sigma$ uniformly and hence pointwise, we have $w^* \geq v_\sigma$. Since σ is arbitrary it follows that $w^* \geq v^*$. (Why?) Therefore $w^* = v^*$. $\qquad\square$

Our next task is to verify the claim that policies are optimal if and only if they are v^*-greedy.

Lemma 10.1.21 *A policy $\sigma \in \Sigma$ is optimal if and only if it is v^*-greedy.*

Proof. Recall that σ is v^*-greedy if and only if it satisfies (10.8), which in operator notation becomes $v^* = T_\sigma v^*$. This is equivalent to the statement $v_\sigma = v^*$, since v_σ is the unique fixed point of T_σ. But $v_\sigma = v^*$ says precisely that σ is optimal. $\qquad\square$

The last claim in theorem 10.1.11 is that at least one optimal policy exists. This now follows from lemma 10.1.10.

10.2   Numerical Methods

Numerical solution of dynamic programming problems is challenging and, at the same time, of great practical significance. In earlier chapters we considered techniques for solving SDPs numerically, such as value iteration and policy iteration. In this section we look more deeply at the theory behind these iterative methods. The algorithms are shown to converge globally to optimal solutions.

10.2.1   Value Iteration

Consider the SDP defined in §10.1. The fact that the Bellman operator is a uniform contraction on bcS for which v^* is the fixed point gives us a natural way to approximate v^*: Pick any $v_0 \in bcS$ and iterate the Bellman operator until $T^n v_0$ is close to v^*. This suggests the algorithm for computing (approximately) optimal policies presented in algorithm 10.1.

Algorithm 10.1 is essentially the same as the earlier value iteration algorithms presented on page 103, although we have added the index n in order to keep track of

---

Algorithm 10.1: Value iteration algorithm

---

read in initial $v_0 \in bcS$ and set $n = 0$
repeat
   | set $n = n + 1$
   | set $v_n = Tv_{n-1}$, where T is the Bellman operator
until *a stopping rule is satisfied*
solve for a v_n-greedy policy σ (cf., definition 10.1.9)
return σ

---

the iterates. Since $v_n = T^n v_0$ converges to v^*, after sufficiently many iterations the resulting policy σ should have relatively good properties, in the sense that $v_\sigma \cong v^*$.[8]

Two obvious questions arise: First, what stopping rule should be used in the loop? We know that $v_n \to v^*$, but v^* is not observable. How can we measure the distance between v^* and v_n for given n? Second, for given v_n, how close to being optimal is the v_n-greedy policy σ that the algorithm produces? These questions are answered in the next theorem.

Theorem 10.2.1 *Let $v_0 \in bcS$. Fix $n \in \mathbb{N}$, and let $v_n := T^n v_0$, where T is the Bellman operator. If $\sigma \in \Sigma$ is v_n-greedy, then*

$$\|v^* - v_\sigma\|_\infty \leq \frac{2\rho}{1-\rho} \|v_n - v_{n-1}\|_\infty \tag{10.13}$$

The next corollary follows directly (the proof is an exercise).

Corollary 10.2.2 *Let $(v_n)_{n \geq 0}$ be as in theorem 10.2.1. If $(\sigma_n)_{n \geq 0}$ is a sequence in Σ such that σ_n is v_n-greedy for each $n \geq 0$, then $\|v^* - v_{\sigma_n}\|_\infty \to 0$ as $n \to \infty$.*

The proof of theorem 10.2.1 is given at the end of this section. Before turning to it, let us make some comments on the theorem.

First, theorem 10.2.1 bounds the deviation between the v_n-greedy policy σ and the optimal policy σ^* in terms of their *value*. The value of σ^* is given by v_{σ^*}, which by definition is equal to v^*. As a result we can say that given any initial condition x,

$$v_{\sigma^*}(x) - v_\sigma(x) = |v_{\sigma^*}(x) - v_\sigma(x)| \leq \|v^* - v_\sigma\|_\infty \leq \frac{2\rho}{1-\rho} \|v_n - v_{n-1}\|_\infty$$

Of course one can also seek to bound some kind of geometric deviation between σ and σ^*, but in applications this is usually less important than bounding the difference between their values.

---

[8]Of course if v_n is exactly equal to v^*, then σ is optimal, and $v_\sigma = v^*$.

Second, the usefulness of this theorem comes from the fact that $\|v_n - v_{n-1}\|_\infty$ is observable. In particular, it can be measured at each iteration of the algorithm. This provides a natural stopping rule: Iterate until $\|v_n - v_{n-1}\|_\infty$ is less than some tolerance ϵ, and then compute a v_n-greedy policy σ. The policy satisfies $\|v^* - v_\sigma\|_\infty \leq 2\rho\epsilon/(1 - \rho)$.

Third, the bound (10.13) is often quite conservative. From this perspective, theorem 10.2.1 might best be viewed as a guarantee that the output of the algorithm converges to the true solution—such a guarantee is indispensable for numerical algorithms in scientific work.

Finally, on a related point, if you wish to supply bounds for a particular solution you have computed, then *relative* optimality bounds are easier to interpret. A relative bound establishes that an approximate optimal policy earns at least some fraction (say, 95%) of maximum value. Here is an example: Suppose that the reward function is nonnegative, so v_σ and v^* are nonnegative on S.[9] Suppose further that we choose v_0 to satisfy $0 \leq v_0 \leq v^*$, in which case $0 \leq v_n \leq v^*$ for all $n \in \mathbb{N}$ by monotonicity of T. By (10.13) and the fact that $v_n \leq v^*$, we have

$$\|v_n - v_{n-1}\|_\infty \leq \eta \implies \frac{v^*(x) - v_\sigma(x)}{v^*(x)} \leq \frac{2\rho}{1 - \rho} \cdot \frac{\eta}{v_n(x)} =: \alpha_n(\eta, x) \qquad (10.14)$$

In other words, if one terminates the value iteration at $\|v_n - v_{n-1}\|_\infty \leq \eta$, then the resulting policy σ obtains at least $(1 - \alpha_n(\eta, x)) \times 100\%$ of the total value available when the initial condition is x.

Let's finish with the proof of theorem 10.2.1:

Proof of theorem 10.2.1. Note that

$$\|v^* - v_\sigma\|_\infty \leq \|v^* - v_n\|_\infty + \|v_n - v_\sigma\|_\infty \qquad (10.15)$$

First let's bound the first term on the right-hand side of (10.15). Using the fact that v^* is a fixed point of T, we get

$$\|v^* - v_n\|_\infty \leq \|v^* - Tv_n\|_\infty + \|Tv_n - v_n\|_\infty \leq \rho\|v^* - v_n\|_\infty + \rho\|v_n - v_{n-1}\|_\infty$$

$$\therefore \quad \|v^* - v_n\|_\infty \leq \frac{\rho}{1 - \rho}\|v_n - v_{n-1}\|_\infty \qquad (10.16)$$

Now consider the second term on the right-hand side of (10.15). Since σ is v_n-greedy, we have $Tv_n = T_\sigma v_n$, and

$$\|v_n - v_\sigma\|_\infty \leq \|v_n - Tv_n\|_\infty + \|Tv_n - v_\sigma\|_\infty \leq \|Tv_{n-1} - Tv_n\|_\infty + \|T_\sigma v_n - T_\sigma v_\sigma\|_\infty$$

---

[9]Since r is already assumed to be bounded, there is no loss of generality in taking r as nonnegative, in the sense that adding a constant to r produces a monotone transformation of the objective function $v_\sigma(x)$, and hence does not alter the optimization problem.

$$\therefore \quad \|v_n - v_\sigma\|_\infty \leq \rho \|v_{n-1} - v_n\|_\infty + \rho \|v_n - v_\sigma\|_\infty$$

$$\therefore \quad \|v_n - v_\sigma\|_\infty \leq \frac{\rho}{1-\rho} \|v_n - v_{n-1}\|_\infty \tag{10.17}$$

Together, (10.15), (10.16), and (10.17) give us (10.13). □

10.2.2 Policy Iteration

Aside from value function iteration, there is another iterative procedure called policy iteration, which we first met in §5.1.3. In this section we describe policy iteration, and some of its convergence properties. The basic algorithm is presented in algorithm 10.2.

Algorithm 10.2: Policy iteration algorithm

read in an initial policy $\sigma_0 \in \Sigma$
set $n = 0$
repeat
    evaluate $v_{\sigma_n} := \sum_{t=0}^{\infty} \rho^t \mathbf{M}_{\sigma_n}^t r_{\sigma_n}$
    compute a v_{σ_n}-greedy policy $\sigma_{n+1} \in \Sigma$
    set $n = n + 1$
until *a stopping rule is satisfied*
return σ_n

It turns out that the sequence of functions v_{σ_n} produced by algorithm 10.2 converges to v^*, and, with a sensible stopping rule, the resulting policy is approximately optimal. Let's clarify these ideas, starting with the following observation:

Lemma 10.2.3 *If $(\sigma_n)_{n \geq 0}$ is a sequence in Σ generated by the policy iteration algorithm, then $v_{\sigma_n} \leq v_{\sigma_{n+1}}$ holds pointwise on S for all n.*

Proof. Pick any $x \in S$ and $n \in \mathbb{N}$. By definition,

$$\sigma_{n+1}(x) \in \underset{u \in \Gamma(x)}{\mathrm{argmax}} \left\{ r(x,u) + \rho \int v_{\sigma_n}[F(x,u,z)]\phi(dz) \right\}$$

From this and the fact that $v_\sigma = T_\sigma v_\sigma$ for all $\sigma \in \Sigma$, we have

$$v_{\sigma_n}(x) = r(x,\sigma_n(x)) + \rho \int v_{\sigma_n}[F(x,\sigma_n(x),z)]\phi(dz)$$

$$\leq r(x,\sigma_{n+1}(x)) + \rho \int v_{\sigma_n}[F(x,\sigma_{n+1}(x),z)]\phi(dz)$$

Rewriting in operator notation, this inequality becomes $v_{\sigma_n} \leq T_{\sigma_{n+1}} v_{\sigma_n}$. Since $T_{\sigma_{n+1}}$ is monotone (lemma 10.1.17), iteration with $T_{\sigma_{n+1}}$ yields $v_{\sigma_n} \leq T_{\sigma_{n+1}}^k v_{\sigma_n}$ for all $k \in \mathbb{N}$.

Taking limits, and using the fact that $T^k_{\sigma_{n+1}} v_{\sigma_n} \to v_{\sigma_{n+1}}$ uniformly and hence pointwise on S, we obtain the conclusion of the lemma.   □

It turns out that just as the value iteration algorithm is globally convergent, so too is the policy iteration algorithm.

Theorem 10.2.4 *If $(\sigma_n)_{n\geq 0} \subset \Sigma$ is a sequence generated by the policy iteration algorithm, then $\|v_{\sigma_n} - v^*\|_\infty \to 0$ as $n \to \infty$.*

Proof. Let $w_n := T^n v_{\sigma_0}$, where T is the Bellman operator, and, as usual, T^0 is the identity map. Since $v_{\sigma_n} \leq v^*$ for all $n \geq 0$ (why?), it is sufficient to prove that $w_n \leq v_{\sigma_n}$ for all $n \geq 0$. (Why?) The claim is true for $n = 0$ by definition. Suppose that it is true for arbitrary n. Then it is true for $n + 1$, since

$$w_{n+1} = Tw_n \leq Tv_{\sigma_n} = T_{\sigma_{n+1}} v_{\sigma_n} \leq T_{\sigma_{n+1}} v_{\sigma_{n+1}} = v_{\sigma_{n+1}}$$

You should not have too much trouble verifying these statements.   □

The rest of this section focuses mainly on policy iteration in the finite case, which we treated previously in §5.1.3. It is proved that when S and A are finite, the exact optimal policy is obtained in finite time.

Lemma 10.2.3 tells us that the value of the sequence (σ_n) is nondecreasing. In fact, if σ_{n+1} cannot be chosen as equal to σ_n, then the value increase is strict. On the other hand, if σ_{n+1} can be chosen as equal to σ_n, then we have found an optimal policy. The next lemma makes these statements precise.

Lemma 10.2.5 *Let (σ_n) be a sequence of policies generated by the policy iteration algorithm. If σ_{n+1} cannot be chosen as equal to σ_n, in the sense that there exists an $x \in S$ such that*

$$\sigma_n(x) \notin \operatorname*{argmax}_{u \in \Gamma(x)} \left\{ r(x,u) + \rho \int v_{\sigma_n}[F(x,u,z)]\phi(dz) \right\}$$

then $v_{\sigma_{n+1}}(x) > v_{\sigma_n}(x)$. Conversely, if σ_{n+1} can be chosen as equal to σ_n, in the sense that

$$\sigma_n(x) \in \operatorname*{argmax}_{u \in \Gamma(x)} \left\{ r(x,u) + \rho \int v_{\sigma_n}[F(x,u,z)]\phi(dz) \right\} \qquad \forall x \in S$$

then $v_{\sigma_n} = v^*$ and σ_n is an optimal policy.

Proof. Regarding the first assertion, let x be a point in S with

$$r(x,\sigma_{n+1}(x)) + \rho \int v_{\sigma_n}[F(x,\sigma_{n+1}(x),z)]\phi(dz)$$

$$> r(x,\sigma_n(x)) + \rho \int v_{\sigma_n}[F(x,\sigma_n(x),z)]\phi(dz)$$

Writing this in operator notation, we have $T_{\sigma_{n+1}} v_{\sigma_n}(x) > T_{\sigma_n} v_{\sigma_n}(x) = v_{\sigma_n}(x)$. But lemma 10.2.3 and the monotonicity of $T_{\sigma_{n+1}}$ now yield

$$v_{\sigma_{n+1}}(x) = T_{\sigma_{n+1}} v_{\sigma_{n+1}}(x) \geq T_{\sigma_{n+1}} v_{\sigma_n}(x) > v_{\sigma_n}(x)$$

Regarding the second assertion, suppose that

$$\sigma_n(x) \in \operatorname*{argmax}_{u \in \Gamma(x)} \left\{ r(x, u) + \rho \int v_{\sigma_n}[F(x, u, z)] \phi(dz) \right\} \qquad \forall x \in S$$

It follows that

$$v_{\sigma_n}(x) = r(x, \sigma_n(x)) + \rho \int v_{\sigma_n}[F(x, \sigma_n(x), z)] \phi(dz)$$

$$= \max_{u \in \Gamma(x)} \left\{ r(x, u) + \rho \int v_{\sigma_n}[F(x, u, z)] \phi(dz) \right\}$$

for every $x \in S$. In other words, v_{σ_n} is the fixed point of the Bellman operator. In which case $v_{\sigma_n} = v^*$, and the proof is done. □

Algorithm 10.3: Policy iteration, finite case

---

read in initial $\sigma_0 \in \Sigma$
set $n = 0$
repeat
    | set $n = n + 1$
    | evaluate $v_{\sigma_{n-1}} = \sum_{t=0}^{\infty} \rho^t \mathbf{M}^t_{\sigma_{n-1}} r_{\sigma_{n-1}}$
    | taking $\sigma_n = \sigma_{n-1}$ if possible, compute a $v_{\sigma_{n-1}}$-greedy policy σ_n
until $\sigma_n = \sigma_{n-1}$
return σ_n

Algorithm 10.3 adds a stopping rule to algorithm 10.2, which is suitable for the finite case. The algorithm works well in the finite state/action case because it always terminates in finite time at an optimal policy. This is the content of our next theorem.

Theorem 10.2.6 *If S and A are finite, then the policy iteration algorithm always terminates after a finite number of iterations, and the resulting policy is optimal.*

Proof. First note that if the algorithm terminates at n with $\sigma_{n+1} = \sigma_n$, then this policy is optimal by the second part of lemma 10.2.5. Next suppose that the algorithm never terminates, generating an infinite sequence of policies (σ_n). At each stage n the stopping rule implies that σ_{n+1} cannot be chosen as equal to σ_n, and the first part of lemma 10.2.5 applies. Thus $v_{\sigma_n} < v_{\sigma_{n+1}}$ for all n (i.e., $v_{\sigma_n} \leq v_{\sigma_{n+1}}$ and $v_{\sigma_n} \neq v_{\sigma_{n+1}}$). But the set of maps from S to A is clearly finite, and hence so is the set of functions $\{v_\sigma : \sigma \in \Sigma\}$. As a result such an infinite sequence is impossible, and the algorithm always terminates. □

10.2.3 Fitted Value Iteration

In §6.2.2 we began our discussion of fitted value iteration and presented the main algorithm. Recall that to approximate the image Tv of a function v we evaluated Tv at finite set of grid points $(x_i)_{i=1}^k$ and then used these samples to construct an approximation to Tv. In doing so, we decomposed $\hat{T}$ into the two operators L and T: First T is applied to v at each of the grid points, and then an approximation operator L sends the result into a function $w = \hat{T}v = L(Tv)$. Thus $\hat{T} := L \circ T :=: LT$. We saw that LT is uniformly contracting whenever L is nonexpansive with respect to d_∞.

In general, we will consider a map $L \colon b\mathscr{B}(S) \to \mathscr{F} \subset b\mathscr{B}(S)$ where, for each $v \in b\mathscr{B}(S)$, the approximation $Lv \in \mathscr{F}$ is constructed based on a sample $(v(x_i))_{i=1}^k$ on grid points $(x_i)_{i=1}^k$. In addition, L is chosen to be nonexpansive:

$$\|Lv - Lw\|_\infty \leq \|v - w\|_\infty \qquad \forall\, v, w \in b\mathscr{B}(S) \tag{10.18}$$

Example 10.2.7 (Piecewise constant approximation) Let $(P_i)_{i=1}^k$ be a partition of S, in that $P_m \cap P_n = \varnothing$ when $m \neq n$ and $S = \cup_{i=1}^k P_i$. Each P_i contains a single grid point x_i. For any $v \colon S \to \mathbb{R}$ we define $v \mapsto Mv$ by

$$Mv(x) = \sum_{i=1}^k v(x_i)\mathbb{1}_{P_i}(x) \qquad (x \in S)$$

Exercise 10.2.8 Show that for any $w, v \in b\mathscr{B}(S)$ and any $x \in S$ we have

$$|Mw(x) - Mv(x)| \leq \sup_{1 \leq i \leq k} |w(x_i) - v(x_i)|$$

Using this result, show that the operator M is nonexpansive on $(b\mathscr{B}(S), d_\infty)$.

Example 10.2.9 (Continuous piecewise linear interpolation) Let's focus on the one-dimensional case. Let $S = [a, b]$, and let the grid points be increasing:

$$x_1 < \ldots < x_k, \quad x_1 = a \text{ and } x_k = b$$

Let N be the operator that maps $w \colon S \to \mathbb{R}$ into its continuous piecewise affine interpolant defined by the grid. That is to say, if $x \in [x_i, x_{i+1}]$ then

$$Nw(x) = \lambda w(x_i) + (1 - \lambda)w(x_{i+1}) \quad \text{where} \quad \lambda := \frac{x_{i+1} - x}{x_{i+1} - x_i}$$

Exercise 10.2.10 Show that for any $w, v \in b\mathscr{B}(S)$ and any $x \in S$ we have

$$|Nw(x) - Nv(x)| \leq \sup_{1 \leq i \leq k} |w(x_i) - v(x_i)|$$

Using this result, show that N is nonexpansive.

Algorithm 10.4: FVI algorithm

---

read in initial $v_0 \in b\mathscr{B}(S)$ and set $n = 0$
repeat
    | set $n = n + 1$
    | sample Tv_{n-1} at a finite set of grid points
    | compute $\hat{T}v_{n-1} = LTv_{n-1}$ from the samples
    | set $v_n = \hat{T}v_{n-1}$
until *the deviation* $\|v_n - v_{n-1}\|_\infty$ *falls below some tolerance*
solve for a v_n-greedy policy σ

Our algorithm for fitted value iteration is given in algorithm 10.4. It terminates in finite time for any strictly positive tolerance, since $\hat{T}$ is a uniform contraction. The policy it produces is approximately optimal, with the deviation given by the following theorem. (The proof is given in the appendix to this chapter.)

Theorem 10.2.11 *Let* $v_0 \in \mathscr{F}$ *and let* $v_n := \hat{T}^n v_0$, *where* $\hat{T} := LT$ *and* $L\colon b\mathscr{B}(S) \to \mathscr{F}$ *is nonexpansive. If* $\sigma \in \Sigma$ *is* v_n-greedy, then

$$\|v^* - v_\sigma\|_\infty \leq \frac{2}{(1-\rho)^2} \times (\rho \|v_n - v_{n-1}\|_\infty + \|Lv^* - v^*\|_\infty)$$

Most of the comments given after theorem 10.2.1 (page 239) apply to theorem 10.2.11. In particular, the bound is conservative, but it shows that the value of σ can be made as close to that of σ^* as desired, provided that L can be chosen so that $\|Lv^* - v^*\|_\infty$ is arbitrarily small. In the case of continuous piecewise linear interpolation on $S = [a, b]$ this is certainly possible.[10]

---

[10] This is due to continuity of v^*, which always holds under our assumptions. When v^* is continuous on $[a, b]$ it is also uniformly continuous, which is to say that for any $\epsilon > 0$ there is a $\delta > 0$ such that $|v^*(x) - v^*(y)| < \epsilon$ whenever $|x - y| < \delta$. From this property it is not too difficult to show that given any $\epsilon > 0$, a sufficiently fine grid will yield a continuous piecewise affine interpolant Lv^* such that $\|v^* - Lv^*\|_\infty < \epsilon$. See, for example, Bartle and Sherbet (1992).

10.3   Commentary

A first-rate theoretical treatment of stochastic dynamic programming can be found in the two monographs of Hernández-Lerma and Lasserre (1996, 1999). Also recommended are Bertsekas (1995) and Puterman (1994). From the economics literature see, for example, Stokey and Lucas (1989) or Le Van and Dana (2003). All these sources contain extensive references to further applications.

Additional discussion of fitted value iteration with nonexpansive approximators can be found in Gordon (1995) and Stachurski (2008). For alternative discussions of value iteration, see, for example, Santos and Vigo-Aguiar (1998) or Grüne and Semmler (2004).

Value iteration and policy iteration are two of many algorithms proposed in the literature. Other popular techniques include projection methods, Taylor series approximation (e.g., linearization), and parameterized expectations. See, for example, Marcet (1988), Tauchen and Hussey (1991), Judd (1992), Den Haan and Marcet (1994), Rust (1996), Judd (1998), Christiano and Fisher (2000), McGrattan (2001), Uhlig (2001), Maliar and Maliar (2005), or Canova (2007). Marimon and Scott (2001) is a useful survey, while Santos (1999) and Aruoba et al. (2006) provide numerical comparisons.

Dynamic programming has many interesting applications in economics that we will have little chance to discuss. Two unfortunate omissions are industrial organization (e.g., Green and Porter 1984, Hopenhayn 1992, Ericson and Pakes 1995, or Pakes and McGuire 2001) and search theory (see McCall 1970 for an early contribution and Rogerson et al. 2005 for a recent survey). For some earlier classics the macroeconomics literature, try Lucas and Prescott (1971), Hall (1978), Lucas (1978), Kydland and Prescott (1982), Brock (1982), or Mehra and Prescott (1985). Dechert and O'Donnell (2006) provide a nice application of dynamic programming to environmental economics.

Chapter 11

Stochastic Dynamics

It's now time to give a treatment of stability for general Markov chains on uncountably infinite state spaces. Although the stability theory we used to study the finite case (chapter 4) and the density case (chapter 8) does not survive the transition without some modification, the underlying ideas are similar, and connections are drawn at every opportunity. Throughout this chapter we take S to be a Borel subset of $\mathbb{R}^n$.

11.1 Notions of Convergence

Before considering the dynamics of general state Markov chains, we need to develop notions of convergence that apply to the measure (as opposed to finite probability or density) setting. Along the way we will cover some fundamental results in asymptotic probability, including the weak and strong laws of large numbers for IID sequences.

11.1.1 Convergence of Sample Paths

Recall that an S-valued stochastic process is a tuple $(\Omega, \mathscr{F}, \mathbb{P}, (X_t)_{t \in \mathbb{T}})$ where $(\Omega, \mathscr{F})$ is a measurable space, $\mathbb{P}$ is a probability on $(\Omega, \mathscr{F})$, $\mathbb{T}$ is an index set such as $\mathbb{N}$ or $\mathbb{Z}$, and $(X_t)_{t \in \mathbb{T}}$ is a family of S-valued random variables on $(\Omega, \mathscr{F})$. Various notions of convergence exist for stochastic processes. We begin with almost sure convergence.

Definition 11.1.1 Let $(\Omega, \mathscr{F}, \mathbb{P}, (X_t)_{t \geq 1})$ be an S-valued stochastic process and let X be an S-valued random variable on $(\Omega, \mathscr{F}, \mathbb{P})$. We say that $(X_t)_{t \geq 1}$ *converges almost surely* (alternatively, *with probability one*) to X if

$$\mathbb{P} \left\{ \lim_{t \to \infty} X_t = X \right\} := \mathbb{P} \left\{ \omega \in \Omega \; : \; \lim_{t \to \infty} X_t(\omega) = X(\omega) \right\} = 1$$

In the language of §7.3.2, almost sure convergence is just convergence $\mathbb{P}$-almost everywhere.

Almost sure convergence plays a vital role in probability theory, although you may wonder why we don't just require that convergence occurs for *every* $\omega \in \Omega$, instead of just those ω in a set of probability one. The reason is that convergence for every path is too strict: In almost all random systems, aberrations can happen that cause such convergence to fail. Neglecting probability zero events allows us to obtain much more powerful conclusions.

Exercise 11.1.2 Expectations are not always very informative measures of stochastic processes dynamics. For example, consider a stochastic process with probability space $((0,1), \mathscr{B}(0,1), \lambda)$ and random variables $X_n := n^2 \mathbb{1}_{(0,1/n)}$. Show that $X_n \to 0$ almost surely, while $\mathbb{E}X_n \uparrow \infty$.

Here is another important notion of convergence of random variables:

Definition 11.1.3 Let $(\Omega, \mathscr{F}, \mathbb{P}, (X_t)_{t \geq 1})$ and X be as above. We say that $(X_t)_{t \geq 1}$ *converges in probability* to X if, for all $\epsilon > 0$,

$$\lim_{t \to \infty} \mathbb{P}\{\|X_t - X\| \geq \epsilon\} := \lim_{t \to \infty} \mathbb{P}\{\omega \in \Omega : \|X_t(\omega) - X(\omega)\| \geq \epsilon\} = 0$$

Convergence in probability is weaker than almost sure convergence:

Lemma 11.1.4 *If $X_t \to X$ almost surely, then $X_t \to X$ in probability.*

Proof. Fix $\epsilon > 0$. Since $X_t \to X$ almost surely, $\mathbb{1}\{\|X_t - X\| \geq \epsilon\} \to 0$ $\mathbb{P}$-almost everywhere on Ω. (Why?) An application of the dominated convergence theorem (page 182) now gives the desired result. $\qquad \square$

The converse is not true. It is an optional (and nontrivial) exercise for you to construct a stochastic process that converges to some limit in probability, and yet fails to converge to that same limit almost surely.

Exercise 11.1.5 The sets in definition 11.1.3 are measurable. Show, for example, that if $S = \mathbb{R}$ and $\epsilon > 0$, then $\{|X_n - X| \geq \epsilon\} \in \mathscr{F}$.

Let's discuss some applications of almost sure convergence and convergence in probability. Let $(X_t)_{t \geq 1}$ be a real-valued stochastic process on $(\Omega, \mathscr{F}, \mathbb{P})$ with common expectation m. Define $\bar{X}_n := n^{-1} \sum_{t=1}^{n} X_t$. The (weak) law of large numbers (WLLN) states that under suitable assumptions the sample mean $\bar{X}_n$ converges in probability to m as $n \to \infty$. We begin with the case $m = 0$.

Exercise 11.1.6 We will make use of the following identity, the proof of which is left as an exercise (use induction):

$$\text{for all } (a_1, \ldots, a_n) \in \mathbb{R}^n, \quad \left(\sum_{i=1}^{n} a_i \right)^2 = \sum_{1 \leq i,j \leq n} a_i a_j = \sum_{j=1}^{n} \sum_{i=1}^{n} a_i a_j$$

Exercise 11.1.7 Suppose the real, zero-mean sequence $(X_t)_{t \geq 1}$ satisfies

1. $\text{Cov}(X_i, X_j) = 0$ for all $i \neq j$, and

2. $\text{Cov}(X_i, X_i) = \text{Var}(X_i) = \mathbb{E}X_i^2 \leq M$ for all $i \in \mathbb{N}$.

Show that $\text{Var}(\bar{X}_n) \leq M/n$ for all $n \in \mathbb{N}$. (Regarding the first property, we usually say that the sequence is *pairwise uncorrelated*.)

Exercise 11.1.8 Now prove that the WLLN holds for this sequence. As a first step, show that $\mathbb{P}\{|\bar{X}_n| \geq \epsilon\} \leq M/(n\epsilon^2)$.[1] Conclude that $\bar{X}_n$ converges to zero in probability as $n \to \infty$.

This result can be extended to the case $m \neq 0$ by considering the zero-mean sequence $Y_t := X_t - m$. This gives us

Theorem 11.1.9 *Let* $(\Omega, \mathscr{F}, \mathbb{P}, (X_t)_{t \geq 1})$ *be a real-valued stochastic process with common mean* m. *If* $(X_t)_{t \geq 1}$ *is pairwise uncorrelated and* $\text{Var}(X_t) \leq M$ *for all* t, *then* $\bar{X}_n \to m$ *in probability. In particular,*

$$\mathbb{P}\{|\bar{X}_n - m| \geq \epsilon\} \leq \frac{M}{n\epsilon^2} \qquad (\epsilon > 0, \, n \in \mathbb{N}) \tag{11.1}$$

When the sequence is independent a stronger result holds:

Theorem 11.1.10 *Let* $(\Omega, \mathscr{F}, \mathbb{P}, (X_t)_{t \geq 1})$ *be a real-valued stochastic process. If* $(X_t)_{t \geq 1}$ *is* IID *and* $\mathbb{E}|X_1| < \infty$, *then* $\bar{X}_n \to m$ *almost surely.*

Theorem 11.1.10 is called the strong law of large numbers (SLLN). A version of this theorem was stated previously on page 94. For a proof, see Dudley (2002, thm. 8.3.5).

How about stochastic processes that are correlated, rather than IID? We have already presented some generalizations of the SLLN that apply to Markov chains (theorem 4.3.33 on page 94 and theorem 8.2.15 on page 206). Although the proofs of these strong LLNs are beyond the scope of this book, let's take a look at the proof of a weak LLN for correlated processes. (Readers keen to progress can skip to the next section without loss of continuity.) We will see that the correlations $\text{Cov}(X_i, X_{i+k})$ must converge to zero in k sufficiently quickly. The following lemma will be useful:

---

[1]Hint: Use the Chebychev inequality (see page 214).

Lemma 11.1.11 *Let $(\beta_k)_{k\geq 1}$ be a sequence in $\mathbb{R}_+$, and let $(\Omega, \mathscr{F}, \mathbb{P}, (X_t)_{t\geq 1})$ be a real-valued stochastic process such that $\mathrm{Cov}(X_i, X_{i+k}) \leq \beta_k$ for all $i \geq 1$. If $(\beta_k)_{k\geq 0}$ satisfies $\sum_{k\geq 1} \beta_k < \infty$, then $\mathrm{Var}(\bar{X}_n) \to 0$ as $n \to \infty$.*

Proof. We have

$$\mathrm{Var}\left(\frac{1}{n}\sum_{i=1}^n X_i\right) = \frac{1}{n^2} \sum_{1\leq i,j\leq n} \mathrm{Cov}(X_i, X_j)$$

$$= \frac{1}{n^2}\sum_{i=1}^n \mathrm{Cov}(X_i, X_i) + \frac{2}{n^2}\sum_{1\leq i<j\leq n}\mathrm{Cov}(X_i, X_j)$$

$$\leq \frac{2}{n^2}\sum_{1\leq i\leq j\leq n}\mathrm{Cov}(X_i, X_j) = \frac{2}{n^2}\sum_{k=0}^{n-1}\sum_{i=1}^{n-k}\mathrm{Cov}(X_i, X_{i+k})$$

$$\therefore \quad \mathrm{Var}\left(\frac{1}{n}\sum_{i=1}^n X_i\right) \leq \frac{2}{n^2}\sum_{k=0}^{n-1}(n-k)\beta_k \leq \frac{2}{n}\sum_{k=0}^{n-1}\beta_k \leq \frac{2}{n}\sum_{k=0}^{\infty}\beta_k \to 0$$

$\square$

Now we can give a weak LLN for correlated, non-identically distributed random variables:

Theorem 11.1.12 *Let $(\Omega, \mathscr{F}, \mathbb{P}, (X_t)_{t\geq 1})$ be a real-valued stochastic process. If*

1. $\mathrm{Cov}(X_i, X_{i+k}) \leq \beta_k$ *for all $i \geq 1$ where $\sum_{k\geq 0}\beta_k < \infty$, and*

2. $\mathbb{E}X_n \to m \in \mathbb{R}$ *as $n \to \infty$,*

then $\bar{X}_n \to m$ in probability as $n \to \infty$.

Proof. Note that $\mathbb{E}X_n \to m$ implies $\mathbb{E}\bar{X}_n \to m$. Now fix $\epsilon > 0$ and choose $N \in \mathbb{N}$ such that $|\mathbb{E}\bar{X}_n - m| \leq \epsilon/2$ whenever $n \geq N$. If $n \geq N$, then

$$\{|\bar{X}_n - m| \geq \epsilon\} \subset \{|\bar{X}_n - \mathbb{E}\bar{X}_n| + |\mathbb{E}\bar{X}_n - m| \geq \epsilon\} \subset \{|\bar{X}_n - \mathbb{E}\bar{X}_n| \geq \epsilon/2\}$$

$$\therefore \quad \mathbb{P}\{|\bar{X}_n - m| \geq \epsilon\} \leq \mathbb{P}\{|\bar{X}_n - \mathbb{E}\bar{X}_n| \geq \epsilon/2\} \qquad (n \geq N)$$

This is sufficient because $|\bar{X}_n - \mathbb{E}\bar{X}_n| \to 0$ in probability when $n \to \infty$ as a result of the Chebychev inequality

$$\mathbb{P}\{|\bar{X}_n - \mathbb{E}\bar{X}_n| \geq \epsilon\} \leq \frac{\mathrm{Var}(\bar{X}_n)}{\epsilon^2}$$

and the fact that $\mathrm{Var}(\bar{X}_n) \to 0$ by lemma 11.1.11. $\square$

Theorem 11.1.12 can be used to prove a weak version of the SLLN stated in theorem 4.3.33 (page 94). Let S be a finite set, let p be a stochastic kernel on S such that the Dobrushin coefficient $\alpha(p)$ is strictly positive, let ψ^* be the unique stationary distribution for p, and let $h\colon S \to \mathbb{R}$ be any function.

Exercise 11.1.13 Show that if T is a uniform contraction with modulus γ and fixed point x^* on metric space (U,ρ), then for any given $x \in U$ we have $\rho(T^k x, x^*) \leq \gamma^k \rho(x, x^*)$. Conclude that there exist two constants $M < \infty$ and $\gamma \in [0,1)$ satisfying

$$\sum_{y \in S} |p^k(x,y) - \psi^*(y)| \leq M\gamma^k \text{ for any } k \in \mathbb{N} \text{ and any } x \in S$$

Exercise 11.1.14 Let $m := \sum_{y \in S} h(y)\psi^*(y)$ be the mean of h with respect to ψ^*. Using exercise 11.1.13, show that there are constants $K < \infty$ and $\gamma \in [0,1)$ such that

$$\left| \sum_{y \in S} h(y)p^k(x,y) - m \right| \leq K\gamma^k \text{ for any } k \in \mathbb{N} \text{ and any } x \in S$$

Now consider a Markov-(p, ψ^*) chain $(X_t)_{t \geq 0}$, where the initial condition has been set to the stationary distribution in order to simplify the proof. In particular, we have $\mathbb{E}h(X_t) = m$ for all t. Regarding the covariances,

$$\begin{aligned}
\text{Cov}(h(X_i), h(X_{i+k})) &= \sum_{x \in S} \sum_{y \in S} [h(x) - m][h(y) - m]\mathbb{P}\{X_{i+k} = y, X_i = x\} \\
&= \sum_{x \in S} \sum_{y \in S} [h(x) - m][h(y) - m]p^k(x,y)\psi^*(x) \\
&= \sum_{x \in S} [h(x) - m]\psi^*(x) \sum_{y \in S} [h(y) - m]p^k(x,y)
\end{aligned}$$

where the second equality is due to

$$\mathbb{P}\{X_{i+k} = y, X_i = x\} = \mathbb{P}\{X_{i+k} = y \mid X_i = x\}\mathbb{P}\{X_i = x\}$$

Exercise 11.1.15 Using these calculations and the result in exercise 11.1.14, show that there are constants $J < \infty$ and $\gamma \in [0,1)$ such that

$$|\text{Cov}(h(X_i), h(X_{i+k}))| \leq J\gamma^k \text{ for any } i, k \geq 0$$

Exercise 11.1.16 Show that the process $(h(X_t))_{t \geq 0}$ satisfies the conditions of theorem 11.1.12.

11.1.2 Strong Convergence of Measures

Now let's turn to another kind of convergence: convergence of the *distribution* of X_n to the distribution of X. Since distributions are measures, what we are seeking here is a metric (and hence a concept of convergence) defined on spaces of measures. In this section we discuss so-called strong convergence (or total variation convergence) of measures. The next section discusses weak convergence.

When defining strong convergence, it is useful to consider not only standard non-negative measures but also signed measures, which are countably additive set functions taking both positive and negative values.

Definition 11.1.17 Let S be a Borel measurable subset of $\mathbb{R}^n$. A (Borel) *signed measure* μ on a S is a countably additive set function from $\mathscr{B}(S)$ to $\mathbb{R}$: Given any pairwise disjoint sequence $(B_n) \subset \mathscr{B}(S)$, we have $\mu(\cup_n B) = \sum_n \mu(B_n)$. The set of all signed measures on S will be denoted by $b\mathscr{M}(S)$.[2]

Exercise 11.1.18 Show that for any $\mu \in b\mathscr{M}(S)$ we have $\mu(\varnothing) = 0$.

Addition and scalar multiplication of signed measures is defined setwise, so $(\alpha\mu + \beta\nu)(B) = \alpha\mu(B) + \beta\nu(B)$ for $\mu, \nu \in b\mathscr{M}(S)$ and scalars α, β. Notice that the difference $\mu - \nu$ of any two finite (nonnegative) measures is a signed measure. It turns out that every signed measure can be represented in this way:

Theorem 11.1.19 (Hahn–Jordan) *For each $\mu \in b\mathscr{M}(S)$ there exists sets S^- and S^+ in $\mathscr{B}(S)$ with $S^- \cap S^+ = \varnothing$, $S^- \cup S^+ = S$,*

- $\mu(B) \geq 0$ *whenever $B \in \mathscr{B}(S)$ and $B \subset S^+$, and*

- $\mu(B) \leq 0$ *whenever $B \in \mathscr{B}(S)$ and $B \subset S^-$.*

As a result, μ can be expressed as the difference $\mu^+ - \mu^-$ of two finite nonnegative measures μ^+ and μ^-, where

- $\mu^+(B) := \mu(B \cap S^+)$ *for all $B \in \mathscr{B}(S)$, and*

- $\mu^-(B) := -\mu(B \cap S^-)$ *for all $B \in \mathscr{B}(S)$.*

The set S^+ is called a *positive set* for μ and S^- is called a *negative set*. They are unique in the sense that if A and B are both positive (negative) for μ, then the set of points in A or B but not both has zero μ-measure. The first part of the theorem (decomposition of S) is called the Hahn decomposition, while the second (decomposition of μ) is called the Jordan decomposition. The proof of the Hahn decomposition is not overly difficult and can be found in almost every text on measure theory. The Jordan decomposition follows from the Hahn decomposition is a straightforward way:

---

[2]Notice that our signed measures are required to take values in $\mathbb{R}$. For this reason some authors call $b\mathscr{M}(S)$ the set of *finite* (or *bounded*) signed measures.

Exercise 11.1.20 Show that $\mu = \mu^+ - \mu^-$ holds setwise on $\mathscr{B}(S)$.

Exercise 11.1.21 Verify that μ^+ and μ^- are measures on $(S, \mathscr{B}(S))$. Show that the equalities $\mu(S^+) = \max_{B \in \mathscr{B}(S)} \mu(B)$ and $\mu(S^-) = \min_{B \in \mathscr{B}(S)} \mu(B)$ both hold.

Exercise 11.1.22 Show that if S^+ (resp., S^-) is positive (resp., negative) for μ, then S^- (resp., S^+) is positive (resp., negative) for $-\mu$.

Exercise 11.1.23 Let $f \in m\mathscr{B}(S)$ with $\lambda(|f|) < \infty$. Let $\mu(B) := \lambda(\mathbb{1}_B f)$. Show that $\mu \in b\mathscr{M}(S)$. Show that if $S^+ := \{x \in S : f(x) \geq 0\}$ and $S^- := \{x \in S : f(x) < 0\}$, then S^+ and S^- form a Hahn decomposition of S with respect to μ; and that $\mu^+(B) = \lambda(\mathbb{1}_B f^+)$ and $\mu^-(B) = \lambda(\mathbb{1}_B f^-)$. Show that the L_1 norm of f is equal to $\mu^+(S) + \mu^-(S)$.

The final result of the last exercise suggests the following generalization of L_1 distance from functions to measures:

Definition 11.1.24 The *total variation norm* of $\mu \in b\mathscr{M}(S)$ is defined as

$$\|\mu\|_{TV} := \mu^+(S) + \mu^-(S) = \mu(S^+) - \mu(S^-)$$

where S^+ and S^- are as in theorem 11.1.19. The function

$$d_{TV}(\mu, \nu) := \|\mu - \nu\|_{TV} \qquad (\mu, \nu \text{ in } b\mathscr{M}(S))$$

is a metric on $b\mathscr{M}(S)$, and $(b\mathscr{M}(S), d_{TV})$ is a metric space.

Exercise 11.1.25 Prove that for $\mu \in b\mathscr{M}(S)$, the norm $\|\mu\|_{TV} = \max_{\pi \in \Pi} \sum_{A \in \pi} |\mu(A)|$, where Π is the set of finite measurable partitions of S.[3]

Exercise 11.1.26 Verify that d_{TV} is a metric on $b\mathscr{M}(S)$.

One of the nice things about the Jordan decomposition is that we can define integrals with respect to signed measures with no extra effort:

Definition 11.1.27 Let $h \in m\mathscr{B}(S)$, let $\mu \in b\mathscr{M}(S)$ and let μ^+, μ^- be the Jordan decomposition of μ. We set $\mu(h) := \mu^+(h) - \mu^-(h)$ whenever at least one term is finite.

For given stochastic kernel P, we can define the Markov operator as a map over the space of signed measures in the obvious way:

$$\mu\mathbf{M}(B) = \int P(x, B)\mu(dx) = \int P(x, B)\mu^+(dx) - \int P(x, B)\mu^-(dx)$$

where $\mu \in b\mathscr{M}(S)$ and $B \in \mathscr{B}(S)$. The next lemma can be used to show that $\mathbf{M}$ is nonexpansive on $(b\mathscr{M}(S), d_{TV})$.

---

[3]Hint: The maximizer is the partition $\pi := \{S^+, S^-\}$.

Lemma 11.1.28 *We have $\|\mu\mathbf{M}\|_{TV} \leq \|\mu\|_{TV}$ for any $\mu \in b\mathscr{M}(S)$.*

Proof. If S^+ and S^- are the positive and negative sets for $\mu\mathbf{M}$, then

$$\|\mu\mathbf{M}\|_{TV} = \mu\mathbf{M}(S^+) - \mu\mathbf{M}(S^-) = \int P(x,S^+)\mu(dx) - \int P(x,S^-)\mu(dx)$$

By the definition of the integral with respect to μ, this becomes

$$\int P(x,S^+)\mu^+(dx) - \int P(x,S^+)\mu^-(dx) - \int P(x,S^-)\mu^+(dx) + \int P(x,S^-)\mu^-(dx)$$

$$\therefore \quad \|\mu\mathbf{M}\|_{TV} \leq \int P(x,S^+)\mu^+(dx) + \int P(x,S^-)\mu^-(dx)$$

The last term is dominated by $\mu^+(S) + \mu^-(S) =: \|\mu\|_{TV}$. $\qquad\square$

Total variation distance seems rather abstract, but for probabilities it has a very concrete alternative expression.

Lemma 11.1.29 *If ϕ and ψ are probability measures, then*

$$d_{TV}(\phi,\psi) := \|\phi - \psi\|_{TV} = 2 \sup_{B \in \mathscr{B}(S)} |\phi(B) - \psi(B)|$$

This equivalence makes the total variation distance particularly suitable for quantitative work. The proof is in the appendix to this chapter.

Theorem 11.1.30 *The metric spaces $(b\mathscr{M}(S), d_{TV})$ and $(\mathscr{P}(S), d_{TV})$ are both complete.*

Proof. See, for example, Stokey and Lucas (1989, lem. 11.8). $\qquad\square$

11.1.3 Weak Convergence of Measures

Although total variation distance is pleasingly quantitative and important for the analysis of Markov chains, it does not cover all bases. We will also consider another type of convergence, known to probabilists as weak convergence. The next exercise starts the ball rolling.

Exercise 11.1.31 Let $S = \mathbb{R}$, let $\phi_n = \delta_{1/n}$ and let $\phi = \delta_0$. Show that for each $n \in \mathbb{N}$ we have $\sup_{B \in \mathscr{B}(S)} |\phi_n(B) - \phi(B)| = 1$. Conclude that $d_{TV}(\phi_n, \phi) \to 0$ fails for this example.

This negative result is somewhat unfortunate in that $\delta_{1/n}$ seems to be converging to δ_0. To make the definition of convergence more accommodating, we can abandon uniformity, and simply require that $\phi_n(B) \to \phi(B)$ for all $B \in \mathscr{B}(S)$. This is usually known as *setwise convergence*. However, a little thought will convince you that the sequence treated in exercise 11.1.31 still fails to converge setwise.

Thus we need to weaken the definition further by requiring that $\phi_n(B) \to \phi(B)$ holds only for a certain restricted class of sets $B \in \mathscr{B}(S)$: We say that $\phi_n \to \phi$ weakly if $\phi_n(B) \to \phi(B)$ for all sets $B \in \mathscr{B}(S)$ such that $\phi(\text{cl } B \setminus \text{int } B) = 0$. In fact this is equivalent to the following—more convenient—definition.

Definition 11.1.32 The sequence $(\phi_n) \subset \mathscr{P}(S)$ is said to converge to $\phi \in \mathscr{P}(S)$ *weakly* if $\phi_n(h) \to \phi(h)$ for every $h \in bcS$, the bounded continuous functions on S. If (X_n) is a sequence of random variables on S, then $X_n \to X$ weakly (or, in distribution; or, in law) means that the distribution of X_n converges weakly to that of X.

Weak convergence is more in tune with the topology of S than setwise convergence. The next exercise illustrates.

Exercise 11.1.33 Show that $\delta_{1/n} \to \delta_0$ holds for weak convergence.

Exercise 11.1.34 Convergence in probability implies convergence in distribution.[4] Show that the converse is not true.[5]

Actually, when testing weak convergence, we don't in fact need to test convergence at every $h \in bcS$. There are various "convergence determining classes" of functions, convergence over each element of which is sufficient for weak convergence. One example is $ibcS$, the increasing functions in bcS:

Theorem 11.1.35 Let (ϕ_n) and ϕ be elements of $\mathscr{P}(S)$. We have $\phi_n \to \phi$ weakly if and only if $\phi_n(h) \to \phi(h)$ for all $h \in ibcS$.[6]

Readers may be concerned that limits under weak convergence are not unique, in the sense that there may exist a sequence $(\phi_n) \subset \mathscr{P}(S)$ with $\phi_n \to \phi$ and $\phi_n \to \phi'$ weakly, where $\phi \neq \phi'$. In fact this is not possible, as can be shown using the following result:

Theorem 11.1.36 Let $\phi, \psi \in \mathscr{P}(S)$. The following statements are equivalent:

1. $\phi = \psi$

2. $\phi(h) = \psi(h)$ for all $h \in bcS$

3. $\phi(h) = \psi(h)$ for all $h \in ibcS$

---

[4]The proof is not trivial. See, for example, Dudley (2002, prop. 9.3.5).
[5]Hint: What happens in the case of an IID sequence?
[6]See Torres (1990, cor. 6.5).

If it can be shown that the third statement implies the first, then the rest of the theorem is easy. A proof that such an implication holds can be found in Torres (1990, thms. 3.7 and 5.3).

Exercise 11.1.37 Using theorem 11.1.36, show that if $(\phi_n) \subset \mathscr{P}(S)$ is a sequence with $\phi_n \to \phi$ and $\phi_n \to \phi'$ weakly, then $\phi = \phi'$.

When $S = \mathbb{R}$, weak convergence implies convergence of distribution functions in the following sense:

Theorem 11.1.38 *Let (ϕ_n) and ϕ be elements of $\mathscr{P}(\mathbb{R})$, and let (F_n) and F be their respective distribution functions. If $\phi_n \to \phi$ weakly, then $F_n(x) \to F(x)$ in $\mathbb{R}$ for every $x \in \mathbb{R}$ such that F is continuous at x.*[7]

One reason weak convergence is so important in probability theory is the magnificent central limit theorem, which concerns the asymptotic distribution of the sample mean $\bar{X}_n := n^{-1} \sum_{t=1}^n X_t$.[8]

Theorem 11.1.39 *Let $(X_t)_{t \geq 1}$ be an IID sequence of real-valued random variables. If $\mu := \mathbb{E}X_1$ and $\sigma^2 := \operatorname{Var} X_1$ are both finite, then $n^{1/2}(\bar{X}_n - \mu)$ converges weakly to $N(0, \sigma^2)$.*

While not obvious from the definition, it turns out that weak convergence on $\mathscr{P}(S)$ can be metrized, in the sense that there exists a metric ρ on $\mathscr{P}(S)$ with the property that $\phi_n \to \phi$ weakly if and only if $\rho(\phi_n, \phi) \to 0$. Actually there are several, and all are at least a little bit complicated. We will now described one such metric, known as the *Fortet–Mourier distance*.

Recall that a function $h: S \to \mathbb{R}$ is called *Lipschitz* if there exists a $K \in \mathbb{R}$ such that $|h(x) - h(y)| \leq K d_2(x, y)$ for all $x, y \in S$. Let $b\ell S$ be the collection of bounded Lipschitz functions on S, and set

$$\|h\|_{b\ell} := \sup_{x \in S} |h(x)| + \sup_{x \neq y} \frac{|h(x) - h(y)|}{d_2(x, y)} \tag{11.2}$$

The Fortet–Mourier distance between ϕ and ψ in $\mathscr{P}(S)$ is defined as

$$d_{FM}(\phi, \psi) := \sup\{|\phi(h) - \psi(h)| : h \in b\ell S, \ \|h\|_{b\ell} \leq 1\} \tag{11.3}$$

It can be shown that d_{FM} so constructed is indeed a metric, and does metrize weak convergence as claimed.[9]

---

[7]For a proof see, for example, Taylor (1997, prop. 6.2.3). The converse is not true in general, but if $F_n(x) \to F(x)$ in $\mathbb{R}$ for every $x \in \mathbb{R}$ such that F is continuous at x then $\phi_n(h) \to \phi(h)$ for every $h \in bcS$ such that $h(S \setminus K) = \{0\}$ for some compact $K \subset S$.

[8]For a proof see, for example, Taylor (1997, thm. 6.7.4.).

[9]See, for example, Dudley (2002, thm. 11.3.3).

There is a large and elegant theory of weak convergence, most of which would take us too far afield. We will content ourselves with stating Prohorov's theorem. It turns out that $\mathscr{P}(S)$ is d_{FM}-compact if and only if S is compact. Prohorov's theorem can be used to prove this result, and also provides a useful condition for (pre)compactness of subsets of $\mathscr{P}(S)$ when S is not compact.

Definition 11.1.40 A subset $\mathscr{M}$ of $\mathscr{P}(S)$ is called *tight* if, for each $\epsilon > 0$, there is a compact $K \subset S$ such that $\phi(S \setminus K) \leq \epsilon$ for all $\phi \in \mathscr{M}$.[10]

Theorem 11.1.41 (Prohorov) *The following statements are equivalent:*

1. *$\mathscr{M} \subset \mathscr{P}(S)$ is tight.*

2. *$\mathscr{M}$ is a precompact subset of the metric space $(\mathscr{P}(S), d_{FM})$.*

For a proof, see Pollard (2002, p. 185) or Dudley (2002, ch. 11).

11.2   Stability: Analytical Methods

We are now ready to tackle some stability results for general state Markov chains. In this section we will focus on analytical techniques related to the metric space theory of chapter 3. In §11.3 we turn to more probabilistic methods.

11.2.1   Stationary Distributions

When we discussed distribution dynamics for a stochastic kernel p on a finite state space, we regarded the Markov operator $\mathbf{M}$ corresponding to p (see page 75) as providing a dynamical system of the form $(\mathscr{P}(S), \mathbf{M})$, where $\mathscr{P}(S)$ was the set of distributions on S. The interpretation was that if $(X_t)_{t \geq 0}$ is Markov-(p, ψ), then $\psi \mathbf{M}^t$ is the distribution of X_t. A fixed point of $\mathbf{M}$ was called a stationary distribution.

A similar treatment was given for the density case (chapter 8), where we considered the dynamical system $(D(S), \mathbf{M})$. Trajectories correspond to sequences of marginal densities, and a fixed point of $\mathbf{M}$ in $D(S)$ is called a stationary density.

For the general (i.e., measure) case, where S is a Borel subset of $\mathbb{R}^n$, P is an arbitrary stochastic kernel and $\mathbf{M}$ is the corresponding Markov operator (see page 224), we consider the dynamical system $(\mathscr{P}(S), \mathbf{M})$, with $\mathscr{P}(S)$ denoting the Borel probability measures on S. The metric imposed on $\mathscr{P}(S)$ is either d_{TV} or d_{FM} (see §11.1.2 and §11.1.3 respectively). A distribution $\psi^* \in \mathscr{P}(S)$ is called *stationary* if $\psi^* \mathbf{M} = \psi^*$; equivalently

$$\int P(x, B) \psi^*(dx) = \psi^*(B) \qquad (B \in \mathscr{B}(S))$$

---

[10]This is a generalization of the definition given for densities on page 203.

Exercise 11.2.1 Consider the deterministic model $X_{t+1} = X_t$. Show that for this model every $\psi \in \mathscr{P}(S)$ is stationary.[11]

In this section we focus on existence of stationary distributions using continuity and compactness conditions. Regarding continuity,

Definition 11.2.2 Let P be a stochastic kernel on S, and let $\mathbf{M}$ be the corresponding Markov operator. We say that P has the *Feller property* if $\mathbf{M}h \in bcS$ whenever $h \in bcS$.

The Feller property is usually easy to check in applications. To illustrate, recall our canonical SRS defined in (9.7) on page 219.

Lemma 11.2.3 *If $x \mapsto F(x, z)$ is continuous on S for all $z \in Z$, then P is Feller.*

Proof. Recall from (9.17) on page 226 that for any $h \in b\mathscr{B}(S)$, and in particular for $h \in bcS$, we have

$$\mathbf{M}h(x) = \int h[F(x,z)]\phi(dz) \qquad (x \in S)$$

So fix any $h \in bcS$. We wish to show that $\mathbf{M}h$ is a continuous bounded function. Verifying boundedness is left to the reader. Regarding continuity, fix $x_0 \in S$ and take some $x_n \to x_0$. For each $z \in Z$, continuity of $x \mapsto F(x,z)$ and h gives us $h[F(x_n, z)] \to h[F(x_0, z)]$ as $n \to \infty$. Since h is bounded the conditions of the dominated convergence theorem (page 182) are all satisfied. Therefore

$$\mathbf{M}h(x_n) := \int h[F(x_n, z)]\phi(dz) \to \int h[F(x_0, z)]\phi(dz) =: \mathbf{M}h(x_0)$$

As x_0 was arbitrary, $\mathbf{M}h$ is continuous on all of S. $\qquad\qquad\qquad\square$

The Feller property is equivalent to continuity of $\psi \mapsto \psi\mathbf{M}$ in $(\mathscr{P}(S), d_{FM})$:

Lemma 11.2.4 *A stochastic kernel P with Markov operator $\mathbf{M}$ is Feller if and only if $\psi \mapsto \psi\mathbf{M}$ is weakly continuous as a map from $\mathscr{P}(S)$ to $\mathscr{P}(S)$.*

Proof. Suppose first that $\mathbf{M}$ is Feller. Take any $(\psi_n) \subset \mathscr{P}(S)$ with $\psi_n \to \psi \in \mathscr{P}(S)$ weakly. We must show that $\psi_n\mathbf{M}(h) \to \psi\mathbf{M}(h)$ for every $h \in bcS$. Pick any such h. Since $\mathbf{M}h \in bcS$, theorem 9.2.15 (page 226) gives

$$\psi_n\mathbf{M}(h) = \psi_n(\mathbf{M}h) \to \psi(\mathbf{M}h) = \psi\mathbf{M}(h) \qquad (n \to \infty)$$

The reverse implication is left as an exercise.[12] $\qquad\qquad\qquad\square$

We can now state the well-known Krylov–Bogolubov existence theorem, the proof of which is given in the appendix to this chapter.

---

[11]Hint: See example 9.2.5 on page 221.
[12]Hint: Try another application of theorem 9.2.15.

Theorem 11.2.5 (Krylov–Bogolubov) *Let P be a stochastic kernel on S, and let $\mathbf{M}$ be the corresponding Markov operator. If P has the Feller property and $(\psi\mathbf{M}^t)_{t\geq 0}$ is tight for some $\psi \in \mathscr{P}(S)$, then P has at least one stationary distribution.*[13]

Remark 11.2.6 If S is compact, then every subset of $\mathscr{P}(S)$ is tight (why?), and hence every kernel with the Feller property on S has at least one stationary distribution. This is theorem 12.10 in Stokey and Lucas (1989).

There are many applications of theorem 11.2.5 in economic theory, particularly for the case where the state space is compact. In fact it is common in economics to assume that the shocks perturbing a given model are bounded above and below, precisely because the authors wish to obtain a compact state space. (Actually such strict restrictions on the shocks are usually unnecessary: we can deal with unbounded shocks using drift conditions as discussed below.)

Example 11.2.7 Consider again the commodity pricing model, which was shown to be stable when the shock is lognormal in §8.2.4. The law of motion for the model is $X_{t+1} = \alpha I(X_t) + W_{t+1}$ with shock distribution $\phi \in \mathscr{P}(Z)$. The Feller property holds because I is continuous (see the definition on page 146). Suppose now that $Z := [a, b]$ for positive constants $a \leq b$.[14] Define $S := [a, b/(1 - \alpha)]$. It is an exercise to show that if $x \in S$ and $z \in Z$, then $\alpha I(x) + z \in S$. Hence the compact set S can be chosen as a state space for the model, and, by the Krylov–Bogolubov theorem, at least one stationary distribution ψ^* exists. It satisfies

$$\psi^*(B) = \int \left[\int \mathbb{1}_B[\alpha I(x) + z]\phi(dz) \right] \psi^*(dx) \qquad (B \in \mathscr{B}(S)) \qquad (11.4)$$

If S is not compact, then to establish existence via the Krylov–Bogolubov theorem, we need to show that at least one trajectory of $\mathbf{M}$ is tight. Fortunately we already know quite a bit about finding tight trajectories, at least in the density case. For example, under geometric drift to the center every trajectory is tight (proposition 8.2.12, page 204). In the general (i.e., measure rather than density) case a similar result applies:

Lemma 11.2.8 *Let $\mathscr{M}$ be a subset of $\mathscr{P}(S)$. If there exists a norm-like function[15] w on S such that $\sup_{\psi \in \mathscr{M}} \psi(w) < \infty$, then $\mathscr{M}$ is tight.*[16]

Proof. Let $M := \sup_{\psi \in \mathscr{M}} \psi(w)$, and fix $\epsilon > 0$. Pick any $\psi \in \mathscr{M}$. We have

$$\psi\{x \in S : w(x) > k\} \leq \frac{\psi(w)}{k} \leq \frac{M}{k} \qquad \forall k \in \mathbb{N}$$

---

[13] That is, $\mathbf{M}$ has at least one fixed point in $\mathscr{P}(S)$.

[14] Since Z is compact, one often says that ϕ has *compact support*.

[15] Recall that a function $w \colon S \to \mathbb{R}_+$ is called norm-like when all sublevel sets are precompact. See definition 8.2.9 on page 204.

[16] Actually the converse is also true. See Meyn and Tweedie (1993, lem. D.5.3).

where the first inequality follows from $w \geq k\mathbb{1}\{x \in S : w(x) > k\}$. Since ψ is arbitrary,

$$\sup_{\psi \in \mathscr{M}} \psi\{x \in S : w(x) > k\} \leq \frac{M}{k} \qquad \forall k \in \mathbb{N}$$

For sufficiently large k the left-hand side is less than ϵ. Defining $C := w^{-1}([0,k])$, we can write this as $\sup_{\psi \in \mathscr{M}} \psi(C^c) < \epsilon$. Since w is norm-like, we know that C is precompact. By exercise 3.2.17 on page 47, there is a compact set K with $C \subset K$, or $K^c \subset C^c$. But then $\psi(K^c) \leq \psi(C^c) \leq \epsilon$ for all $\psi \in \mathscr{M}$. □

The easiest way to apply lemma 11.2.8 is via a drift condition:

Lemma 11.2.9 *Let P be a stochastic kernel on S with Markov operator* **M**, *and let $\psi \in \mathscr{P}(S)$. If there exists a norm-like function w on S and constants $\alpha \in [0,1)$ and $\beta \in \mathbb{R}_+$ with*

$$\mathbf{M}w(x) \leq \alpha w(x) + \beta \qquad (x \in S) \tag{11.5}$$

then there exists a $\psi \in \mathscr{P}(S)$ such that the trajectory $(\psi_t) := (\psi\mathbf{M}^t)$ is tight.

The intuition is similar to that for proposition 8.2.12 (page 204).

Proof. Let $\psi := \delta_x$ for some $x \in S$. Using theorem 9.2.15 (page 226) and then monotonicity property M4 of the integral, we have

$$\psi_t(w) = (\psi_{t-1}\mathbf{M})(w) = \psi_{t-1}(\mathbf{M}w) \leq \psi_{t-1}(\alpha w + \beta) = \alpha\psi_{t-1}(w) + \beta$$

From this bound one can verify (use induction) that

$$\psi_t(w) \leq \alpha^t \psi(w) + \frac{\beta}{1-\alpha} = \alpha^t w(x) + \frac{\beta}{1-\alpha} \qquad \forall t \in \mathbb{N}$$

Tightness of (ψ_t) now follows from lemma 11.2.8. □

11.2.2 Testing for Existence

Let's investigate how one might use lemma 11.2.9 in applications. First we can repackage the results of the previous section as a corollary that applies to the canonical SRS given in (9.7) on page 219. The proof is an exercise.

Corollary 11.2.10 *If $x \mapsto F(x,z)$ is continuous on S for each $z \in Z$, and there exists a norm-like function w on S and constants $\alpha \in [0,1)$ and $\beta \in \mathbb{R}_+$ with*

$$\int w[F(x,z)]\phi(dz) \leq \alpha w(x) + \beta \qquad (x \in S) \tag{11.6}$$

then at least one stationary distribution exists.

Example 11.2.11 Let $S = Z = \mathbb{R}^n$, and let $F(x,z) = Ax + b + z$, where A is an $n \times n$ matrix and b is an $n \times 1$ vector. Let $\phi \in \mathscr{P}(Z)$. Suppose that for some norm $\|\cdot\|$ on S we have $\lambda := \sup\{\|Ax\| : \|x\| = 1\} < 1$, and in addition $\int \|z\| \phi(dz) < \infty$. Using exercise 4.1.17 (page 60), we have

$$\int \|Ax + \|b\| + z\| \phi(dz) \le \|Ax\| + \|b\| + \int \|z\| \phi(dz) \le \lambda \|x\| + \|b\| + \int \|z\| \phi(dz)$$

Setting $\alpha := \lambda$ and $\beta := \|b\| + \int \|z\| \phi(dz)$ gives the drift condition (11.6). It is left as an exercise to show that the Feller property holds. Since $\|\cdot\|$ is norm-like on S (see example 8.2.10), a stationary distribution exists.

Example 11.2.12 Let $S = Z = \mathbb{R}^n$, and let $F(x,z) = G(x) + z$, where $G: \mathbb{R}^n \to \mathbb{R}^n$ is a continuous function with the property that for some $M < \infty$ and some $\alpha < 1$ we have $\|G(x)\| \le \alpha \|x\|$ whenever $\|x\| > M$. In other words, when x is sufficiently far from the origin, $G(x)$ is closer to the origin than x. Also note that $L := \sup_{\|x\| \le M} \|G(x)\|$ is finite because continuous functions map compact sets into compact sets. By considering the two different cases $\|x\| \le M$ and $\|x\| > M$, you should be able to show that $\|G(x)\| \le \alpha \|x\| + L$ for every $x \in \mathbb{R}^n$. As a result

$$\int \|G(x) + z\| \phi(dz) \le \|G(x)\| + \int \|z\| \phi(dz) \le \alpha \|x\| + L + \int \|z\| \phi(dz)$$

If $\int \|z\| \phi(dz) < \infty$, then the drift condition (11.6) holds. As the Feller property clearly holds a stationary distribution must exist.

Exercise 11.2.13 Show that example 11.2.11 is a special case of example 11.2.12.

Exercise 11.2.14 Consider the log-linear Solow–Swan model $k_{t+1} = s k_t^\alpha W_{t+1}$. Set $S = Z = (0, \infty)$, and assume $\alpha < 1$ and $\mathbb{E}|\ln W_t| < \infty$. One way to show existence of a stationary distribution is by way of taking logs and converting our log-linear system into a linear one. Example 11.2.11 then applies. However, this still leaves the task of showing that existence of a stationary distribution for the linear model implies existence of a stationary distribution for the original model. Instead of log-linearizing, prove the existence of a stationary distribution directly, by applying corollary 11.2.10.[17]

Finally, let's treat the problem of existence in a more involved application. The application requires an extension of lemma 11.2.9 on page 260. A proof can be found in Meyn and Tweedie (1993, thm. 12.1.3).

---

[17]Hint: Use $w(x) = |\ln x|$ as the norm-like function.

Lemma 11.2.15 *Let P be a stochastic kernel on S with Markov operator $\mathbf{M}$, and let $\psi \in \mathscr{P}(S)$. If P has the Feller property, and in addition there exists a norm-like function w on S and constants $\alpha \in [0, 1)$ and $\beta \in \mathbb{R}_+$ with*

$$\mathbf{M}^t w(x) \leq \alpha w(x) + \beta \qquad (x \in S) \tag{11.7}$$

for some $t \in \mathbb{N}$, then P has at least one stationary distribution.

As an application, consider the SRS with correlated shocks defined by

$$X_{t+1} = g(X_t) + \xi_{t+1} \quad \text{and} \quad \xi_{t+1} = A\xi_t + W_{t+1} \tag{11.8}$$

Here X_t and ξ_t both take values in $\mathbb{R}^k$, and $(W_t)_{t \geq 1}$ is an $\mathbb{R}^k$ valued IID sequence with distribution $\phi \in \mathscr{P}(\mathbb{R}^k)$. The matrix A is $k \times k$, while $g \colon \mathbb{R}^k \to \mathbb{R}^k$ is Borel measurable. Although $(X_t)_{t \geq 0}$ is not generally Markovian, the joint process $(X_t, \xi_t)_{t \geq 0}$ is Markovian in $S := \mathbb{R}^k \times \mathbb{R}^k$. The associated stochastic kernel is

$$P((x, \xi), B) = \int \mathbb{1}_B[g(x) + A\xi + z, A\xi + z]\phi(dz) \qquad ((x, \xi) \in S, \ B \in \mathscr{B}(S))$$

If g is a continuous function, then in view of lemma 11.2.15 a stationary distribution will exist for P whenever the drift condition (11.7) holds. That this drift condition does hold under some restrictions is the content of the next proposition.

Proposition 11.2.16 *Let $\| \cdot \|$ be any norm on $\mathbb{R}^k$. Let ρ be a constant such that A satisfies $\|Ax\| \leq \rho\|x\|$ for all $x \in \mathbb{R}^k$, and let w be the norm-like function $w(x, \xi) = \|x\| + \|\xi\|$. Set $\mu := \mathbb{E}\|W_1\|$. If $\mu < \infty$, $\rho < 1$, and there exists constants $\lambda \in [0, 1)$ and $L \in \mathbb{R}_+$ such that*

$$\|g(x)\| \leq \lambda\|x\| + L \qquad (x \in \mathbb{R}^k)$$

then there exists constants $t \in \mathbb{N}$, $\alpha \in [0, 1)$ and $\beta \in \mathbb{R}_+$ such that (11.7) holds.

Proof. Consider the joint process $(X_t, \xi_t)_{t \geq 0}$ from constant initial condition $(x_0, \xi_0) \in S$. From the definition of the SRS and the growth condition on g, we have

$$\mathbb{E}\|X_{t+1}\| \leq \lambda\mathbb{E}\|X_t\| + L + \mathbb{E}\|\xi_{t+1}\| \quad \text{and} \quad \mathbb{E}\|\xi_{t+1}\| \leq \rho\mathbb{E}\|\xi_t\| + \mu$$

From these bounds one can see (use induction) that for any $t \geq 0$,

$$\mathbb{E}\|X_t\| \leq \lambda^t\|x_0\| + \frac{L}{1 - \lambda} + \sum_{i=0}^{t-1} \lambda^i \mathbb{E}\|\xi_{t-i}\| \tag{11.9}$$

and

$$\mathbb{E}\|\xi_t\| \leq \rho^t\|\xi_0\| + \frac{\mu}{1 - \rho} \tag{11.10}$$

Substituting (11.10) into (11.9) and rearranging gives

$$\mathbb{E}\|X_t\| \leq \lambda^t \|x_0\| + \frac{L}{1-\lambda} + \frac{\mu}{(1-\lambda)(1-\rho)} + \sum_{i=0}^{t-1} \lambda^i \rho^{t-i} \|\xi_0\| \tag{11.11}$$

Adding (11.10) and (11.11), we obtain

$$\mathbb{E}\|X_t\| + \mathbb{E}\|\xi_t\| \leq \lambda^t \|x_0\| + \rho^t \|\xi_0\| + \sum_{i=0}^{t-1} \lambda^i \rho^{t-i} \|\xi_0\| + \beta$$

where β is a constant. Since $\lim_{t\to\infty} \sum_{i=0}^{t-1} \lambda^i \rho^{t-i} = 0$, we can choose a $t \in \mathbb{N}$ such that

$$\rho^t + \sum_{i=0}^{t-1} \lambda^i \rho^{t-i} < 1$$

Letting α be the maximum of this term and λ^t, we obtain

$$\mathbb{E}\|X_t\| + \mathbb{E}\|\xi_t\| \leq \alpha\|x_0\| + \alpha\|\xi_0\| + \beta$$

This inequality is equivalent to the claim in the proposition. $\square$

11.2.3 The Dobrushin Coefficient, Measure Case

Now let's turn to the problem of uniqueness and stability of stationary distributions. As a first step we discuss a contraction mapping approach based on the Dobrushin coefficient. As we saw in §8.2.2, for unbounded state spaces this approach is not always successful. Nevertheless, it provides a useful departure point, and the basic ideas will later be extended to handle more general models.

Let S be a Borel subset of $\mathbb{R}^n$. For stochastic kernel P with a density representation p (i.e., $P(x, dy) = p(x, y)dy$ for all $x \in S$), the Dobrushin coefficient was defined in §8.2.2 as

$$\alpha(p) := \inf\left\{ \int p(x,y) \wedge p(x',y)dy : (x,x') \in S \times S \right\} \tag{11.12}$$

The corresponding Markov operator is a uniform contraction of modulus $1 - \alpha(p)$ on $(D(S), d_1)$ whenever $\alpha(p) > 0$.

The concept of the Dobrushin coefficient can be extended to kernels without density representation. To do so we need a notion equivalent to the affinity measure $\int f \wedge g$ between densities f and g used in (11.12). Since $f \wedge g$ is the largest function less than both f and g, it is natural to extend this idea by considering the largest measure less than two given measures μ and ν.[18] Such a measure is called the *infimum* of μ and ν, and denoted by $\mu \wedge \nu$.

---

[18] The ordering is setwise: $\mu \leq \nu$ if $\mu(B) \leq \nu(B)$ for all $B \in \mathscr{B}(S)$.

Things are not quite as simple as the density case. In particular, it is *not* correct to define $\mu \wedge \nu$ as the set function $m\colon B \mapsto \mu(B) \wedge \nu(B)$. The reason is that m is not always additive, and hence fails to be a measure (example?) However, the infimum of two measures does always exists:

Lemma 11.2.17 *If $\mu \in b\mathcal{M}(S)$ and $\nu \in b\mathcal{M}(S)$, then there exists a unique element of $b\mathcal{M}(S)$, denoted here by $\mu \wedge \nu$, such that*

1. *both $\mu \wedge \nu \leq \mu$ and $\mu \wedge \nu \leq \nu$, and*

2. *if $\kappa \in b\mathcal{M}(S)$ and both $\kappa \leq \mu$ and $\kappa \leq \nu$, then $\kappa \leq \mu \wedge \nu$.*

Proof. Let S^+ be a positive set for $\mu - \nu$, and let S^- be a negative set (for definitions, see page 252). It follows that if $B \in \mathcal{B}(S)$ and $B \subset S^+$, then $\mu(B) \geq \nu(B)$, while if $B \subset S^-$, then $\nu(B) \geq \mu(B)$. Now set

$$(\mu \wedge \nu)(B) := \mu(B \cap S^-) + \nu(B \cap S^+)$$

Evidently $\mu \wedge \nu$ is countably additive. That $\mu \wedge \nu \leq \mu$ is immediate:

$$\mu(B) = \mu(B \cap S^-) + \mu(B \cap S^+) \geq \mu(B \cap S^-) + \nu(B \cap S^+)$$

The proof that $\mu \wedge \nu \leq \nu$ is similar. To check the second claim, let $\kappa \in b\mathcal{M}(S)$ with $\kappa \leq \mu$ and $\kappa \leq \nu$. Then

$$\kappa(B) = \kappa(B \cap S^-) + \kappa(B \cap S^+) \leq \mu(B \cap S^-) + \nu(B \cap S^+)$$

Hence $\kappa \leq \mu \wedge \nu$, as was to be shown. $\square$

Exercise 11.2.18 Show that if μ is a probability with density f, and ν has density g, then $\mu \wedge \nu$ has density $f \wedge g$.

For probabilities μ and ν, the value $(\mu \wedge \nu)(S)$ is sometimes called the *affinity* between μ and ν, and is a measure of similarity.

Exercise 11.2.19 Show that $(\mu \wedge \nu)(S) = \min_\pi \sum_{A \in \pi} \mu(A) \wedge \nu(A)$ for any μ and ν in $\mathscr{P}(S)$, where the minimum is over all finite measurable partitions π of S.[19] Show that the affinity between μ and ν has a maximum value of 1, which is attained if and only if $\mu = \nu$.

We can now define the Dobrushin coefficient for general kernel P.

---

[19]Here $\mu(A) \wedge \nu(A)$ is a simple infimum in $\mathbb{R}$, rather than in $b\mathcal{M}(S)$. Hint: For the minimizer consider $\pi = \{S^+, S^-\}$, where S^+ and S^- are as in the proof of lemma 11.2.17.

Definition 11.2.20 Let P be a stochastic kernel on S. Writing $P_x \wedge P_{x'}$ for the infimum measure $P(x, dy) \wedge P(x', dy)$, the *Dobrushin coefficient* of P is defined as

$$\alpha(P) := \inf \{ (P_x \wedge P_{x'})(S) : (x, x') \in S \times S \}$$

When $P(x, dy) = p(x, y)dy$, this reduces to (11.12) above, while if S is finite and P is defined by a finite kernel p (i.e., $P(x, B) = \sum_{y \in B} p(x, y)$ for all $x \in S$ and $B \subset S$), then it reduces to the definition given on page 89.

As for the finite and density cases, the Dobrushin coefficient is closely connected to stability. In fact the following theorem holds:

Theorem 11.2.21 *Let P be a stochastic kernel on S with Markov operator **M**. For every pair ϕ, ψ in $\mathscr{P}(S)$ we have*

$$\|\phi\mathbf{M} - \psi\mathbf{M}\|_{TV} \leq (1 - \alpha(P))\|\phi - \psi\|_{TV}$$

Moreover this bound is the best available, in the sense that if $\lambda < 1 - \alpha(P)$, then there exists a pair ϕ, ψ in $\mathscr{P}(S)$ such that $\|\phi\mathbf{M} - \psi\mathbf{M}\|_{TV} > \lambda\|\phi - \psi\|_{TV}$.

This result closely parallels the result in theorem 4.3.17, page 89. The proof is similar to the finite case (i.e., the proof of theorem 4.3.17) and as such is left to the enthusiastic reader as an exercise. The intuition behind the theorem is also similar to the finite case: The affinity $(P_x \wedge P_{x'})(S)$ is a measure of the similarity of the kernels $P(x, dy)$ and $P(x', dy)$. If all the kernels are identical then $\alpha(P) = 1$ and $\mathbf{M}$ is a constant map—the ultimate in global stability. More generally, high values of $\alpha(P)$ correspond to greater similarity across the kernels, and hence more stability.

The first half of theorem 11.2.21 says that if $\alpha(P) > 0$, then $\mathbf{M}$ is a uniform contraction with modulus $1 - \alpha(P)$ on $\mathscr{P}(S)$. Since $(\mathscr{P}(S), d_{TV})$ is a complete metric space (theorem 11.1.30), it follows that $(\mathscr{P}(S), \mathbf{M})$ is globally stable whenever $\alpha(P^t) > 0$ for some $t \in \mathbb{N}$.[20] Let us record these findings as a corollary.

Corollary 11.2.22 *Let P be a stochastic kernel on S, and let $\mathbf{M}$ be the associated Markov operator. If $\alpha(P^t) > 0$ for some $t \in \mathbb{N}$, then $(\mathscr{P}(S), \mathbf{M})$ is globally stable with unique stationary distribution ψ^*. Moreover if $h \colon S \to \mathbb{R}$ is a measurable function satisfying $\psi^*|h| < \infty$ and $\psi \in \mathscr{P}(S)$, then any Markov-(P, ψ) chain $(X_t)_{t \geq 0}$ satisfies*

$$\frac{1}{n} \sum_{t=1}^{n} h(X_t) \to \psi^*(h) \quad \text{with probability one as } n \to \infty \tag{11.13}$$

The second part of the corollary, which is a law of large numbers for Markov chains, follows from global stability. The proof is omitted, but interested readers can consult Meyn and Tweedie (1993, ch. 17).

---

[20]This follows from Banach's fixed point theorem, nonexpansiveness of $\mathbf{M}$ with respect to d_{TV} (see lemma 11.1.28 on page 254) and lemma 4.1.21 on page 61.

Example 11.2.23 Consider the well-known stability condition

$$\exists\, m \in \mathbb{N},\ \nu \in \mathscr{P}(S),\ \epsilon > 0 \quad \text{such that} \quad P^m(x, dy) \geq \epsilon \nu \quad \forall x \in S \qquad (11.14)$$

Here $P^m(x, dy) \geq \epsilon \nu$ means that $P^m(x, B) \geq \epsilon \nu(B)$ for all $B \in \mathscr{B}(S)$. If this condition holds, then $\alpha(P^m) > 0$ and $(\mathscr{P}(S), \mathbf{M})$ is globally stable, as the next exercise asks you to confirm.

Exercise 11.2.24 Show that if (11.14) holds, then $\alpha(P^m) \geq \epsilon$.

The condition (11.14) is particularly easy to verify when $P^m(x, dy)$ puts uniformly positive probability mass on a point $z \in S$, in the sense that

$$\exists\, z \in S,\ \gamma > 0 \quad \text{such that} \quad P^m(x, \{z\}) \geq \gamma \quad \forall x \in S \qquad (11.15)$$

It turns out that if (11.15) holds, then (11.14) holds with $\nu := \delta_z$ and $\epsilon := \gamma$. To see this, pick any $x \in S$ and any $B \in \mathscr{B}(S)$. If $z \in B$, then

$$P^m(x, B) \geq P^m(x, \{z\}) \geq \gamma = \epsilon \nu(B)$$

On the other hand, if $z \notin B$, then $P^m(x, B) \geq 0 = \epsilon \nu(B)$. Thus $P^m(x, B) \geq \epsilon \nu(B)$ for all $x \in S$ and $B \in \mathscr{B}(S)$ as claimed.

Example 11.2.25 Stokey and Lucas (1989, p. 348) use the following condition for stability, which they refer to as condition M: There exists a $m \in \mathbb{N}$ and an $\epsilon > 0$ such that for any $A \in \mathscr{B}(S)$, either $P^m(x, A) \geq \epsilon$ for all $x \in S$ or $P^m(x, A^c) \geq \epsilon$ for all $x \in S$. This condition is stricter than $\alpha(P^m) > 0$, as the next exercise asks you to confirm.

Exercise 11.2.26 Show that if condition M holds, then $\alpha(P^m) \geq \epsilon$.[21]

We saw in §8.2.2 that when the state space is unbounded, existence of a $t \in \mathbb{N}$ such that $\alpha(P^t) > 0$ often fails (recall the discussion of the AR(1) model in that section). In that case the method for establishing global stability discussed in this section cannot be applied. However, as we will see, the basic ideas can be extended to a wide variety of problems—including those on unbounded spaces.

11.2.4   Application: Credit-Constrained Growth

In this section we apply the stability condition in corollary 11.2.22 to a model of economic growth under credit constraints due to Matsuyama (2004), which tries to reconcile classical and structuralist views on the effects of global financial market integration. The classical view is that such integration fosters growth of developing countries

---

[21]Hint: For any $x, x' \in S$ we have $(P_x \wedge P_{x'})(S) = P^m(x, S^-) + P^m(x', S^+)$, where S^- and S^+ are negative and positive for $P^m(x, dy) - P^m(x', dy)$ respectively. Now use the fact that $S^+ = (S^-)^c$.

by giving them access to scarce capital. Structuralists argue that poorer economies would not be able to compete with richer countries in global financial markets, and that the gap between rich and poor may even be magnified.

We will not delve into the many economic ideas that are treated in the paper. Instead our focus will be on technical issues, in particular, on analyzing the dynamics of a small open economy model constructed by Matsuyama. The model has been slightly modified to better suit our purposes.

To begin, consider a small open economy populated by agents who live for two periods. Agents supply one unit of labor when young and consume only when old. Each successive generation has unit mass. At the start of time t a shock ξ_t is realized and production takes place, combining the current aggregate stock of capital k_t supplied by the old with the unit quantity of labor supplied by the young.[22] The resulting output is $y_t = f(k_t)\xi_t$, where the production function $f\colon \mathbb{R}_+ \to \mathbb{R}_+$ is increasing, strictly concave, differentiable, and $f'(x) \uparrow \infty$ as $x \downarrow 0$. The shocks $(\xi_t)_{t\geq 0}$ are IID on $\mathbb{R}_+$. For convenience we set $\mathbb{E}\xi_t = 1$.[23]

Factor markets are competitive, paying young workers the wage $w_t := w(k_t)\xi_t$, where

$$w(k) := f(k) - kf'(k) \qquad (k \in \mathbb{R}_+)$$

and a gross return on capital given by $f'(k_t)\xi_t$. Since the old supply k_t units of capital to production, the sum of factor payments exhausts aggregate income.[24]

After production and the distribution of factor payments, the old consume and disappear from the model, while the young take their wage earnings and invest them. In doing so, the young have two choices:

1. a loan to international investors at the risk free gross world interest rate R, or

2. an indivisible project, which takes one unit of the consumption good and returns in the next period Q units of the capital good.

The gross rate of return on the second option, measured in units of the consumption good, is $Qf'(k_{t+1})\xi_{t+1}$. In this expression, k_{t+1} is the outcome of investment in the project by the young. Factors of production are not internationally mobile, and FDI is ruled out.

Agents are assumed to be risk neutral, and as a result they invest in the project until the expected rate of return $Qf'(k_{t+1})$ is equal to the risk-free rate. Thus k_{t+1} is determined by the equation

$$R = Qf'(k_{t+1}) \tag{11.16}$$

---

[22]Capital depreciates fully between periods, so capital stock is equal to investment.

[23]If the mean differs from one then this constant can be absorbed into the production function. Thus $\mathbb{E}\xi_t = 1$ is essentially a finite first-moment assumption.

[24]That is, $k_t f'(k_t)\xi_t + w(k_t)\xi_t = y_t$.

We assume that $Qf'(Q) \leq R$ so that if every young agent starts a project, then the return to projects is driven below that of the risk-free rate.

We have already constructed a dynamic model, with $(k_t)_{t \geq 0}$ converging immediately to the (constant) solution to (11.16). To make matters more interesting, however, let's investigate how the model changes when capital markets are imperfect. The imperfection we consider is a constraint on borrowing, where lending is dependent on the provision of collateral.

When $w_t < 1$, young agents who start projects must borrow $1 - w_t$ at the risk free rate R. As a result their obligation at $t + 1$ is given by $R(1 - w_t)$. Against this obligation, borrowers can only credibly pledge a fraction $\lambda \in [0, 1]$ of their expected earnings $Qf'(k_{t+1})$. This results in the borrowing constraint

$$R(1 - w_t) \leq \lambda Qf'(k_{t+1})$$

This constraint is binding only if $1 - w_t > \lambda$, or, as a restriction on w_t, if $w_t < 1 - \lambda$; otherwise, agents are able to choose the unconstrained equilibrium value of k_{t+1} defined in (11.16). If the constraint is in fact binding, then it holds with equality

$$R = \frac{\lambda}{1 - w_t} Qf'(k_{t+1})$$

Combining this with (11.16), we can write the equation that determines k_{t+1} as

$$R = \theta(w_t)Qf'(k_{t+1}), \qquad \theta(w) := \begin{cases} \lambda/(1-w) & \text{if } w < 1 - \lambda \\ 1 & \text{otherwise} \end{cases} \tag{11.17}$$

The function $w \mapsto \theta(w)$ is shown in figure 11.1. It is monotone increasing and takes values in the interval $[\lambda, 1]$. The determination of next period's capital stock k_{t+1} is depicted in figure 11.2 as the value of k at which the curve $k \mapsto \theta(w_t)Qf'(k)$ intersects the horizontal line R. This value is the solution to (11.17).

Let $g := (f')^{-1}$ be the inverse function of f'. The constants a and b in the figure are defined by

$$a := g\left(\frac{R}{\lambda Q}\right) \quad \text{and} \quad b := g\left(\frac{R}{Q}\right)$$

The lower bound a is the quantity of domestic capital at $t + 1$ when $w_t = 0$. In this case the entire cost of the project must be financed by borrowing, and k_{t+1} solves $R = \lambda Qf'(k_{t+1})$. The upper bound b is the unconstrained solution (11.16). As is clear from these two figures, higher wages increases θ, which increases k_{t+1}.

Using g, we can write the stochastic law of motion for $(k_t)_{t \geq 0}$ as the SRS

$$k_{t+1} = g\left(\frac{R}{\theta(w(k_t)\xi_t)Q}\right) \tag{11.18}$$

A suitable state space for this SRS is provided by the interval $S = [a, b]$:

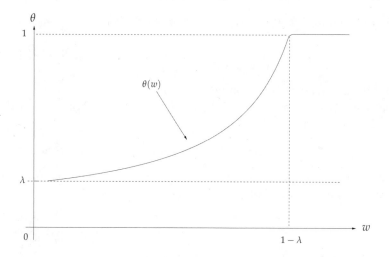

Figure 11.1 The function $\theta(w)$

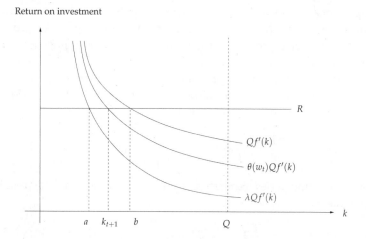

Figure 11.2 Determination of k_{t+1}

Exercise 11.2.27 Show that if $k_t \in S$ and $\xi_t \in \mathbb{R}_+$, then $k_{t+1} \in S$.

The corresponding stochastic kernel P on S is defined by

$$P(x, B) = \int \mathbb{1}_B \left[g \left(\frac{R}{\theta(w(x)z)Q} \right) \right] \phi(dz) \qquad (x \in S,\ B \in \mathscr{B}(S))$$

where ϕ is the common distribution of the shocks $(\xi_t)_{t \geq 0}$.

Global stability holds whenever $\alpha(P) > 0$ (corollary 11.2.22). To show that $\alpha(P) > 0$, we can use the condition

$$\exists z \in S,\ \gamma > 0 \quad \text{such that} \quad P(x, \{z\}) \geq \gamma \quad \forall x \in S$$

presented in (11.15) on page 266. For the z in this expression we use b, the upper bound of $S = [a, b]$. For γ we set

$$\gamma := \mathbb{P}\{w(a)\xi_t \geq 1 - \lambda\} = \phi \left\{ z \in \mathbb{R}_+ : z \geq \frac{1 - \lambda}{f(a) - af'(a)} \right\}$$

We assume ϕ is such that $\gamma > 0$. It remains to be shown that $P(x, \{b\}) \geq \gamma$ for all $x \in S$. To see that this is the case, note that by monotonicity we have $P(x, \{b\}) \geq P(a, \{b\})$ for all $x \in S$. The latter quantity $P(a, \{b\})$ is the probability of jumping from the lowest state a to the highest in state b (the unconstrained equilibrium) in one period. This occurs whenever $\theta(w(a)\xi_t) = 1$, which in turn holds when $w(a)\xi_t \geq 1 - \lambda$. The probability of this event is precisely γ, and $P(x, \{b\}) \geq \gamma$ for all x is now verified.

Although global stability holds, this general result masks important differences in the dynamics that occur when the parameters are varied. To gain some understanding of these differences let us compute the stationary distribution ψ^* and see how it is affected by variation in parameters.

The stationary distribution is not a density (it puts positive mass on b) and hence one cannot use the look-ahead estimator (6.12) introduced on page 131. Instead we use the empirical cumulative distribution function

$$F_n^*(x) := \frac{1}{n} \sum_{t=1}^{n} \mathbb{1}\{k_t \leq x\} = \frac{1}{n} \sum_{t=1}^{n} \mathbb{1}_{[0,x]}(k_t) \qquad (x \in S)$$

where $(k_t)_{t \geq 0}$ is a time series generated from (11.18). From the LLN result (11.13) on page 265, we have

$$\lim_{n \to \infty} F_n^*(x) = \int \mathbb{1}_{[0,x]}(y)\psi^*(dy) = \psi^*\{y : y \leq x\} =: F^*(x)$$

with probability one, where the far right-hand side function F^* is defined to be the cumulative distribution corresponding to the probability measure ψ^*. Thus the estimator F_n^* of F^* converges at each point in the state space with probability one.

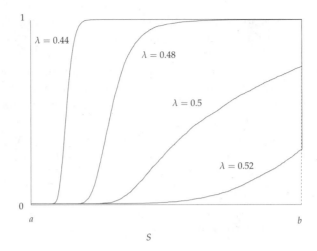

Figure 11.3 Estimates of F^* for different λ

Figure 11.3 shows four observations of F_n^*, each generated with individual time series of length 5,000. The four observations correspond to different values of the credit constraint parameter λ, as shown in the figure.[25] Notice that the stationary distributions are very sensitive to λ, with probability mass rapidly shifting toward the unconstrained equilibrium state b as λ increases.

Exercise 11.2.28 Use the LLN to estimate $\psi^*(\{b\})$ for different values of λ. Graph these values against λ as λ varies over $[0.4, 0.6]$.[26]

11.3   Stability: Probabilistic Methods

So far our techniques for treating stability of Markov chains have been mainly analytical. Now it's time to treat probabilistic methods, which are perhaps even more fundamental to modern Markov chain theory. As you will see, the flavor of the proofs is quite different. This unusual taste has limited the diffusion of probabilistic methods into economics. However, a bit of study will illustrate how powerful—and beautiful— these ideas can be.

---

[25]The model is otherwise defined by $f(k) = k^\alpha$, $\alpha = 0.6$, $R = 1$, $Q = 2$ and $\ln \xi_t \sim N(\mu, \sigma^2)$, where $\sigma = 0.1$ and $\mu = -\sigma^2/2$ (which gives $\mathbb{E}\xi_t = 1$).

[26]In view of the LLN, $\psi^*(\{b\})$ can be interpreted as the fraction of time that $(k_t)_{t \geq 0}$ spends at the unconstrained equilibrium b over the long run.

Underlying most probabilistic methods is the notion of coupling. Coupling is used to make assertions about a collection of distributions by constructing random variables on a common probability space that (1) have these distributions, and (2) also have certain properties useful for establishing the assertion one wishes to prove. In the case of Markov chains the distributions in question are usually the stationary distribution ψ^* and the marginal distribution $\psi \mathbf{M}^t$, and the assertion one seeks to prove is that $\|\psi \mathbf{M}^t - \psi^*\| \to 0$ as $t \to \infty$.

At this stage you might like to review the material in §5.2.3, which uses coupling to prove the stability of finite state Markov chains. However, the next section can be read independently of §5.2.3, and despite the infinite state space, is perhaps a little easier to follow.

11.3.1 Coupling with Regeneration

Let us begin by reconsidering stability of the commodity pricing model. This model has a regenerative structure making it particularly well suited to illustrating the fundamentals of coupling. The commodity pricing model was shown to be globally stable in §8.2.4 when the harvest (i.e., shock) process is lognormally distributed. Let us now prove that global stability holds when the harvest is distributed according to a general Borel probability measure ϕ (as opposed to a density).

In §6.3.1 we assumed that the shock is bounded away from zero in order to set up a contraction mapping argument in a space of bounded functions. Now we assume that W_t has compact support $[0, b]$. The contraction mapping arguments of §6.3.1 can be maintained if $P(0) < \infty$, where P is the inverse demand function. Assume this to be the case.

The law of motion for the commodity pricing model is of the form

$$X_{t+1} = \alpha I(X_t) + W_{t+1}, \quad (W_t)_{t \geq 1} \overset{\text{IID}}{\sim} \phi, \quad X_0 \sim \psi$$

where I is the equilibrium investment function in (6.27), page 146. As usual, we assume that X_0 is independent of $(W_t)_{t \geq 1}$. The shocks $(W_t)_{t \geq 1}$ and the initial condition X_0 are random variables on some probability space $(\Omega, \mathscr{F}, \mathbb{P})$. The stochastic kernel P is given by

$$P(x, B) := \int \mathbb{1}_B[\alpha I(x) + z] \phi(dz) \qquad (x \in S, B \in \mathscr{B}(S))$$

and the Markov operator $\mathbf{M}$ is defined in the usual way. In example 11.2.7 (page 259) we saw that $S := [0, \bar{s}]$ is a valid state space for this model when $\bar{s} := b/(1 - \alpha)$, and that a stationary distribution $\psi^* \in \mathscr{P}(S)$ always exists. When discussing stability, we will use the metric

$$\|\mu - \nu\| := \sup_{B \in \mathscr{B}(S)} |\mu(B) - \nu(B)| \tag{11.19}$$

on $\mathscr{P}(S)$, which is proportional to total variation distance (see page 254).

Theorem 11.3.1 *If $\phi(z_0) > 0$ whenever $z_0 > 0$, then $(\mathscr{P}(S), \mathbf{M})$ is globally stable. In fact there exists a $k \in \mathbb{N}$ and a $\delta < 1$ such that*

$$\|\psi \mathbf{M}^{t \times k + 1} - \psi^*\| \leq \delta^t \qquad \text{for all } \psi \in \mathscr{P}(S), \ t \in \mathbb{N} \tag{11.20}$$

It is an exercise to show that (11.20) implies global stability of $(\mathscr{P}(S), \mathbf{M})$.[27] The proof of (11.20) is based on the following inequality, a version of which was previously discussed in lemma 5.2.5 (page 115):

Lemma 11.3.2 *If X and Y are two random variables on $(\Omega, \mathscr{F}, \mathbb{P})$ with $X \sim \psi_X \in \mathscr{P}(S)$ and $Y \sim \psi_Y \in \mathscr{P}(S)$, then*

$$\|\psi_X - \psi_Y\| \leq \mathbb{P}\{X \neq Y\} \tag{11.21}$$

Intuitively, if the probability that X and Y differ is small, then so is the distance between their distributions. The beauty of lemma 11.3.2 is that it holds for *any* X and Y with the appropriate distributions, and careful choice of these random variables can yield a tight bound.

Proof. Pick any $B \in \mathscr{B}(S)$. We have

$$\mathbb{P}\{X \in B\} = \mathbb{P}\{X \in B\} \cap \{X = Y\} + \mathbb{P}\{X \in B\} \cap \{X \neq Y\}, \text{ and}$$

$$\mathbb{P}\{Y \in B\} = \mathbb{P}\{Y \in B\} \cap \{X = Y\} + \mathbb{P}\{Y \in B\} \cap \{X \neq Y\}$$

Since $\{X \in B\} \cap \{X = Y\} = \{Y \in B\} \cap \{X = Y\}$, we have

$$\mathbb{P}\{X \in B\} - \mathbb{P}\{Y \in B\} = \mathbb{P}\{X \in B\} \cap \{X \neq Y\} - \mathbb{P}\{Y \in B\} \cap \{X \neq Y\}$$

$$\therefore \quad \mathbb{P}\{X \in B\} - \mathbb{P}\{Y \in B\} \leq \mathbb{P}\{X \in B\} \cap \{X \neq Y\} \leq \mathbb{P}\{X \neq Y\}$$

Reversing the roles of X and Y gives

$$|\mathbb{P}\{X \in B\} - \mathbb{P}\{Y \in B\}| \leq \mathbb{P}\{X \neq Y\}$$

Since B is arbitrary, we have established (11.21). □

Our strategy for proving theorem 11.3.1 is as follows: Given the harvest process $(W_t)_{t \geq 1}$, let $(X_t)_{t \geq 0}$ and $(X_t^*)_{t \geq 0}$ be defined by

$$X_{t+1} = \alpha I(X_t) + W_{t+1}, \ X_0 \sim \psi \quad \text{and} \quad X_{t+1}^* = \alpha I(X_t^*) + W_{t+1}, \ X_0^* \sim \psi^*$$

---

[27] Hint: Use nonexpansiveness. See lemma 11.1.28 on page 254.

Notice that $X_t^* \sim \psi^*$ for all t. Hence, by (11.21), we have

$$\|\psi \mathbf{M}^t - \psi^*\| \leq \mathbb{P}\{X_t \neq X_t^*\} \qquad (t \in \mathbb{N}) \tag{11.22}$$

Thus to bound $\|\psi \mathbf{M}^t - \psi^*\|$, it is sufficient to bound $\mathbb{P}\{X_t \neq X_t^*\}$. In other words, we need to show that the probability X_t and X_t^* remain distinct converges to zero in t—or, conversely, that X_t and X_t^* are eventually equal with high probability.

Critical to the proof are two facts. One is that $(X_t)_{t \geq 0}$ and $(X_t^*)_{t \geq 0}$ are driven by the *same* sequence of harvests $(W_t)_{t \geq 1}$. As a result, if the two processes meet, then they remain equal: if $X_j = X_j^*$ for some j, then $X_t = X_t^*$ for all $t \geq j$. Second, there exists an $x_b > 0$ such that $I(x) = 0$ for all $x \leq x_b$ (see lemma 11.3.3 below). As a result, if both $X_j \leq x_b$ and $X_j^* \leq x_b$, then $I(X_j) = I(X_j^*) = 0$, in which case $X_{j+1} = X_{j+1}^* = W_{j+1}$.

As a consequence of these two properties, for $X_t = X_t^*$ to hold, *it is sufficient that both $X_j \leq x_b$ and $X_j^* \leq x_b$ for some $j < t$*. Moreover this will occur whenever there is a sufficiently long sequence of sufficiently small harvests. We will show that the probability such a sequence has occurred at least once prior to t converges to one as $t \to \infty$; and hence $\mathbb{P}\{X_t \neq X_t^*\} \to 0$.

An illustration of the coupling of $(X_t)_{t \geq 0}$ and $(X_t^*)_{t \geq 0}$ is given in figure 11.4, which shows simulated paths for these two time series.[28] At $t = 4$ both X_t and X_t^* are below the threshold x_b at which investment becomes zero. As a result $X_t = X_t^*$ for all $t \geq 5$. The two processes are said to couple at $t = 5$, and that date is referred as the coupling time.

Now let's turn to details, beginning with the following lemma. (To maintain continuity, the proof is given in the appendix to this chapter.)

Lemma 11.3.3 *There exists an $x_b > 0$ such that $x \leq x_b$ implies $I(x) = 0$.*

The processes $(X_t)_{t \geq 0}$ and $(X_t^*)_{t \geq 0}$ couple at $j + 1$ if $\{X_j \leq x_b\} \cap \{X_j^* \leq x_b\}$ occurs. To check for occurrence of this event, it is convenient to define a third process that acts as an upper bound for $(X_t)_{t \geq 0}$ and $(X_t^*)_{t \geq 0}$:

$$X_{t+1}' = \alpha X_t' + W_{t+1}, \quad X_0' = \bar{s}$$

Exercise 11.3.4 Show by induction that the inequalities $X_j \leq X_j'$ and $X_j^* \leq X_j'$ hold pointwise on Ω, and as a consequence

$$\{X_j' \leq x_b\} \subset \{X_j \leq x_b\} \cap \{X_j^* \leq x_b\}$$

for all $j \geq 0$. (We are, of course, referring to subsets of Ω here.)

---

[28]In the simulation, the primitives are $\alpha = 0.9$, $\xi \sim cU$ where $c = 4$ and U is Beta(5,5), and $P(x) = \bar{s}e^{-x}$. As above, $\bar{s} = b(1 - \alpha)^{-1}$. Since $U \leq 1$, we have $b = 4$, and $\bar{s} = 40$.

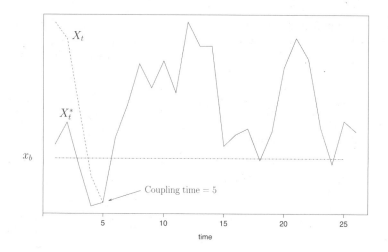

Figure 11.4 Coupling of $(X_t)_{t\geq 0}$ and $(X_t^*)_{t\geq 0}$ at $t = 5$

Given that if $(X_t)_{t\geq 0}$ and $(X_t^*)_{t\geq 0}$ meet they remain equal, and given that $X_j' \leq x_b$ implies $X_{j+1} = X_{j+1}^*$, we have

$$X_j' \leq x_b \text{ for some } j \leq t \implies X_{t+1} = X_{t+1}^*$$

In terms of subsets of Ω, this can be stated as

$$\cup_{j\leq t}\{X_j' \leq x_b\} \subset \{X_{t+1} = X_{t+1}^*\}$$

$$\therefore \quad \mathbb{P}\cup_{j\leq t}\{X_j' \leq x_b\} \leq \mathbb{P}\{X_{t+1} = X_{t+1}^*\}$$

$$\therefore \quad \mathbb{P}\{X_{t+1} \neq X_{t+1}^*\} \leq \mathbb{P}\cap_{j\leq t}\{X_j' > x_b\}$$

The probability of the event $\cap_{j\leq t}\{X_j' > x_b\}$ can be bounded relatively easily. Indeed suppose that harvests W_{n+1} to W_{n+k} are all below z_0, where k and z_0 are chosen to satisfy

$$\alpha^k \bar{s} + z_0 \frac{1 - \alpha^k}{1 - \alpha} \leq x_b \tag{11.23}$$

Since the harvests are all below z_0, we have

$$X_j' \leq \alpha X_{j-1}' + z_0 \qquad j = n+1, \ldots, n+k$$

Combining these k inequalities gives

$$X_{n+k}' \leq \alpha^k X_n' + z_0 \frac{1 - \alpha^k}{1 - \alpha} \leq \alpha^k \bar{s} + z_0 \frac{1 - \alpha^k}{1 - \alpha} \leq x_b$$

where the last inequality follows from (11.23). Thus a sequence of k harvests below z_0 forces $(X'_t)_{t\geq 0}$ below x_b by the end of the sequence. For dates of the form $t \times k$ there are t nonoverlapping sequences of k consecutive harvests prior to $t \times k$. Let E_i be the event that the i-th of these t sequences has all harvests below z_0:

$$E_i = \cap_{j=k\times(i-1)+1}^{i\times k}\{W_j \leq z_0\} \qquad i = 1,\ldots,t$$

If X'_j never falls below x_b in the period up to $t \times k$, then none of the events E_i has occurred.

$$\therefore \quad \cap_{j\leq t\times k}\{X'_j > x_b\} \subset \cap_{i=1}^t E_i^c$$

$$\therefore \quad \mathbb{P}\{X_{t\times k+1} \neq X^*_{t\times k+1}\} \leq \mathbb{P} \cap_{j\leq t\times k} \{X'_j > x_b\} \leq \mathbb{P} \cap_{i=1}^t E_i^c$$

Since the sequences of harvests that make up each E_i are nonoverlapping, these events are independent, and $\mathbb{P} \cap_{i=1}^t E_i^c = \prod_{i=1}^t (1 - \mathbb{P}(E_i))$.

$$\therefore \quad \mathbb{P}\{X_{t\times k+1} \neq X^*_{t\times k+1}\} \leq \prod_{i=1}^t (1 - \mathbb{P}(E_i)) = (1 - \phi(z_0)^k)^t$$

We have now established (11.20) and hence theorem 11.3.1.

11.3.2   Coupling and the Dobrushin Coefficient

Let $S \in \mathscr{B}(\mathbb{R}^n)$, and let P be an arbitrary stochastic kernel on S. In the previous section S was a subset of $\mathbb{R}$, and P had the attractive property that for any x, x' in $[0, x_b] \subset S, P(x, dy) = P(x', dy)$. Such a set is called an *atom* of P. Existence of an atom to which the state returns regularly makes coupling particularly simple: Whenever the two chains $(X_t)_{t\geq 0}$ and $(X^*_t)_{t\geq 0}$ in §11.3.1 enter the atom simultaneously they can be coupled.

Unfortunately, most Markov chains studied in economic applications fail to have this structure. However, it turns out that with a little bit of trickery one can construct couplings for many chains without using atoms. In this section we discuss the case where P has a positive Dobrushin coefficient. As we show, positivity of the Dobrushin coefficient is very closely connected with the possibility of successful coupling.

Understanding the connection between coupling and the Dobrushin coefficient is satisfying because the latter plays such an important role in analytical proofs of stability. Coupling will shine a light on the role of the Dobrushin coefficient from a new angle. More importantly, the basic idea behind the coupling proof we use here can be generalized to a large number of different models.

As in §11.3.1, we will endow $\mathscr{P}(S)$ with the metric defined by (11.19), which is proportional to total variation distance. Our main result is as follows:

Proposition 11.3.5 *Let $\psi, \psi' \in \mathscr{P}(S)$. If $\alpha(P)$ is the Dobrushin coefficient for P and $\mathbf{M}$ is the corresponding Markov operator, then*

$$\|\psi\mathbf{M}^t - \psi'\mathbf{M}^t\| \leq (1 - \alpha(P))^t \qquad (t \in \mathbb{N})$$

If $\alpha(P) = 0$, then this inequality is trivial. On the other hand, if $\alpha(P) > 0$, then for any initial conditions ψ and ψ', we have $\|\psi\mathbf{M}^t - \psi'\mathbf{M}^t\| \to 0$ at a geometric rate. In particular, if $\psi' = \psi^*$ is stationary, then $\|\psi\mathbf{M}^t - \psi^*\| \to 0$. Note also that we are proving nothing new here. Theorem 11.2.21 yields the same result. What is new is the proof—which is radically different.

Since the proposition is trivial when $\alpha(P) = 0$, we assume in all of what follows that $\alpha(P)$ is strictly positive. To make the proof, we will build two processes $(X_t)_{t \geq 0}$ and $(X_t')_{t \geq 0}$ such that the distribution ψ_t of X_t is $\psi\mathbf{M}^t$ and the distribution ψ_t' of X_t' is $\psi'\mathbf{M}^t$. In view of lemma 11.3.2 (page 273), we then have

$$\|\psi\mathbf{M}^t - \psi'\mathbf{M}^t\| \leq \mathbb{P}\{X_t \neq X_t'\} \tag{11.24}$$

Given (11.24) it is sufficient to show that $\mathbb{P}\{X_t \neq X_t'\} \leq (1 - \alpha(P))^t$. The trick to the proof is constructing $(X_t)_{t \geq 0}$ and $(X_t')_{t \geq 0}$ in a rather special way. In particular, we build these processes so that there is an independent $\alpha(P)$ probability of meeting at every step, and, moreover, once the processes meet (i.e., $X_j = X_j'$ for some j) they remain coupled (i.e., $X_t = X_t'$ for all $t \geq j$). It then follows that if $X_t \neq X_t'$, then the two processes have never met, and the probability of this is less than $(1 - \alpha(P))^t$.

While $(X_t)_{t \geq 0}$ and $(X_t')_{t \geq 0}$ are constructed in a special way, care must be taken that $\psi_t = \psi\mathbf{M}^t$ and $\psi_t' = \psi'\mathbf{M}^t$ remains valid. This requires that $X_0 \sim \psi$, $X_0' \sim \psi'$, and, recursively, $X_{t+1} \sim P(X_t, dy)$ and $X_{t+1}' \sim P(X_t', dy)$. That such is the case does not appear obvious from the construction, but at the end of our proof we will verify that it is.

In order to construct the processes $(X_t)_{t \geq 0}$ and $(X_t')_{t \geq 0}$, it is convenient to introduce some additional notation. First, let

$$\gamma(x, x') := (P_x \wedge P_{x'})(S) \qquad ((x, x') \in S \times S)$$

be the affinity between $P(x, dy)$ and $P(x', dy)$, so $\alpha(P) = \inf_{x,x'} \gamma(x, x')$. Evidently $\gamma(x, x') \geq \alpha(P) > 0$ for every x and x'.

Next let us define three new functions from $S \times S \times \mathscr{B}(S)$ to $[0, 1]$ by

$$\nu(x, x', B) := \frac{(P_x \wedge P_{x'})(B)}{\gamma(x, x')}$$

$$\mu(x, x', B) := \frac{P(x, B) - (P_x \wedge P_{x'})(B)}{1 - \gamma(x, x')}$$

$$\mu'(x, x', B) := \frac{P(x', B) - (P_x \wedge P_{x'})(B)}{1 - \gamma(x, x')}$$

Crucially, $\nu(x, x', dy)$, $\mu(x, x', dy)$ and $\mu'(x, x', dy)$ are all probability measures on S. In particular, $\nu(x, x', B)$, $\mu(x, x', B)$ and $\mu'(x, x', B)$ are nonnegative for all $B \in \mathscr{B}(S)$; and $\nu(x, x', S) = \mu(x, x', S) = \mu'(x, x', S) = 1$. The next exercise asks you to show that this is the case.

Exercise 11.3.6 Prove that $\nu(x, x', dy)$, $\mu(x, x', dy)$ and $\mu'(x, x', dy)$ are probability measures on S for every $(x, x') \in S \times S$ such that $\gamma(x, x') < 1$.[29]

Algorithm 11.1: Coupling algorithm

---

draw $X_0 \sim \psi$ and $X_0' \sim \psi^*$
set $t = 0$
while True **do**
    if $X_t = X_t'$ **then**
        | draw $Z \sim P(X_t, dy)$ and set $X_{t+1} = X_{t+1}' = Z$
    else
        draw U_{t+1} independently from Uniform$(0, 1)$
        if $U_{t+1} \leq \gamma(X_t, X_t')$ **then**   `// with probability` $\gamma(X_t, X_t')$
        | draw $Z \sim \nu(X_t, X_t', dy)$ and set $X_{t+1} = X_{t+1}' = Z$
        else   `// with probability` $1 - \gamma(X_t, X_t')$
            draw $X_{t+1} \sim \mu(X_t, X_t', dy)$
            draw $X_{t+1}' \sim \mu'(X_t, X_t', dy)$
        end
    end
    set $t = t + 1$
end

We are now ready to build the processes $(X_t)_{t \geq 0}$ and $(X_t')_{t \geq 0}$. This is done in algorithm 11.1. The while loop in the algorithm is an infinite loop. If at time t we have $X_t \neq X_t'$, then a uniform random variable U_{t+1} is drawn to determine the next action. The probability that $U_{t+1} \leq \gamma(X_t, X_t')$ is $\gamma(X_t, X_t')$, and if this occurs, then both X_{t+1} and X_{t+1}' are set equal to a single draw from $\nu(X_t, X_t', dy)$. Otherwise, they are drawn from $\mu(X_t, X_t', dy)$ and $\mu'(X_t, X_t', dy)$ respectively. All random variables are assumed to live on probability space $(\Omega, \mathscr{F}, \mathbb{P})$.

Our first claim is that for the processes $(X_t)_{t \geq 0}$ and $(X_t')_{t \geq 0}$ we have

$$\mathbb{P}\{X_t \neq X_t'\} \leq (1 - \alpha(P))^t \qquad (t \in \mathbb{N}) \tag{11.25}$$

---
[29]Note that $\gamma(x, x') > 0$ by assumption.

The proof is not difficult: Fix any $t \in \mathbb{N}$. Observe that if $U_j \leq \gamma(X_{j-1}, X'_{j-1})$ for just one $j \leq t$, then the two processes couple and $X_t = X'_t$. So, if $X_t \neq X'_t$, then we must have $U_j > \gamma(X_{j-1}, X'_{j-1})$ for all $j \leq t$. Since $\gamma(X_{j-1}, X'_{j-1}) \geq \alpha(P)$, and since $(U_t)_{t \geq 1}$ is IID and uniformly distributed on $(0,1)$, the probability of this event is no more than $(1 - \alpha(P))^t$. As t is arbitrary, the proof of (11.25) is now done.

In view of (11.25), proposition 11.3.5 will be established if we can verify (11.24); in other words, we need to show that

$$\|\psi \mathbf{M}^t - \psi' \mathbf{M}^t\| \leq \mathbb{P}\{X_t \neq X'_t\}$$

As discussed above, this is implied by lemma 11.3.2 (page 273), provided that the distributions of X_t and X'_t are $\psi \mathbf{M}^t$ and $\psi' \mathbf{M}^t$ respectively. Since $X_0 \sim \psi$ and $X'_0 \sim \psi'$ certainly hold, we need only show that $X_{t+1} \sim P(X_t, dy)$ and $X'_{t+1} \sim P(X'_t, dy)$ at each step.

To see that this is the case, suppose that at time t we have $(X_t, X'_t) = (x, x')$. We claim that after the next iteration of algorithm 11.1, the probability that $X_{t+1} \in B$ is $P(x, B)$, while the probability that $X'_{t+1} \in B$ is $P(x', B)$. Let's focus for now on the claim that the probability that $X_{t+1} \in B$ is $P(x, B)$.

Suppose first that $x \neq x'$. If $\gamma(x, x') = 1$, then X_{t+1} is drawn from $\nu(x, x', dy)$ with probability one. But $\gamma(x, x') = 1$ implies that $P(x, dy) = P(x', dy)$, and hence $\nu(x, x', dy) = P(x, dy) = P(x', dy)$. In particular, the probability that $X_{t+1} \in B$ is $P(x, B)$. On the other hand, if $\gamma(x, x') < 1$, then X_{t+1} is drawn from $\nu(x, x', dy)$ with probability $\gamma(x, x')$, and from $\mu(x, x', dy)$ with probability $1 - \gamma(x, x')$. As a result the probability that $X_{t+1} \in B$ is

$$\gamma(x, x') \nu(x, x', B) + (1 - \gamma(x, x')) \mu(x, x', B)$$

A look at the definitions of ν and μ confirms that this is precisely $P(x, B)$.

The argument that $X'_{t+1} \in B$ with probability $P(x', B)$ is almost identical.

Finally, suppose that $x = x'$. Then $X_{t+1} \sim P(x, dy)$, so clearly $X_{t+1} \in B$ with probability $P(x, B)$. Also $X'_{t+1} = X_{t+1}$, so $X'_{t+1} \in B$ with probability $P(x, B)$. But since $x' = x$, we have $P(x', B) = P(x, B)$. Hence the probability that $X'_{t+1} \in B$ is $P(x', B)$. The proof is done.

11.3.3   Stability via Monotonicity

Economic models often possess a degree of monotonicity in the laws of motion. If a nice property such as monotonicity holds, then we would like to exploit it when considering stability. The question of when monotone stochastic systems are stable is the topic of this section.

Consider again our canonical SRS introduced on page 219. To reiterate, the state space S is a Borel subset of $\mathbb{R}^n$ and the shock space Z is a Borel subset of $\mathbb{R}^k$. The SRS is described by a (Borel) measurable function $F \colon S \times Z \to S$ and a distribution $\phi \in \mathscr{P}(Z)$, where

$$X_{t+1} = F(X_t, W_{t+1}), \quad (W_t)_{t \geq 1} \overset{\text{IID}}{\sim} \phi, \quad X_0 \sim \psi \in \mathscr{P}(S) \tag{11.26}$$

The sequence of shocks $(W_t)_{t \geq 1}$ and the initial condition X_0 live on a common probability space $(\Omega, \mathscr{F}, \mathbb{P})$ and are mutually independent. The stochastic kernel for (11.26) is given by

$$P(x, B) = \int \mathbb{1}_B[F(x, z)]\phi(dz) \quad (x \in S, B \in \mathscr{B}(S)) \tag{11.27}$$

Let $\mathbf{M}$ be the corresponding Markov operator, so ψ^* is stationary for (11.26) if and only if $\psi^* \mathbf{M} = \psi^*$.

Definition 11.3.7 The SRS (11.26) is said to be *monotone increasing* if, for all $z \in Z$,[30]

$$F(x, z) \leq F(x', z) \text{ whenever } x \leq x' \tag{11.28}$$

Example 11.3.8 Consider the one-dimensional linear system

$$X_{t+1} = \alpha X + W_{t+1}, \quad (W_t)_{t \geq 1} \overset{\text{IID}}{\sim} \phi \in \mathscr{P}(\mathbb{R}) \tag{11.29}$$

where $S = Z = \mathbb{R}$ and $F(x, z) = \alpha x + z$. This SRS is monotone increasing if and only if $\alpha \geq 0$.

Exercise 11.3.9 Suppose that (11.26) is monotone increasing, and $h \colon S \to \mathbb{R}$. Show that if $h \in ibS$, then $\mathbf{M}h \in ibS$.[31]

Exercise 11.3.10 A set $B \subset S$ is called an *increasing set* if $x \in B$ and $x' \in S$ with $x \leq x'$ implies $x' \in B$. Show that $B \subset S$ is an increasing set if and only if $\mathbb{1}_B \in ibS$. (Since $P(x, B) = \mathbf{M}\mathbb{1}_B(x)$, it follows that $x \mapsto P(x, B)$ is increasing whenever B is an increasing set. What is the intuition?)

Returning to the general SRS (11.26), let's assume that at least one stationary distribution ψ^* exists, and see what conditions we might need to obtain uniqueness and stability of ψ^* under the hypothesis of monotonicity. In this section stability of ψ^* will have the slightly nonstandard definition

$$\forall \psi \in \mathscr{P}(S), \quad (\psi \mathbf{M}^t)(h) \to \psi^*(h) \text{ as } t \to \infty \text{ for all } h \in ibS \tag{11.30}$$

---

[30] Our order on $\mathbb{R}^n$ is the usual one: $(x_i)_{i=1}^n \leq (y_i)_{i=1}^n$ if $x_i \leq y_i$ for all i with $1 \leq i \leq n$.
[31] As usual, *ibS* denotes the increasing bounded functions $h \colon S \to \mathbb{R}$.

To understand condition (11.30), recall that a sequence $(\mu_n) \subset \mathscr{P}(S)$ converges to $\mu \in \mathscr{P}(S)$ weakly if $\mu_n(h) \to \mu(h)$ as $n \to \infty$ for all $h \in bcS$. Moreover, to verify weak convergence, it suffices to check $\mu_n(h) \to \mu(h)$ only for $h \in ibcS$ (theorem 11.1.35, page 255). Since $ibcS \subset ibS$, the convergence in (11.30) is stronger than $\psi \mathbf{M}^t \to \psi^*$ weakly. Finally, (11.30) implies uniqueness of ψ^*, as follows from the next exercise.

Exercise 11.3.11 Show that if $\psi^{**} \in \mathscr{P}(S)$ satisfies $\psi^{**}\mathbf{M} = \psi^{**}$ and (11.30) holds, then ψ^{**} and ψ^* must be equal.[32]

Now let $(W_t)_{t \geq 1}$ and $(W'_t)_{t \geq 1}$ be jointly independent IID processes on Z, both distributed according to ϕ and defined on the common probability space $(\Omega, \mathscr{F}, \mathbb{P})$. Using these two independent shock processes, we can introduce a condition that is sufficient for stability of monotone systems.

Definition 11.3.12 The SRS (11.26) is called *order mixing* if, given any two independent initial conditions X_0 and X'_0, the processes $(X_t)_{t \geq 0}$ and $(X'_t)_{t \geq 0}$ defined by $X_{t+1} = F(X_t, W_{t+1})$ and $X'_{t+1} = F(X'_t, W'_{t+1})$ satisfy

$$\mathbb{P} \cup_{t \geq 0} \{X_t \leq X'_t\} = \mathbb{P} \cup_{t \geq 0} \{X'_t \leq X_t\} = 1 \tag{11.31}$$

Exercise 11.3.13 Prove: $\mathbb{P} \cup_{t \geq 0} \{X_t \leq X'_t\} = 1$ iff $\lim_{T \to \infty} \mathbb{P} \cap_{t \leq T} \{X_t \not\leq X'_t\} = 0$.[33]

Paired with monotonicity, order mixing is sufficient for stability. In particular, we have the following result:

Theorem 11.3.14 *Suppose that (11.26) has at least one stationary distribution $\psi^* \in \mathscr{P}(S)$. If it is monotone increasing and order mixing, then ψ^* is the only stationary distribution, and moreover ψ^* is globally stable in the sense of (11.30).*

The proof is given at the end of §11.3.4. For now let us consider how to apply this result. To do so, it is necessary to develop sufficient conditions for order mixing that are easy to check in applications. One set of conditions that implies order mixing is that introduced by Razin and Yahav (1979), and extended and popularized by Stokey and Lucas (1989) and Hopenhayn and Prescott (1992). Following Stokey and Lucas (1989, assumption 12.1), suppose that $S = [a,b] := \{x \in \mathbb{R}^n : a \leq x \leq b\}$ for fixed $a, b \in \mathbb{R}^n$, and that

$$\exists \, m \in \mathbb{N}, \, c \in S, \, \epsilon > 0 \text{ such that } P^m(a, [c,b]) \geq \epsilon \text{ and } P^m(b, [a,c]) \geq \epsilon \tag{11.32}$$

Let's consider this condition in our setting, where P is the kernel (11.27) and (11.26) is monotone increasing.

---

[32] Hint: See theorem 11.1.36 on page 255.
[33] Hint: Use exercise 7.1.25 on page 167. Note that all sets are subsets of Ω.

Exercise 11.3.15 Show that under (11.32) the kernel P satisfies $P^m(x, [c, b]) \geq \epsilon$ and $P^m(x, [a, c]) \geq \epsilon$ for all $x \in S$.[34]

If (11.26) satisfies (11.32), then it is order mixing. To get a feel for the proof, suppose that (11.32) holds with $m = 1$, and consider the probability that there exists a $t \geq 0$ with $X_t \leq X_t'$. In light of exercise 11.3.13, to show this probability is one, it is sufficient to show that $\lim_{T \to \infty} \mathbb{P}(E_T) = 0$, where $E_T := \cap_{t \leq T}\{X_t \not\leq X_t'\}$. The probability of $X_t \leq X_t'$ given $(X_{t-1}, X_{t-1}') = (x, x')$ is

$$\mathbb{P}\{F(x, W_t) \leq F(x', W_t')\} \geq \mathbb{P}\{F(x, W_t) \leq c\} \cap \{F(x', W_t') \geq c\}$$
$$= \mathbb{P}\{F(x, W_t) \leq c\}\mathbb{P}\{F(x', W_t') \geq c\}$$
$$= P(x, [a, c])P(x', [c, b]) \geq \epsilon^2$$

Hence for each t the probability of $X_t \leq X_t'$ occurring is at least ϵ^2, independent of the lagged value of the state. One can then show that the probability $\mathbb{P}(E_T)$ of this event never occurring prior to T is $\leq (1 - \epsilon^2)^T \to 0$.[35] From exercise 11.3.13 we then have $\mathbb{P} \cup_{t \geq 0} \{X_t \leq X_t'\} = 1$. A similar argument establishes $\mathbb{P} \cup_{t \geq 0} \{X_t \geq X_t'\} = 1$, and hence order mixing.

Condition (11.32) can be restrictive, as the state space must be of the form $\{x \in \mathbb{R}^n : a \leq x \leq b\}$. To devise a weaker set of sufficient conditions, we introduce the concept of order inducing sets and order norm-like functions.

Definition 11.3.16 A $C \in \mathscr{B}(S)$ is called *order inducing* for kernel P if there exists a $c \in S$ and an $m \in \mathbb{N}$ such that

$$\inf_{x \in C} P^m(x, \{z : z \leq c\}) > 0 \quad \text{and} \quad \inf_{x \in C} P^m(x, \{z : z \geq c\}) > 0$$

A measurable function $v \colon S \to \mathbb{R}_+$ is called *order norm-like* for P if every sublevel set of v is order inducing for P.[36]

Example 11.3.17 Continuing with the model (11.29) in example 11.3.8, suppose that $\mathbb{P}\{W_t \leq d\} > 0$ and $\mathbb{P}\{W_t \geq d\} > 0$ for all $d \in S = \mathbb{R}$. Then every set of the form $[-b, b]$ is order inducing with $m = 1$ and $c = 0$. To see this, pick any $b \geq 0$. For $x \in [-b, b]$ we have

$$P(x, \{z : z \leq 0\}) = \mathbb{P}\{\alpha x + W \leq 0\} = \mathbb{P}\{W \leq -\alpha x\} \geq \mathbb{P}\{W \leq -\alpha b\}$$

which is positive. The proof that $\inf_{x \in C} P(x, \{z : z \geq 0\}) > 0$ is similar.

Since all sets of the form $[-b, b]$ are order inducing, it follows that $v(x) = |x|$ is order norm-like for this model. (Why?)

---

[34]Hint: Use exercise 11.3.10.

[35]A complete proof can be made along the lines of of proposition 5.2.6, or using conditional expectations. See also Kamihigashi and Stachurski (2008).

[36]Recall that the sublevel sets of v are sets of the form $\{x \in S : v(x) \leq a\}$ for $a \in \mathbb{R}_+$.

Exercise 11.3.18 Show that all measurable subsets of order inducing sets are order inducing. Show that $v \colon S \to \mathbb{R}_+$ is order norm-like if and only if there exists a $K \in \mathbb{R}_+$ such that $\{x \in S : v(x) \le a\}$ is order inducing for all $a \ge K$.

We can now state the following sufficient condition for order mixing:

Theorem 11.3.19 *If there exists an order norm-like function v for (11.26) and constants $\alpha \in [0, 1)$ and $\beta \in \mathbb{R}_+$ such that*

$$\int v[F(x, z)]\phi(dz) \le \alpha v(x) + \beta \qquad (x \in S) \tag{11.33}$$

then (11.26) is order mixing.

The intuition is that under (11.33) there is drift back to (sufficiently large) sublevel sets of v, which are order inducing. When two independent chains $(X_t)_{t \ge 0}$ and $(X_t')_{t \ge 0}$ enter such an order inducing set, there is a positive probability of the orderings $X_t \le X_t'$ and $X_t \ge X_t'$ occurring within m periods. Repeated over an infinite horizon, these orderings eventually occur with probability one. For a proof, see Kamihigashi and Stachurski (2008).

Example 11.3.20 Continuing with example 11.3.17, we saw that $v(x) = |x|$ is order norm-like for this model. If $\mathbb{E}|W_t| < \infty$ and $|\alpha| < 1$, then the model is order mixing, since for any $x \in S$,

$$\int v[F(x, z)]\phi(dz) = \int |\alpha x + z|\phi(dz) \le |\alpha|v(x) + \int |z|\phi(dz)$$

11.3.4   More on Monotonicity

Let's now turn to a less trivial example of how monotonicity and order mixing can be used to establish stability.[37] Recall the stochastic Solow–Swan growth model discussed in example 9.2.4, page 220, with law of motion

$$k_{t+1} = sf(k_t)W_{t+1} + (1 - \delta)k_t \tag{11.34}$$

and increasing production function $f \colon \mathbb{R}_+ \to \mathbb{R}_+$ with $f(k) > 0$ for all $k > 0$. The productivity shock W_t and the capital stock k_t take values in $Z := S := (0, \infty)$, while the parameters satisfy $s \in (0, 1)$ and $\delta \in (0, 1]$. The model (11.34) is monotone increasing in the sense of definition 11.3.7.

Let's look for conditions under which (11.34) is order mixing. For simplicity we assume that $\delta = 1$.[38] We suppose further that $\lim_{k \to 0} f(k)/k = \infty$, and $\lim_{k \to \infty} f(k)/k =$

---

[37]See also §12.1.3, which treats the optimal growth model using monotonicity.
[38]The case $\delta < 1$ can be accommodated at the cost of a longer proof.

0. (More traditional Inada conditions can be used if f is assumed to be differentiable.) Regarding the distribution ϕ, we assume that both $\mathbb{E}W_t$ and $\mathbb{E}(1/W_t)$ are finite, and that $\mathbb{P}\{W_t \leq x\}$ and $\mathbb{P}\{W_t \geq x\}$ are strictly positive for every $x \in S$.[39]

First let's show that all closed intervals $[a, b] \subset S$ are order inducing for this model. To do so, pick any $a \leq b$. Fix any $c \in S$. It suffices to show that $\inf_{a \leq x \leq b} P(x, [c, \infty)) > 0$ and likewise for $(0, c]$. Given our assumptions on ϕ, for any $x \in [a, b]$ we have

$$
\begin{aligned}
P(x, [c, \infty)) &= \mathbb{P}\{sf(x)W_{t+1} \geq c\} \\
&= \mathbb{P}\{W_{t+1} \geq c/(sf(x))\} \geq \mathbb{P}\{W_{t+1} \geq c/(sf(a))\} > 0
\end{aligned}
$$

A similar argument gives $\inf_{a \leq x \leq b} P(x, (0, c]) > 0$.

Now consider the function $v(x) := 1/x + x$. Since sublevel sets of v are closed intervals in S, the function v is order norm-like on S.

Exercise 11.3.21 Suppose that f is unbounded, and $f(0) = 0$. Show that neither $x \mapsto x$ nor $x \mapsto 1/x$ are order norm-like for (11.34).

To complete the proof of stability, we must show that the drift condition (11.33) holds for v. That is, we need to establish the existence of an $\alpha \in [0, 1)$ and a $\beta \in \mathbb{R}_+$ such that

$$
\mathbf{M}v \leq \alpha v + \beta \quad \text{when} \quad v(x) = x + \frac{1}{x} \tag{11.35}
$$

where $\mathbf{M}$ is the Markov operator defined by $\mathbf{M}h(x) = \int h[sf(x)z]\phi(dz)$. To this end, suppose that we can establish the existence of $\alpha_1, \alpha_2 \in [0, 1)$ and $\beta_1, \beta_2 \in \mathbb{R}_+$ with

$$
\mathbf{M}v_1 \leq \alpha_1 v_1 + \beta_1 \quad \text{and} \quad \mathbf{M}v_2 \leq \alpha_2 v_2 + \beta_2 \tag{11.36}
$$

where $v_1(x) = x$ and $v_2(x) = 1/x$. Then since $v = v_1 + v_2$, adding across the two inequalities in (11.36) gives the desired inequality (11.35) because

$$
\mathbf{M}v = \mathbf{M}(v_1 + v_2) = \mathbf{M}v_1 + \mathbf{M}v_2 \leq \alpha_1 v_1 + \beta_1 + \alpha_2 v_2 + \beta_2 \leq \alpha v + \beta
$$

when $\alpha := \max\{\alpha_1, \alpha_2\}$ and $\beta := \beta_1 + \beta_2$. In other words, to establish the drift condition (11.35) and hence order mixing, it is sufficient to establish separately the two drift conditions in (11.36). Intuitively, the drift condition on v_1 prevents divergence to $+\infty$, which the condition on v_2 prevents divergence to zero.

Let's attempt to establish the two inequalities in (11.36), starting with the left-hand side (i.e., the inequality $\mathbf{M}v_1 \leq \alpha_1 v_1 + \beta_1$).

---

[39]We require a lot of mixing for global stability because f is not assumed to be concave. In fact f can be rather arbitrary, and the deterministic model (i.e., $W_t \equiv 1$ for all t) can have many fixed points.

Exercise 11.3.22 Fix any constant $\alpha_1 \in (0,1)$. Using the assumption $\lim_{x\to\infty} f(x)/x = 0$, establish the existence of a $\gamma \in S$ satisfying

$$sf(x)\mathbb{E}W_1 \leq \alpha_1 x \qquad \forall x > \gamma$$

Using monotonicity of f, show further that there is a $\beta_1 \in \mathbb{R}_+$ with

$$sf(x)\mathbb{E}W_1 \leq \beta_1 \qquad \forall x \leq \gamma$$

Based on these two inequalities, conclude that

$$\mathbf{M}v_1(x) = sf(x)\mathbb{E}W_1 \leq \alpha_1 x + \beta_1 = \alpha_1 v_1(x) + \beta_1 \qquad \forall x \in S$$

The last exercise establishes the first of the two inequalities in (11.36). The next exercise establishes the second inequality.

Exercise 11.3.23 Fix any constant $\alpha_2 \in (0,1)$. Using the assumption $\lim_{x\to 0} f(x)/x = \infty$, establish the existence of a $\gamma \in S$ satisfying

$$\mathbb{E}\left[\frac{1}{sf(x)W_1}\right] \leq \alpha_2 \frac{1}{x} \qquad \forall x < \gamma$$

Using monotonicity of f, show there exists a $\beta_2 \in \mathbb{R}_+$ with

$$\mathbb{E}\left[\frac{1}{sf(x)W_1}\right] \leq \beta_2 \qquad \forall x \geq \gamma$$

Based on these two inequalities, conclude that

$$\mathbf{M}v_2(x) = \mathbb{E}\left[\frac{1}{sf(x)W_1}\right] \leq \alpha_2 \frac{1}{x} + \beta_2 = \alpha_2 v_2(x) + \beta_2 \qquad \forall x \in S$$

Thus the second inequality in (11.36) also holds, and theorem 11.3.19 implies that under our assumptions the stochastic Solow–Swan growth model is order mixing. Since it is also monotone increasing, theorem 11.3.14 implies that should any stationary distribution $\psi^* \in \mathscr{P}(S)$ exist, that stationary distribution would be unique and globally stable in the sense of (11.30). If f is continuous, then since v is also norm-like (as well as order norm-like, see lemma 8.2.17 on page 209), existence of ψ^* follows immediately from corollary 11.2.10.

To complete this section, let's give the proof of theorem 11.3.14. Using the two independent series $(W_t)_{t\geq 1}$ and $(W_t')_{t\geq 1}$ in the definition of order mixing, algorithm 11.2 defines four processes, denoted by $(X_t)_{t\geq 0}$, $(X_t')_{t\geq 0}$, $(X_t^L)_{t\geq 0}$ and $(X_t^U)_{t\geq 0}$.[40] The process $(X_t)_{t\geq 0}$ is the original SRS in (11.26), with initial condition ψ. The process $(X_t')_{t\geq 0}$

---

[40]Formally, $\omega \in \Omega$ is drawn at the start of time according to $\mathbb{P}$, thereby determining the initial conditions $X_0(\omega)$ and $X_0'(\omega)$, and the shock realizations $(W_t(\omega))_{t\geq 1}$ and $(W_t'(\omega))_{t\geq 1}$. These in turn determine the realizations of all other random variables.

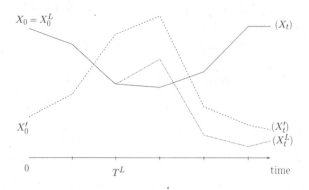

Figure 11.5 The process $(X_t^L)_{t \geq 0}$

has the same law of motion, is driven by independent shocks $(W_t')_{t \geq 1}$, and starts at the stationary distribution ψ^*. Clearly, $X_t' \sim \psi^*$ for all $t \geq 0$. The process $(X_t^L)_{t \geq 0}$ starts off equal to X_0, and updates with shock W_{t+1}' if $X_t^L \leq X_t'$ and with W_{t+1} otherwise. The process $(X_t^U)_{t \geq 0}$ also starts off equal to X_0, and updates with W_{t+1}' if $X_t^U \geq X_t'$ and with W_{t+1} otherwise. An illustration of the process $(X_t^L)_{t \geq 0}$ is given in figure 11.5.

To help clarify the algorithm, we introduce two random variables:

$$T^L := \min\{t \geq 0 : X_t \leq X_t'\} \quad \text{and} \quad T^U := \min\{t \geq 0 : X_t \geq X_t'\}$$

with the usual convention that $\min \varnothing = \infty$. Three properties of $(X_t^L)_{t \geq 0}$ and $(X_t^U)_{t \geq 0}$ are pertinent. The first is that $(X_t^L)_{t \geq 0}$ and $(X_t^U)_{t \geq 0}$ are identical to $(X_t)_{t \geq 0}$ until T^L and T^U respectively. This just follows from the logic of the algorithm. Second, for $t \geq T^L$ we have $X_t^L \leq X_t'$; and $X_t^U \geq X_t'$ for $t \geq T^U$. To see this, consider the case of $(X_t^L)_{t \geq 0}$. Note that $X_{T^L}^L = X_{T^L} \leq X_{T^L}'$ by the definition of T^L, and then

$$X_{T^L+1}^L = F(X_{T^L}^L, W_{T^L+1}') \leq F(X_{T^L}', W_{T^L+1}') = X_{T^L+1}'$$

by monotonicity. Continuing in this way, we have $X_t^L \leq X_t'$ for all $t \geq T^L$. The argument for $(X_t^U)_{t \geq 0}$ is similar. A third property is that both X_t^L and X_t^U have the same distribution $\psi \mathbf{M}^t$ as X_t for *all* t. The reason is that X_0^L and X_0^U are drawn from ψ, and both processes are then updated by the same SRS as the original process $(X_t)_{t \geq 0}$, even though the source of shocks switches from $(W_t)_{t \geq 1}$ to $(W_t')_{t \geq 1}$ at T^L and T^U respectively.[41]

---

[41]A more formal argument can be made via the so-called strong Markov property. See also Kamihigashi and Stachurski (2008).

Algorithm 11.2: Four (F, ϕ) processes

---

generate independent draws $X_0 \sim \psi$ and $X_0' \sim \psi^*$
set $X_0^L = X_0^U = X_0$
for $t \geq 0$ **do**
  draw $W_{t+1} \sim \phi$ and $W_{t+1}' \sim \phi$ independently
  set $X_{t+1} = F(X_t, W_{t+1})$ and $X_{t+1}' = F(X_t', W_{t+1}')$
  if $X_t^L \leq X_t'$ **then**
    | set $X_{t+1}^L = F(X_t^L, W_{t+1}')$
  else
    | set $X_{t+1}^L = F(X_t^L, W_{t+1})$
  end
  if $X_t^U \geq X_t'$ **then**
    | set $X_{t+1}^U = F(X_t^U, W_{t+1}')$
  else
    | set $X_{t+1}^U = F(X_t^U, W_{t+1})$
  end
end

Now to the proof. Pick any $h \in ibS$. We wish to show that $(\psi \mathbf{M}^t)(h) \to \psi^*(h)$ as $t \to \infty$. Order mixing tells us precisely that $\mathbb{P}\{T^L < \infty\} = 1$, or $\mathbb{P}\{T^L \leq t\} \to 1$ as $t \to \infty$. Note that $T^L \leq t$ implies $X_t^L \leq X_t'$, and since h is increasing, this implies $h(X_t^L) \leq h(X_t')$. Therefore

$$h(X_t^L)\mathbb{1}\{T^L \leq t\} \leq h(X_t')\mathbb{1}\{T^L \leq t\}$$

$$\therefore \quad \mathbb{E}h(X_t^L)\mathbb{1}\{T^L \leq t\} \leq \mathbb{E}h(X_t')\mathbb{1}\{T^L \leq t\} \tag{11.37}$$

Using $\mathbb{P}\{T^L \leq t\} \to 1$ and taking the limit superior now gives

$$\limsup_{t \to \infty} \mathbb{E}h(X_t^L) \leq \limsup_{t \to \infty} \mathbb{E}h(X_t') \tag{11.38}$$

where the precise derivation of (11.38) is left until the end of the proof. To continue, since $X_t' \sim \psi^*$ for all t, the right-hand side is just $\psi^*(h)$. And since $X_t^L \sim \psi \mathbf{M}^t$ for all t, we have proved that

$$\limsup_{t \to \infty}(\psi \mathbf{M}^t)(h) \leq \psi^*(h)$$

By a similar argument applied to X_t^U instead of X_t^L, we obtain

$$\psi^*(h) \leq \liminf_{t \to \infty}(\psi \mathbf{M}^t)(h)$$

It now follows that $\lim_{t\to\infty}(\psi \mathbf{M}^t)(h) = \psi^*(h)$, and since h is an arbitrary element of ibS, the claim in (11.30) is established.

To end the section, let's see how (11.38) is derived. We have

$$\limsup_{t\to\infty} \mathbb{E}h(X_t^L) \leq \limsup_{t\to\infty} \mathbb{E}h(X_t^L)\mathbb{1}\{T^L \leq t\} + \limsup_{t\to\infty} \mathbb{E}h(X_t^L)\mathbb{1}\{T^L > t\}$$

Since h is bounded the last term is zero, and hence

$$\limsup_{t\to\infty} \mathbb{E}h(X_t^L) \leq \limsup_{t\to\infty} \mathbb{E}h(X_t^L)\mathbb{1}\{T^L \leq t\} \leq \limsup_{t\to\infty} \mathbb{E}h(X_t')\mathbb{1}\{T^L \leq t\} = \psi^*(h)$$

where the second inequality is due to (11.37), and the final equality holds because

$$\mathbb{E}h(X_t')\mathbb{1}\{T^L \leq t\} = \mathbb{E}h(X_t') - \mathbb{E}h(X_t')\mathbb{1}\{T^L > t\} = \psi^*(h) - \mathbb{E}h(X_t')\mathbb{1}\{T^L > t\}$$

and boundedness of h gives $\mathbb{E}h(X_t')\mathbb{1}\{T^L > t\} \to 0$ as $t \to \infty$.

11.3.5   Further Stability Theory

In the theory covered so far, we have illustrated how stability problems in unbounded state spaces can be treated using drift conditions. Drift conditions were used in §8.2.3 for existence, uniqueness, and stability in the density case, in §11.2.1 for existence in the general (i.e., measure) case, and in §11.3.3 for stability in the general case under monotonicity. It would be nice to add a stability result suitable for unbounded spaces (i.e., using drift) that requires neither density assumptions nor monotonicity. Let us now address this gap, providing a result for the general case that gives existence, uniqueness, and stability without specifically requiring continuity, monotonicity or densities. While full proofs are beyond the scope of this text, an intuitive explanation based on coupling and the Dobrushin coefficient is provided.

Let P be a stochastic kernel on $S \in \mathscr{B}(\mathbb{R}^n)$. In §11.3.2 we showed how processes on S can be coupled when the Dobrushin coefficient is strictly positive, and how this coupling can be used to prove stability. However, we know that on unbounded state space the Dobrushin coefficient $\alpha(P)$ of P is often zero.[42] The problem is that $\alpha(P)$ is defined as the infimum of the affinities $\gamma(x, x') := (P_x \wedge P_{x'})(S)$ over all (x, x') pairs in $S \times S$. The affinity is close to one when $P(x, dy)$ and $P(x', dy)$ put probability mass in similar areas, and converges to zero as their supports diverge. If S is unbounded, then as the distance between x and x' increases without bound, it is likely that the supports of $P(x, dy)$ and $P(x', dy)$ also diverge from one another, and $\gamma(x, x')$ can be made arbitrarily small. The end result is $\alpha(P) = 0$.

---

[42]Recall the example shown in figure 8.7 on page 202.

If $\alpha(P) = 0$, then the coupling result in §11.3.2 fails. To see this, recall that the proof is based on the bound

$$\|\psi \mathbf{M}^t - \psi' \mathbf{M}^t\| \leq \mathbb{P}\{X_t \neq X_t'\} \tag{11.39}$$

where $(X_t)_{t \geq 0}$ and $(X_t')_{t \geq 0}$ are processes with $X_0 \sim \psi$, $X_0' \sim \psi'$, $X_{t+1} \sim P(X_t, dy)$ and $X_{t+1}' \sim P(X_t', dy)$. To show that $\mathbb{P}\{X_t \neq X_t'\} \to 0$ as $t \to \infty$, we constructed the sequences $(X_t)_{t \geq 0}$ and $(X_t')_{t \geq 0}$ using algorithm 11.1, where the probability of coupling at $t + 1$ is $\gamma(X_t, X_t') \geq \alpha(P)$. This gave the bound $\mathbb{P}\{X_t \neq X_t'\} \leq (1 - \alpha(P))^t$.

Of course, if $\alpha(P) = 0$, then this bound has no bite. However, there is a way to extend our stability result to such cases. The basic idea is as follows: While the infimum of $\gamma(x, x')$ may be zero when taken over all of $S \times S$, it may well be a positive value ϵ when taken over $C \times C$, where C is some bounded subset of S. If such a C exists, then we can modify our strategy by attempting to couple the chains *only when both are in C*. In this case the probability of coupling at $t + 1$ is $\gamma(X_t, X_t') \geq \epsilon$.

This idea is illustrated in algorithm 11.3. (Compare with algorithm 11.1.)

Algorithm 11.3: Coupling with drift

---

draw $X_0 \sim \psi$, $X_0' \sim \psi^*$ and set $t = 0$
while True **do**
   if $X_t = X_t'$ **then**
     | draw $Z \sim P(X_t, dy)$ and set $X_{t+1} = X_{t+1}' = Z$
   else
     if $X_t \in C$ *and* $X_t' \in C$ **then**
       draw U_{t+1} from the uniform distribution on $(0, 1)$
       if $U_{t+1} \leq \gamma(X_t, X_t')$ **then**   // with probability $\gamma(X_t, X_t')$
        | draw $Z \sim \nu(X_t, X_t', dy)$ and set $X_{t+1} = X_{t+1}' = Z$
       else   // with probability $1 - \gamma(X_t, X_t')$
        | draw $X_{t+1} \sim \mu(X_t, X_t', dy)$ and $X_{t+1}' \sim \mu'(X_t, X_t', dy)$
       end
     else
       | draw $X_{t+1} \sim P(X_t, dy)$ and $X_{t+1}' \sim P(X_t', dy)$
     end
   end
   set $t = t + 1$
end

The algorithm can be made to work along the following lines: Suppose that the kernel P and the set C are such that $(X_t)_{t \geq 0}$ and $(X_t')_{t \geq 0}$ return to C infinitely often with probability one. Each time both chains return to C simultaneously, there is an ϵ probability of coupling. From this it can be shown that $\mathbb{P}\{X_t \neq X_t'\} \to 0$ as $t \to \infty$, which implies stability via (11.39).

The formal arguments are not trivial, and rather than attempting such a proof here, we present instead some standard results that can be understood using the same intuition. In particular, we will be looking for (1) a set $C \subset S$ such that the infimum of $\gamma(x, x')$ on $C \times C$ is positive, and (2) some kind of drift condition ensuring the chain returns to C infinitely often. These conditions capture the essence of stability on unbounded state spaces: mixing at the center of the state sufficient to rule out multiple local equilibria, and drift back to the center sufficient to rule out divergence.

Before presenting these standard results, it might be helpful to consider an example for which there exists a set $C \subset S$ such that the infimum of $\gamma(x, x')$ on $C \times C$ is positive:

Example 11.3.24 Consider again the STAR model discussed in §8.2.4, where $Z = S = \mathbb{R}$, and the state evolves according to

$$X_{t+1} = g(X_t) + W_{t+1}, \quad (W_t)_{t \geq 1} \overset{\text{IID}}{\sim} \phi \in D(S) \tag{11.40}$$

The kernel P corresponding to this model is of the form $P(x, dy) = p(x, y)dy$, where $p(x, y) = \phi(y - g(x))$. The affinities are given by

$$\gamma(x, x') = \int p(x, y) \wedge p(x', y)dy = \int \phi(y - g(x)) \wedge \phi(y - g(x'))dy$$

Suppose that ϕ and g are both continuous, and that ϕ is strictly positive on $\mathbb{R}$, in which case

$$S \times S \ni (x, y) \mapsto p(x, y) = \phi(y - g(x)) \in \mathbb{R}$$

is continuous and positive on $S \times S$. It follows that for any compact set C, the infimum of $(x, y) \mapsto p(x, y)$ on $C \times C$ is greater than some $\delta > 0$. But then

$$\gamma(x, x') = \int p(x, y) \wedge p(x', y)dy \geq \int_C p(x, y) \wedge p(x', y)dy \geq \delta\lambda(C) =: \epsilon$$

for any $(x, x') \in C \times C$. If $\lambda(C) > 0$, then $\inf_{(x,x') \in C \times C} \gamma(x, x') \geq \epsilon > 0$.

The set C is called a small set. The traditional definition is as follows:

Definition 11.3.25 Let $\nu \in \mathscr{P}(S)$ and let $\epsilon > 0$. A set $C \in \mathscr{B}(S)$ is called (ν, ϵ)-*small* for P if for all $x \in C$ we have

$$P(x, A) \geq \epsilon\nu(A) \quad (A \in \mathscr{B}(S))$$

C is called *small* for P if it is (ν, ϵ)-*small* for some $\nu \in \mathscr{P}(S)$ and $\epsilon > 0$.

This definition works for us because if C is small, then the infimum of γ on $C \times C$ is strictly positive. In particular, if C is (ν, ϵ)-small, then the infimum is greater than ϵ. The proof is left as an exercise:

Exercise 11.3.26 Show that if C is (ν, ϵ)-small for kernel P, then $\gamma(x, x') := (P_x \wedge P_{x'})(S) \geq \epsilon$ for any x, x' in C.

Exercise 11.3.27 Show that every measurable subset of a small set is small.

Exercise 11.3.28 Show that if $P(x, dy) = p(x, y)dy$ for some density kernel p, then $C \in \mathscr{B}(S)$ is small for P whenever there exists a measurable $g \colon S \to \mathbb{R}_+$ with $\int g(y)dy > 0$ and $p(x, y) \geq g(y)$ for all $x \in C$ and $y \in S$.

Exercise 11.3.29 Consider the stochastic kernel $P(x, dy) = p(x, y)dy = N(ax, 1)$, which corresponds to a linear AR(1) process with normal shocks. Show that every bounded $C \subset \mathbb{R}$ is a small set for this process.

Now let's introduce a suitable drift condition.

Definition 11.3.30 The kernel P satisfies *drift to a small set* if there exists a small set C, a measurable function $v \colon S \to [1, \infty)$ and constants $\lambda \in [0, 1)$ and $L \in \mathbb{R}_+$ such that

$$Mv(x) \leq \lambda v(x) + L\mathbb{1}_C(x) \qquad (x \in S) \tag{11.41}$$

Drift to a small set is not dissimilar to our other drift conditions. In this case, if the current state X_t is at some point $x \notin C$, then since $\lambda < 1$, the next period expectation $\mathbb{E}[v(X_{t+1}) \mid X_t = x]$ of the "Lyapunov function" v is a fraction of the current value $v(x)$. Since $v \geq 1$, this cannot continue indefinitely, and the state tends back toward C.[43]

Sometimes the following condition is easier to check than (11.41).

Lemma 11.3.31 *Suppose that there exists a measurable function* $v \colon S \to \mathbb{R}_+$ *and constants* $\alpha \in [0, 1)$ *and* $\beta \in \mathbb{R}_+$ *such that*

$$Mv(x) \leq \alpha v(x) + \beta \qquad (x \in S) \tag{11.42}$$

If all sublevel sets of v are small, then P satisfies drift to a small set.[44]

Exercise 11.3.32 Prove lemma 11.3.31. Show that we can assume without loss of generality that v in the lemma satisfies $v \geq 1$.[45] Now show that if we pick any $\lambda \in (\alpha, 1)$, set $C := \{x : v(x) \leq \beta/(\lambda - \alpha)\}$ and $L := \beta$, then v, C, λ, and L satisfy the conditions of definition 11.3.30.

---

[43]Condition (11.41) corresponds to (V4) in Meyn and Tweedie (1993, §15.2.2).

[44]Recall that the sublevel sets of v are sets of the form $\{x \in S : v(x) \leq K\}$, $K \in \mathbb{R}_+$.

[45]Show that if the lemma holds for some v, α and β then it also holds for the function $v' := v + 1$ and constants $\alpha' := \alpha$ and $\beta' = \beta + 1$.

Finally, to present the main result, we need two definitions that appear frequently in classical Markov chain theory. The first is aperiodicity:

Definition 11.3.33 A kernel P is called *aperiodic* if P has a (ν, ϵ)-small set C with $\nu(C) > 0$.[46]

Exercise 11.3.34 Continuing exercise 11.3.28, suppose there is a $C \in \mathscr{B}(S)$ and measurable $g: S \to \mathbb{R}_+$ with $\int g(y)dy > 0$ and $p(x,y) \geq g(y)$ for all $x \in C$ and $y \in S$. Show that if $\int_C g(x)dx > 0$, then P is aperiodic.

Definition 11.3.35 Let $\mu \in \mathscr{P}(S)$. The kernel P is called μ-*irreducible* if, for all $x \in S$ and $B \in \mathscr{B}(S)$ with $\mu(B) > 0$, there exists a $t \in \mathbb{N}$ with $P^t(x, B) > 0$. P is called *irreducible* if it is μ-irreducible for some $\mu \in \mathscr{P}(S)$.

Let P be a stochastic kernel on S. Let $\mathbf{M}$ be the corresponding Markov operator, and let $\mathscr{P}(S)$ be endowed with the total variation metric. We can now state the following powerful stability result:

Theorem 11.3.36 *If P is irreducible, aperiodic and satisfies drift to a small set, then the system $(\mathscr{P}(S), \mathbf{M})$ is globally stable with unique stationary distribution $\psi^* \in \mathscr{P}(S)$. Moreover if v is the function in definition 11.3.30 and $h: S \to \mathbb{R}$ is any measurable function satisfying $|h| \leq v$, then*

$$\frac{1}{n} \sum_{t=1}^{n} h(X_t) \to \psi^*(h) \quad \text{with probability one as } n \to \infty \tag{11.43}$$

Finally, if $h^2 \leq v$, then there is a $\sigma \in \mathbb{R}_+$ with

$$\sqrt{n} \left(\frac{1}{n} \sum_{t=1}^{n} h(X_t) - \psi^*(h) \right) \to N(0, \sigma^2) \quad \text{weakly as } n \to \infty \tag{11.44}$$

For a proof of theorem 11.3.36, the reader is referred to Meyn and Tweedie (1993, thms. 16.0.1 and 17.0.1), or Roberts and Rosenthal (2004, thms. 9 and 28). The proofs in the second reference are closer to the coupling intuition provided in algorithm 11.3.

Remark 11.3.37 Under the conditions of theorem 11.3.36 it is known that $\psi^*(v) < \infty$, so $|h| \leq v$ implies $\psi^*(|h|) < \infty$. Actually this last restriction is sufficient for h to satisfy the LLN (11.43).

Remark 11.3.38 Why is aperiodicity required in theorem 11.3.36? To gain some intuition we can refer back to algorithm 11.3. Under the conditions of theorem 11.3.36 we have drift back to a small set, which corresponds to C in algorithm 11.3. Each time

---

[46] Our definition corresponds to what is usually called strong aperiodicity. See Meyn and Tweedie (1993, ch. 5).

the chains both return to C there is an opportunity to couple, and with enough such opportunities the probability of not coupling prior to t converges to zero in t. At issue here is that the chains must not only drift back to C, they must also return to C *at the same time*. With periodic chains this can be problematic because starting at different points in the cycle can lead to infinitely many visits to C by both chains that never coincide.

Let's try to apply theorem 11.3.36 to the STAR model in (11.40). We assume that the function g in (11.40) is continuous and satisfies

$$|g(x)| \leq \alpha|x| + c \qquad (x \in S = \mathbb{R})$$

for some $\alpha < 1$ and $c \in \mathbb{R}_+$, that the density ϕ is continuous and everywhere positive on $\mathbb{R}$, and that $\mathbb{E}|W_1| = \int |z|\phi(z)dz < \infty$.

Exercise 11.3.39 Show that the corresponding kernel P is irreducible on $S = \mathbb{R}$.[47]

Next let's prove that every compact $C \subset S$ is small for P. Since measurable subsets of small sets are themselves small (exercise 11.3.27), we can assume without loss of generality that the Lebesgue measure $\lambda(C)$ of C is strictly positive. (Why?) Now set $\delta := \min\{p(x,y) : (x,y) \in C \times C\}$. The quantity δ is strictly positive. (Why?) Finally, let $g := \delta \mathbb{1}_C$.

Exercise 11.3.40 Show that C and g satisfy the all of the conditions in exercises 11.3.28 and 11.3.34.

From exercise 11.3.40 we conclude that all compact sets are small, and, moreover, that P is aperiodic. To verify theorem 11.3.36, it remains only to check drift to a small set holds. In view of lemma 11.3.31, it is sufficient to exhibit a function $v \colon S \to \mathbb{R}_+$ and constants $\alpha \in [0,1)$, $\beta \in \mathbb{R}_+$ such that all sublevel sets of v are compact and the drift in (11.42) holds. The condition (11.42) has already been shown to hold for $v(x) := |x|$ in §8.2.4. Moreover the sublevel sets of v are clearly compact. The conditions of theorem 11.3.36 are now verified.

11.4   Commentary

The outstanding reference for stability of Markov chains in general state spaces is Meyn and Tweedie (1993). (At the time of writing, a second edition is reportedly in production through Cambridge University Press.) Another good reference on the topic is Hernández-Lerma and Lasserre (2003). Bhattacharya and Majumdar (2007)

---

[47]Hint: Show first that if $x \in S$ and $B \in \mathscr{B}(S)$ has positive Lebesgue measure, then $P(x,B) > 0$. Now let $f \in D(S)$, and let $\mu \in \mathscr{P}(S)$ be defined by $\mu(B) = \int_B f(x)dx$. Show that P is μ-irreducible.

give a general treatment of Markov chains with extensive discussion of stability. For an application of Meyn and Tweedie's ideas to time series models, see Kristensen (2007).

I learned about the Dobrushin coefficient and its role in stability through reading lecture notes of Eric Moulines, and about coupling via Lindvall (1992), Rosenthal (2002), and Roberts and Rosenthal (2004). The idea of using coupling to prove stability of Markov chains is due to the remarkable Wolfgang Doeblin (1938). The link between the Dobrushin coefficient and coupling in §11.3.2 is my own work, extending similar studies by the authors listed above. While I was unable to find this line of argument elsewhere, I certainly doubt that it is new.

The monotonicity-based approach to stability introduced in §11.3.3 is due to Kamihigashi and Stachurski (2008). For other monotonicity-based treatments of stability, see Razin and Yahav (1979), Bhattacharya and Lee (1988), Hopenhayn and Prescott (1992), or Zhang (2007).

In this chapter we gave only a very brief discussion of the central limit theorem for Markov chains. See Meyn and Tweedie (1993) for standard results, and Jones (2004) for a survey of recent theory.

Chapter 12

More Stochastic Dynamic Programming

In this chapter we treat some extensions to the fundamental theory of dynamic programming, as given in chapter 10. In §12.1 we investigate how additional structure can be used to obtain new results concerning the value function and the optimal policy. In §12.2 we show how to modify our earlier optimality results when the reward function is not bounded.

12.1 Monotonicity and Concavity

In many economic models we have more structure at our disposal than continuity and compactness (assumptions 10.1.3–10.1.5, page 231), permitting sharper characterizations of optimal actions and more efficient numerical algorithms. Often this additional structure is in the form of monotonicity or convexity. We begin by discussing monotonicity.

12.1.1 Monotonicity

Let's recall some definitions. Given vectors $x = (x_1, \ldots, x_n)$ and $y = (y_1, \ldots, y_n)$ in $\mathbb{R}^n$, we say that $x \leq y$ if $x_i \leq y_i$ for $1 \leq i \leq n$; and $x < y$ if, in addition, we have $x \neq y$. A function $w \colon \mathbb{R}^n \supset E \to \mathbb{R}$ is called *increasing on E* if, given $x, x' \in E$ with $x \leq x'$, we have $w(x) \leq w(x')$; and strictly increasing if $x < x'$ implies $w(x) < w(x')$. A correspondence Γ from E to any set A is called *increasing on E* if $\Gamma(x) \subset \Gamma(x')$ whenever $x \leq x'$. A set $B \subset E$ is called an *increasing subset of E* if $\mathbb{1}_B$ is an increasing function on E; equivalently, if $x \in B$ and $x' \in E$ with $x \leq x'$ implies $x' \in B$. It is called a *decreasing subset of E* if $x' \in B$ and $x \in E$ with $x \leq x'$ implies $x \in B$. For convex $E \subset \mathbb{R}^n$, a function $w \colon E \to \mathbb{R}$ is called *concave* if,

$$\lambda w(x) + (1 - \lambda)w(y) \leq w(\lambda x + (1 - \lambda)y) \qquad \forall \lambda \in [0, 1] \text{ and } x, y \in E$$

and *strictly concave* if the inequality is strict for all $x \neq y$ and all $\lambda \in (0, 1)$.

In all of this section, $S \in \mathscr{B}(\mathbb{R}^n)$ and $A \in \mathscr{B}(\mathbb{R}^k)$. Furthermore *the set S is assumed to be convex*. We will be working with an arbitrary SDP $(r, F, \Gamma, \phi, \rho)$ that obeys our usual assumptions (page 231). The state space is S, the action space is A, and the shock space is Z. Finally, *ibcS* is the increasing bounded continuous functions on S.

Exercise 12.1.1 Show that *ibcS* is a closed subset of (bcS, d_∞).[1] The same is not true for the *strictly* increasing bounded continuous functions. (Why?)

Our first result gives sufficient conditions for the value function associated with this SDP to be increasing on S.

Theorem 12.1.2 *The value function v^* is increasing on S whenever Γ is increasing on S and, for any $x, x' \in S$ with $x \leq x'$, we have*

1. *$r(x, u) \leq r(x', u)$ for all $u \in \Gamma(x)$, and*

2. *$F(x, u, z) \leq F(x', u, z)$ for all $u \in \Gamma(x), z \in Z$.*

Proof. In the proof of lemma 10.1.20 (page 237) we saw that the Bellman operator T satisfies $T \colon bcS \to bcS$. Since *ibcS* is a closed subset of *bcS* and since v^* is the fixed point of T, we need only show that $T \colon ibcS \to ibcS$. (Recall exercise 4.1.20, page 61.) To do so, take any x and x' in S with $x \leq x'$ and fix $w \in ibcS$. Let σ be w-greedy (definition 10.1.9, page 233) and let $u^* = \sigma(x)$. From $w \in ibcS$ and our hypotheses we obtain

$$Tw(x) = r(x, u^*) + \rho \int w[F(x, u^*, z)]\phi(dz)$$

$$\leq r(x', u^*) + \rho \int w[F(x', u^*, z)]\phi(dz)$$

$$\leq \max_{\Gamma(x')} \left\{ r(x', u) + \rho \int w[F(x', u, z)]\phi(dz) \right\} =: Tw(x')$$

where the second inequality follows from the assumption that Γ is increasing. (Why?) We conclude that $Tw \in ibcS$, and hence so is v^*. $\square$

---

[1]Hint: Recall that if $f_n \to f$ uniformly, then $f_n \to f$ pointwise.

Exercise 12.1.3 Show that if, in addition to the hypotheses of the theorem, $x < x'$ implies $r(x, u) < r(x', u)$, then v^* is strictly increasing.[2]

Exercise 12.1.4 Recall the optimal growth model, with $S = A = \mathbb{R}_+$, $\Gamma(y) = [0, y]$ and $\operatorname{gr} \Gamma = $ all pairs (y, k) with $k \in [0, y]$. The reward function is given on $\operatorname{gr} \Gamma$ by $r(y, k) := U(y - k)$, where $U \colon \mathbb{R}_+ \to \mathbb{R}$. The transition function is $F(y, k, z) = f(k, z)$. The shocks $(W_t)_{t \geq 1}$ are independent and takes values in $Z \subset \mathbb{R}$ according to $\phi \in \mathscr{P}(Z)$. Let the conditions of assumption 10.1.12 be satisfied (see page 235). Show that the value function v^* is increasing whenever U is, and strictly increasing when U is. (Notice that monotonicity of the production function plays no role.)

Next let's consider parametric monotonicity. The question is as follows: Suppose that an objective function has maximizer u. If one now varies a given parameter in the objective function and maximizes again, a new maximizer u' is determined. Does the maximizer always increase when the parameter increases? The connection to dynamic programming comes when the state variable is taken to be the parameter and the corresponding optimal action is the maximizer. We wish to know when the optimal action is monotone in the state. Monotone policies are of interest not only for their economic interpretation but also because they can speed up algorithms for approximating optimal policies.

Definition 12.1.5 Let Γ and $\operatorname{gr} \Gamma$ be as above. A function $g \colon \operatorname{gr} \Gamma \to \mathbb{R}$ satisfies *increasing differences* on $\operatorname{gr} \Gamma$ if, whenever $x, x' \in S$ with $x \leq x'$ and $u, u' \in \Gamma(x) \cap \Gamma(x')$ with $u \leq u'$, we have

$$g(x, u') - g(x, u) \leq g(x', u') - g(x', u) \tag{12.1}$$

The function is said to satisfy *strictly increasing differences* on $\operatorname{gr} \Gamma$ if the inequality (12.1) is strict whenever $x < x'$ and $u < u'$.

Intuitively, the impact of increasing the argument from u to u' has more effect on g when the parameter x is larger. The requirement $u, u' \in \Gamma(x) \cap \Gamma(x')$ ensures that g is properly defined at all the points in (12.1).

Example 12.1.6 Consider the optimal growth model of exercise 12.1.4. If U is strictly concave, then $r(y, k) = U(y - k)$ has strictly increasing differences on $\operatorname{gr} \Gamma$. To see this, pick any $y, y', k, k' \in \mathbb{R}_+$ with $y < y'$, $k < k'$ and $k, k' \in \Gamma(y) \cap \Gamma(y') = [0, y]$. We are claiming that

$$U(y - k') - U(y - k) < U(y' - k') - U(y' - k)$$

or alternatively,

$$U(y - k') + U(y' - k) < U(y' - k') + U(y - k) \tag{12.2}$$

---

[2]Hint: Show that the result $Tw(x) \leq Tw(x')$ in the proof of the theorem can be strengthened to $Tw(x) < Tw(x')$. Hence T sends $ibcS$ into the *strictly* increasing bounded continuous functions. Since $v^* \in ibcS$ it follows that Tv^* is strictly increasing. But then v^* is strictly increasing. (Why?!)

It is left as an exercise for the reader to show that strict concavity of U implies that for each $\lambda \in (0,1)$ and each pair x, x' with $0 \leq x < x'$ we have

$$U(x) + U(x') < U(\lambda x + (1 - \lambda)x') + U(\lambda x' + (1 - \lambda)x)$$

This yields (12.2) when

$$\lambda := \frac{k' - k}{y' - y + k' - k}, \quad x := y - k', \quad x' := y' - k$$

Exercise 12.1.7 Show that if $g \colon \operatorname{gr}\Gamma \to \mathbb{R}$ satisfies strictly increasing differences on $\operatorname{gr}\Gamma$ and $h \colon \operatorname{gr}\Gamma \to \mathbb{R}$ satisfies increasing differences on $\operatorname{gr}\Gamma$, then $g + h$ satisfies strictly increasing differences on $\operatorname{gr}\Gamma$.

We now present a parametric monotonicity result in the case where the action space A is one-dimensional. Although not as general as some other results, it turns out to be useful, and the conditions in the theorem are relatively easy to check in applications. In the statement of the theorem we are assuming that $A \subset \mathbb{R}$, $g \colon \operatorname{gr}\Gamma \to \mathbb{R}$, and $\operatorname{argmax}_{u \in \Gamma(x)} g(x, u)$ is nonempty for each $x \in S$.

Theorem 12.1.8 *Suppose that g satisfies strictly increasing differences on $\operatorname{gr}\Gamma$, that Γ is increasing on S, and that $\Gamma(x)$ is a decreasing subset of A for every $x \in S$. Let $x, x' \in S$ with $x \leq x'$. If u is a maximizer of $a \mapsto g(x, a)$ on $\Gamma(x)$ and u' is a maximizer of $a \mapsto g(x', a)$ on $\Gamma(x')$, then $u \leq u'$.*

Proof. When $x = x'$ the result is trivial, so take $x < x'$. Let u and u' be as in the statement of the theorem. Suppose to the contrary that $u > u'$. Since Γ is increasing, we have $\Gamma(x) \subset \Gamma(x')$, and hence both u and u' are in $\Gamma(x')$. Also, since $u' < u \in \Gamma(x)$, and since $\Gamma(x)$ is a decreasing set, both u and u' are in $\Gamma(x)$. It then follows from strictly increasing differences that

$$g(x', u) - g(x', u') > g(x, u) - g(x, u')$$

However, from $u \in \Gamma(x')$, $u' \in \Gamma(x)$ and the definition of maxima,

$$g(x', u') - g(x', u) \geq 0 \geq g(x, u') - g(x, u)$$

$$\therefore \quad g(x', u) - g(x', u') \leq g(x, u) - g(x, u')$$

Contradiction. $\square$

We can now derive a general parametric monotonicity result for SDPs.

Corollary 12.1.9 *Let* $(r, F, \Gamma, \phi, \rho)$ *define an SDP satisfying the conditions in theorem 12.1.2. If in addition* $\Gamma(x)$ *is a decreasing subset of* A *for every* $x \in S$, r *satisfies strictly increasing differences on* $\operatorname{gr} \Gamma$, *and,* $\forall w \in ibcS$,

$$\operatorname{gr} \Gamma \ni (x, u) \mapsto \int w[F(u, x, z)]\phi(dz) \in \mathbb{R}$$

satisfies increasing differences on $\operatorname{gr} \Gamma$, *then every optimal policy is monotone increasing on S.*

Proof. Let v^* be the value function for this SDP, and set

$$g(x, u) := r(x, u) + \rho \int v^*[F(x, u, z)]\phi(dz)$$

on $\operatorname{gr} \Gamma$. If a policy is optimal, then it maximizes $a \mapsto g(x, a)$ over $\Gamma(x)$ for each $x \in S$. Hence we need only verify the conditions of theorem 12.1.8 for our choice of g and Γ. The only nontrivial assertion is that g satisfies strictly increasing differences on $\operatorname{gr} \Gamma$. This follows from exercise 12.1.7 and the fact that $v^* \in ibcS$ (see theorem 12.1.2). $\qquad \square$

From corollary 12.1.9 it can be shown that investment in the optimal growth model is monotone increasing whenever U is increasing and strictly concave. In particular, no shape restrictions on f are necessary.

Exercise 12.1.10 Verify this claim.

12.1.2 Concavity and Differentiability

Next we consider the role of concavity and differentiability in dynamic programs. This topic leads naturally to the Euler equation, which holds at the optimal policy whenever the solution is interior and all primitives are sufficiently smooth. Although we focus on the growth model, many other models in economics have Euler equations, and they can be derived using steps similar to those shown below. Detailed proofs are provided, although most have been consigned to the appendix to this chapter.

To begin, let $\mathscr{C}ibcS$ denote the set of all concave functions in $ibcS$, where the latter is endowed as usual with the supremum norm d_∞.

Exercise 12.1.11 Show that the set $\mathscr{C}ibcS$ is a closed subset of $ibcS$.

Our first result gives conditions under which the value function v^* is concave. Here v^* corresponds to our canonical SDP $(r, F, \Gamma, \phi, \rho)$ that obeys the standard assumptions (page 231).

Theorem 12.1.12 *Let the conditions of theorem 12.1.2 hold. If, in addition,*

1. gr Γ *is convex,*

2. *r is concave on* gr Γ, *and*

3. $(x, u) \mapsto F(x, u, z)$ *is concave on* gr Γ *for all* $z \in Z$,

then the value function v^* *is concave. In particular, we have* $v^* \in \mathscr{C}ibcS$.

Proof. By theorem 12.1.2, $T : ibcS \to ibcS$ and $v^* \in ibcS$. We wish to show additionally that $v^* \in \mathscr{C}ibcS$. Analogous to the proof of theorem 12.1.2, since $\mathscr{C}ibcS$ is a closed subset of $ibcS$, it suffices to show that T maps $\mathscr{C}ibcS$ into itself. So let $w \in \mathscr{C}ibcS$. Since $Tw \in ibcS$, we need only show that Tw is also concave. Let $x, x' \in S$, and let $\lambda \in [0, 1]$. Set $x'' := \lambda x + (1 - \lambda)x'$. Let σ be a w-greedy policy, let $u := \sigma(x)$, and let $u' := \sigma(x')$. Define $u'' := \lambda u + (1 - \lambda)u'$. Condition 1 implies that $u'' \in \Gamma(x'')$, and hence

$$Tw(x'') \geq r(x'', u'') + \rho \int w[F(x'', u'', z)]\phi(dz)$$

Consider the two terms on the right-hand side. By condition 2,

$$r(x'', u'') \geq \lambda r(x, u) + (1 - \lambda)r(x', u')$$

By condition 3 and $w \in \mathscr{C}ibcS$,

$$\int w[F(x'', u'', z)]\phi(dz) \geq \int w[\lambda F(x, u, z) + (1 - \lambda)F(x', u', z)]\phi(dz)$$

$$\geq \lambda \int w[F(x, u, z)]\phi(dz) + (1 - \lambda)\int w[F(x', u', z)]\phi(dz)$$

$$\therefore \quad Tw(x'') \geq \lambda Tw(x) + (1 - \lambda)Tw(x')$$

Hence Tw is concave on S, $Tw \in \mathscr{C}ibcS$, and v^* is concave. $\square$

Exercise 12.1.13 Show that if, in addition to the hypotheses of the theorem, r is strictly concave on gr Γ, then v^* is strictly concave.[3]

Exercise 12.1.14 Consider again the stochastic optimal growth model (see exercise 12.1.4 on page 297 for notation and assumption 10.1.12 on page 235 for our assumptions on the primitives). Suppose that the utility function U is strictly increasing and strictly concave, in which case v^* is strictly increasing (exercise 12.1.4) and any optimal investment policy is increasing (exercise 12.1.10). Show that if, in addition, $k \mapsto f(k, z)$ is concave on $\mathbb{R}_+$ for each fixed $z \in Z$, then v^* is also strictly concave.

Under the present assumptions one can show that for the optimal growth model the optimal policy is unique. Uniqueness in turn implies continuity. The details are left for you:

---

[3]Hint: The argument is similar to that of exercise 12.1.3.

Exercise 12.1.15 Let $[a, b] \subset \mathbb{R}$, where $a < b$, and let $g \colon [a, b] \to \mathbb{R}$. Show that if g is strictly concave, then g has at most one maximizer on $[a, b]$. Show that under the conditions of exercise 12.1.14, there is one and only one optimal policy for the growth model. Show that in addition it is continuous everywhere on S.[4]

Now let's turn to the Euler equation. First we need to strengthen our assumptions; in particular, we need to ensure that our primitives are smooth.

Assumption 12.1.16 For each $z \in Z$, the function $k \mapsto f(k, z)$ is concave, increasing, and differentiable, while $z \mapsto f(k, z)$ is Borel measurable for each $k \in \mathbb{R}_+$. The utility function U is bounded, strictly increasing, strictly concave and differentiable. Moreover

$$\lim_{k \downarrow 0} f'(k, z) > 0 \quad \forall z \in Z, \quad \text{and} \quad \lim_{c \downarrow 0} U'(c) = \infty$$

Here and below, $f'(k, z)$ denotes the partial derivative of f with respect to k. Since U is bounded we can and do assume that $U(0) = 0$.[5]

Under the conditions of assumption 12.1.16, we know that the value function is strictly concave and strictly increasing, while the optimal policy is unique, increasing, and continuous. More can be said. A preliminary result is as follows.

Proposition 12.1.17 *Let assumption 12.1.16 hold, and let $w \in \mathscr{C}ibcS$. If σ is w-greedy, then $\sigma(y) < y$ for every $y > 0$.*

The proof can be found in the appendix to this chapter. We are now ready to state a major differentiability result.

Proposition 12.1.18 *Let $w \in \mathscr{C}ibcS$ and let σ be w-greedy. If assumption 12.1.16 holds, then Tw is differentiable at every $y \in (0, \infty)$, and moreover*

$$(Tw)'(y) = U'(y - \sigma(y)) \qquad (y > 0)$$

Concavity plays a key role in the proof, which can be found in the appendix to the chapter.[6]

Corollary 12.1.19 *Let σ be the optimal policy. If assumption 12.1.16 holds, then v^* is differentiable and $(v^*)'(y) = U'(y - \sigma(y))$ for all $y > 0$.*

That corollary 12.1.19 follows from proposition 12.1.18 is left as an exercise.

---

[4]Hint: Regarding continuity, see Berge's theorem (page 340).

[5]Adding a constant to an objective function (in this case the function $\sigma \mapsto v_\sigma(y)$) affects the maximum but not the maximizer.

[6]The argument is one of the class of so-called "envelope theorem" results. There is no one envelope theorem that covers every case of interest, so it is worth going over the proof to get a feel for how the bits and pieces fit together.

Exercise 12.1.20 Show using corollary 12.1.19 that optimal consumption is strictly increasing in income.

There is another approach to corollary 12.1.19, which uses the following lemma from convex analysis.

Lemma 12.1.21 *If $g\colon \mathbb{R}_+ \to \mathbb{R}$ is concave, and on some neighborhood N of $y_0 \in (0, \infty)$ there is a differentiable concave function $h\colon N \to \mathbb{R}$ with $h(y_0) = g(y_0)$ and $h \leq g$ on N, then g is differentiable at y_0 and $g'(y_0) = h'(y_0)$.*

Exercise 12.1.22 Prove proposition 12.1.18 using lemma 12.1.21.

We have been working toward a derivation of the Euler (in)equality. In our statement of the result, $\sigma := \sigma^*$ is the optimal policy and $c(y) := y - \sigma(y)$ is optimal consumption.

Proposition 12.1.23 *Let $y > 0$. Under assumption 12.1.16, we have*

$$U' \circ c(y) \geq \rho \int U' \circ c[f(\sigma(y), z)] f'(\sigma(y), z) \phi(dz) \qquad (y > 0) \qquad (12.3)$$

When is the Euler inequality an equality? Here is one answer:

Proposition 12.1.24 *Under the additional assumption $f(0, z) = 0$ for all $z \in Z$, we have $\sigma(y) > 0$ for all $y > 0$, and the Euler inequality always holds with equality. On the other hand, if the inequality is strict at some $y > 0$, then $\sigma(y) = 0$.*

The proofs of these propositions are in the appendix to this chapter.

12.1.3   Optimal Growth Dynamics

Next we consider dynamics of the growth model when agents follow the optimal policy. We would like to know whether the system is globally stable, and, in addition, whether the resulting stationary distribution is nontrivial in the sense that it is not concentrated on zero. The problem is not dissimilar to that for the Solow–Swan model treated in §11.3.4. However, the fact that the savings rate is endogenous and nonconstant means that we will have to work a little harder. In particular, we need to extract any necessary information about savings from the Euler equation.

Throughout this section we will suppose that the conditions of assumption 12.1.16 all hold, and moreover that $f(0, z) = 0$ for all $z \in Z$. Together, these conditions, proposition 12.1.17, and proposition 12.1.24 give us interiority of the optimal policy and the Euler equality

$$U' \circ c(y) = \rho \int U' \circ c[f(\sigma(y), z)] f'(\sigma(y), z) \phi(dz) \qquad (y > 0)$$

We study the process $(y_t)_{t\geq 0}$ generated by the optimal law of motion

$$y_{t+1} = f(\sigma(y_t), W_{t+1}) \quad (W_t)_{t\geq 1} \overset{\text{IID}}{\sim} \phi \in \mathscr{P}(Z) \tag{12.4}$$

For our state space S we will use $(0, \infty)$ rather than $\mathbb{R}_+$. The reason is that when $S = \mathbb{R}_+$ the degenerate measure $\delta_0 \in \mathscr{P}(S)$ is stationary for (12.4). Hence any proof of existence based on a result such as the Krylov–Bogolubov theorem is entirely redundant. Moreover global convergence to a nontrivial stationary distribution is impossible because δ_0 will never converge to such a distribution. Hence global stability never holds. Third, if we take $S := (0, \infty)$, then any stationary distribution we can obtain must automatically be nontrivial.

To permit $(0, \infty)$ to be the state space, we require that $f(k, z) > 0$ whenever $k > 0$ and $z \in Z$. (For example, if $f(k, z) = k^{\alpha} z$ and $Z = (0, \infty)$ then this assumption holds.) Observe that since $\sigma(y) > 0$ for all $y \in S = (0, \infty)$, we then have $f(\sigma(y), z) \in S$ for all $y \in S$ and $z \in Z$. Hence S is a valid state space for the model. Observe also that if we permit $f(k, z) = 0$ independent of k for z in a subset of Z with ϕ-measure $\epsilon > 0$, then $\mathbb{P}\{y_t \neq 0\} \leq (1 - \epsilon)^t$ for all t, and $y_t \to 0$ in probability. (Why?) Under such conditions a nontrivial steady state cannot be supported.

To keep our assumptions clear let's now state them formally.

Assumption 12.1.25 All the conditions of assumption 12.1.16 hold. Moreover for any $z \in Z$, $f(k, z) = 0$ if, and only if, $k = 0$.

Let us begin by considering existence of a (nontrivial) stationary distribution. We will use the Krylov–Bogolubov theorem; in particular, corollary 11.2.10 on page 260. The corollary requires that $y \mapsto f(\sigma(y), z)$ is continuous on S for each $z \in Z$, and that there exists a norm-like function w on S and nonnegative constants α and β with $\alpha < 1$ and

$$\mathbf{M}w(y) = \int w[f(\sigma(y), z)]\phi(dz) \leq \alpha w(y) + \beta \quad \forall y \in S \tag{12.5}$$

Since σ is continuous on S (see exercise 12.1.15 on page 301), $y \mapsto f(\sigma(y), z)$ is continuous on S for each $z \in Z$. Thus it remains only to show that there exists a norm-like function w on S and nonnegative constants α and β with $\alpha < 1$ such that (12.5) holds.

In this connection recall from lemma 8.2.17 on page 209 that $w \colon S \to \mathbb{R}_+$ is norm-like on S if and only if $\lim_{x \to 0} w(x) = \lim_{x \to \infty} w(x) = \infty$. So suppose that we have two nonnegative real-valued functions w_1 and w_2 on S with the properties $\lim_{x \to 0} w_1(x) = \infty$ and $\lim_{x \to \infty} w_2(x) = \infty$. Then the sum $w := w_1 + w_2$ is norm-like on S. (Why?) If, in addition,

$$\mathbf{M}w_1 \leq \alpha_1 w_1 + \beta_1 \text{ and } \mathbf{M}w_2 \leq \alpha_2 w_2 + \beta_2 \text{ pointwise on } S \tag{12.6}$$

for some $\alpha_1, \alpha_2, \beta_1, \beta_2$ with $\alpha_i < 1$ and $\beta_i < \infty$, then $w = w_1 + w_2$ satisfies (12.5), as the next exercise asks you to confirm.

Exercise 12.1.26 Using linearity of **M** and assuming (12.6), show that w satisfies (12.5) for $\alpha := \max\{\alpha_1, \alpha_2\}$ and $\beta := \beta_1 + \beta_2$.

The advantage of decomposing w in this way stems from the fact that we must confront two rather separate problems. We are trying to show that at least one trajectory of distributions is tight, keeping almost all of its mass on a compact $K \subset S$. A typical compact subset of S is a closed interval $[a, b]$ with $0 < a < b < \infty$. Thus we require positive a such that almost all probability mass is above a, and finite b such that almost all mass is less than b. In other words, income must neither collapse toward zero nor drift out to infinity. The existence of a w_1 with $\lim_{x \to 0} w_1(x) = \infty$ and $\mathbf{M}w_1 \leq \alpha_1 w_1 + \beta_1$ prevents drift to zero, and requires that the agent has sufficient incentives to invest at low income levels. The existence of a w_2 with $\lim_{x \to \infty} w_2(x) = \infty$ and $\mathbf{M}w_2 \leq \alpha_2 w_2 + \beta_2$ prevents drift to infinity, and requires sufficiently diminishing returns.

Let's start with the first (and hardest) problem, that is, guaranteeing that the agent has sufficient incentives to invest at low income levels. We need some kind of Inada condition on marginal returns to investment. In the deterministic case (i.e., $f(k, z) = f(k)$) a well-known condition is that $\lim_{k \downarrow 0} \rho f'(k) > 1$, or $\lim_{k \downarrow 0} 1/\rho f'(k) < 1$. This motivates the following assumption:

Assumption 12.1.27 Together, ϕ, ρ and f jointly satisfy

$$\lim_{k \downarrow 0} \int \frac{1}{\rho f'(k, z)} \phi(dz) < 1$$

With this assumption we can obtain a function w_1 with the desired properties, as shown in the next lemma. (The proof is straightforward but technical, and can be found in the appendix of this chapter.)

Lemma 12.1.28 *For $w_1 := (U' \circ c)^{1/2}$ there exist positive constants $\alpha_1 < 1$ and $\beta_1 < \infty$ such that $\mathbf{M}w_1 \leq \alpha_1 w_1 + \beta_1$ pointwise on S.*

Now let's turn to the second problem, which involves bounding probability mass away from infinity via diminishing returns. We assume that

Assumption 12.1.29 There exists constants $a \in [0, 1)$ and $b \in \mathbb{R}_+$ such that

$$\int f(k, z)\phi(dz) \leq ak + b \qquad \forall k \in S \tag{12.7}$$

This is a slightly nonstandard and relatively weak diminishing returns assumption. Standard assumptions forcing marginal returns to zero at infinity can be shown to imply assumption 12.1.29.

We now establish the complementary result for a suitable function w_2.

Lemma 12.1.30 *For $w_2(y) := y$ there exist positive constants $\alpha_2 < 1$ and $\beta_2 < \infty$ such that $Mw_2 \leq \alpha_2 w_2 + \beta_2$ pointwise on S.*

Exercise 12.1.31 Prove lemma 12.1.30 using assumption 12.1.29.

Since $\lim_{x \to 0} w_1(x) = \infty$ and $\lim_{x \to \infty} w_2(x) = \infty$, we have proved the following result:

Proposition 12.1.32 *Under assumptions 12.1.25–12.1.29 there exists a norm-like function w on S and nonnegative constants α and β with $\alpha < 1$ and $\beta < \infty$ such that $Mw \leq \alpha w + \beta$ pointwise on S. As a result the optimal income process (12.4) has at least one nontrivial stationary distribution $\psi^* \in \mathscr{P}(S)$.*

Having established existence, let us now consider the issue of global stability. This property can be obtained quite easily if the shocks are unbounded (for example, multiplicative, lognormal shocks). If the shocks are bounded the proofs are more fiddly, and interested readers should consult the commentary at the end of this chapter.[7]

Assumption 12.1.33 For each $k > 0$ and each $c \in S$ both $\mathbb{P}\{f(k,W) \geq c\}$ and $\mathbb{P}\{f(k,W) \leq c\}$ are strictly positive.

Proposition 12.1.34 *If, in addition to the conditions of proposition 12.1.32, assumption 12.1.33 holds, then the optimal income process is globally stable.*

Proof. We will show that the function w in proposition 12.1.32 is also order norm-like (see definition 11.3.16 on page 282). Since $y \mapsto f(\sigma(y), z)$ is monotone increasing this is sufficient for the proof (see theorems 11.3.14 and 11.3.19 in §11.3.3). As a first step, let's establish that all intervals $[a, b] \subset S$ are order inducing for the growth model. To do so, pick any $a \leq b$. Fix any $c \in S$. In view of assumption 12.1.33,

$$\forall y \in [a,b], \quad \mathbb{P}\{f(\sigma(y), W_{t+1}) \geq c\} \geq \mathbb{P}\{f(\sigma(a), W_{t+1}) \geq c\} > 0$$

$$\therefore \quad \inf_{a \leq y \leq b} P(y, [c, \infty)) > 0$$

A similar argument shows that $\inf_{a \leq y \leq b} P(y, (0, c]) > 0$. Therefore $[a, b]$ is order inducing. Now since any sublevel set of w is contained in a closed interval $[a, b] \subset S$ (why?), and since subsets of order inducing sets are order inducing, it follows that every sublevel set of w is order inducing. Hence w is order norm-like, as was to be shown. $\qquad \square$

---

[7]Don't be afraid of assuming unbounded shocks—this is just a modeling assumption that approximates reality. For example, the normal distribution is often used to model human height, but no one is claiming that a 20 meter giant is going to be born.

12.2   Unbounded Rewards

One weakness of the dynamic programming theory provided in chapter 10 is that the reward function is must be bounded. This constraint is violated in many applications. The problem of a potentially unbounded reward function can sometimes be rectified by compactifying the state space so that the (necessarily continuous) reward function is automatically bounded on the state (despite perhaps being unbounded on a larger domain). In other situations such tricks do not work, or are ultimately unsatisfying in terms of the model they imply.

Unfortunately, there is no really general theory of dynamic programming with unbounded rewards. Different models are tackled in different ways, which is time-consuming and intellectually unrewarding. Below we treat perhaps the most general method available, for programs with reward and value functions that are bounded when "weighted" by some function κ. We travel to the land of weighted supremum norms, finding an elegant technique and the ability to treat quite a large class of models.

Should you seek to use this theory for a given application, you will quickly discover that while the basic ideas are straightforward, the problem of choosing a suitable weighting function can be quite tricky. We give some indication of how to go about this using our benchmark example: the optimal growth model.

12.2.1   Weighted Supremum Norms

To begin, let κ be a function from S to $\mathbb{R}$ such that $\kappa \geq 1$. For any other $v \colon S \to \mathbb{R}$, define the κ-weighted supremum norm

$$\|v\|_\kappa := \sup_{x \in S} \frac{|v(x)|}{\kappa(x)} = \left\| \frac{v}{\kappa} \right\|_\infty \tag{12.8}$$

Definition 12.2.1 Let $b_\kappa S$ be the set of all $v \colon S \to \mathbb{R}$ such that $\|v\|_\kappa < \infty$. We refer to these functions as the κ-*bounded* functions on S. On $b_\kappa S$ define the metric

$$d_\kappa(v, w) := \|v - w\|_\kappa = \|v/\kappa - w/\kappa\|_\infty$$

Define also $b_\kappa \mathscr{B}(S) := b_\kappa S \cap m\mathscr{B}(S)$ and $b_\kappa cS := b_\kappa S \cap cS$. In other words, $b_\kappa \mathscr{B}(S)$ is the κ-bounded functions on S that are also (Borel) measurable, and $b_\kappa cS$ is the κ-bounded functions on S that are also continuous.

Exercise 12.2.2 Show that $v \in b_\kappa S$ if and only if $v/\kappa \in bS$.

Exercise 12.2.3 Show that $b\mathscr{B}(S) \subset b_\kappa \mathscr{B}(S)$ and $bcS \subset b_\kappa cS$.

Exercise 12.2.4 Confirm that $(b_\kappa S, d_\kappa)$ is a metric space.

It is a convenient fact that d_κ-convergence implies pointwise convergence. Precisely, if (w_n) is a sequence in $b_\kappa S$ and $d_\kappa(w_n, w) \to 0$ for some $w \in b_\kappa S$, then $w_n(x) \to w(x)$ for every $x \in S$. To see this, pick any $x \in S$. We have

$$|w_n(x)/\kappa(x) - w(x)/\kappa(x)| \leq \|w_n - w\|_\kappa \to 0$$

$$\therefore \quad |w_n(x) - w(x)| \leq \|w_n - w\|_\kappa \kappa(x) \to 0$$

The next lemma states that the usual pointwise ordering on $b_\kappa S$ is "closed" with respect to the d_κ metric. The proof is an exercise.

Lemma 12.2.5 *If (w_n) is a d_κ-convergent sequence in $b_\kappa S$ with $w_n \leq w \in b_\kappa S$ for all $n \in \mathbb{N}$, then $\lim w_n \leq w$.*

The space $(b_\kappa S, d_\kappa)$ and its closed subspaces would not be of much use to us should they fail to be complete. Fortunately all the spaces of κ-bounded functions are complete under reasonable assumptions.

Theorem 12.2.6 *The space $(b_\kappa S, d_\kappa)$ is a complete metric space.*

Proof. Let (v_n) be a Cauchy sequence in $(b_\kappa S, d_\kappa)$. It is left to the reader to show that (v_n/κ) is then Cauchy in (bS, d_∞). Since the latter space is complete, there exists some function $\hat{v} \in bS$ with $\|v_n/\kappa - \hat{v}\|_\infty \to 0$. We claim that $\hat{v} \cdot \kappa \in b_\kappa S$ and $\|v_n - \hat{v} \cdot \kappa\|_\kappa \to 0$, in which case the completeness of $(b_\kappa S, d_\kappa)$ is established. That $\hat{v}\kappa \in b_\kappa S$ follows from boundedness of $\hat{v}$. Moreover,

$$\|v_n - \hat{v}\kappa\|_\kappa = \|v_n/\kappa - (\hat{v}\kappa)/\kappa\|_\infty = \|v_n/\kappa - \hat{v}\|_\infty \to 0 \qquad (n \to \infty)$$

The completeness of $(b_\kappa S, d_\kappa)$ is now verified. $\qquad\qquad\square$

Exercise 12.2.7 Show that if κ is Borel measurable, then $b_\kappa \mathscr{B}(S)$ is a closed subset of $(b_\kappa S, d_\kappa)$, and that if κ is continuous, then $b_\kappa cS$ is a closed subset of $(b_\kappa S, d_\kappa)$.[8] Now prove the following theorem:

Theorem 12.2.8 *If κ is measurable, then $(b_\kappa \mathscr{B}(S), d_\kappa)$ is complete. If κ is continuous, then $(b_\kappa cS, d_\kappa)$ is complete.*

The following result is a useful extension of Blackwell's sufficient condition, which can be used to establish that a given operator is a uniform contraction on $b_\kappa S$.

Theorem 12.2.9 *Let $T: b_\kappa S \to b_\kappa S$ be a monotone operator, in the sense that $v \leq v'$ implies $Tv \leq Tv'$. If, in addition, there is a $\lambda \in [0, 1)$ such that*

$$T(v + a\kappa) \leq Tv + \lambda a\kappa \quad \text{for all } v \in b_\kappa S \text{ and } a \in \mathbb{R}_+ \tag{12.9}$$

then T is uniformly contracting on $(b_\kappa S, d_\kappa)$ with modulus λ.

---

[8]Hint: Pointwise limits of measurable functions are measurable. Uniform limits of continuous functions are continuous.

Exercise 12.2.10 Prove this result by modifying the proof of theorem 6.3.8 (page 344) appropriately.

12.2.2   Results and Applications

Let S, A, r, F, Γ, and ρ again define an SDP, just as in §10.1. Let gr Γ have its previous definition. However, instead of assumptions 10.1.3–10.1.5 on page 231, we assume the following:

Assumption 12.2.11 The reward function r is continuous on gr Γ.

Assumption 12.2.12 $\Gamma\colon S \to \mathscr{B}(A)$ is continuous and compact valued.

Assumption 12.2.13 gr $\Gamma \ni (x,u) \mapsto F(x,u,z) \in S$ is continuous for all $z \in Z$.

The only difference so far is that r is not required to be bounded. For our last assumption we replace boundedness of r by

Assumption 12.2.14 There exists a continuous function $\kappa\colon S \to [1,\infty)$ and constants $R \in \mathbb{R}_+$ and $\beta \in [1,1/\rho)$ satisfying the conditions

$$\sup_{u \in \Gamma(x)} |r(x,u)| \leq R\kappa(x) \qquad \forall\, x \in S \tag{12.10}$$

$$\sup_{u \in \Gamma(x)} \int \kappa[F(x,u,z)]\phi(dz) \leq \beta\kappa(x) \qquad \forall\, x \in S \tag{12.11}$$

In addition the map $(x,u) \mapsto \int \kappa[F(x,u,z)]\phi(dz)$ is continuous on gr Γ.

Remark 12.2.15 Actually it is sufficient to find a continuous *nonnegative* function κ satisfying the conditions of assumption 12.2.14. The reason is that if κ is such a function, then $\hat\kappa := \kappa + 1$ is a continuous function that is greater than 1 and satisfies the conditions of the assumption with the same constants R and β. You may want to check this claim as an exercise.

In applications the difficulty is in constructing the required function κ. The following example illustrates how this might be done:

Example 12.2.16 Consider again the stochastic optimal growth model, this time satisfying all of the conditions in assumption 10.1.12 (page 235) apart from boundedness of U. Instead U is required to be nonnegative. In addition we assume that

$$\kappa(y) := \sum_{t=0}^{\infty} \delta^t \mathbb{E}U(\hat y_t) < \infty \qquad (y \in S) \tag{12.12}$$

Here δ is a parameter satisfying $\rho < \delta < 1$, and $(\hat{y}_t)_{t\geq 0}$ is defined by

$$\hat{y}_{t+1} = f(\hat{y}_t, W_{t+1}) \quad \text{and} \quad \hat{y}_0 = y \tag{12.13}$$

The process $(\hat{y}_t)$ is an upper bound for income under the set of feasible policies. It is the path for income when consumption is zero in each period.

We claim that the function κ in (12.12) satisfies all the conditions of assumption 12.2.14 for the optimal growth model.[9] In making the argument, it is useful to define $\mathbf{N}$ to be the Markov operator corresponding to (12.13). Hopefully it is clear to you that for this operator we have $\mathbb{E}U(\hat{y}_t) = \mathbf{N}^t U(y)$ for each $t \geq 0$, so κ can be expressed as $\sum_t \delta^t \mathbf{N}^t U$.

Lemma 12.2.17 *The function κ in (12.12) is continuous and increasing on $\mathbb{R}_+$.*

The proof can be found in the appendix to this chapter. Let us show instead that the conditions of assumption 12.2.14 are satisfied, beginning with (12.10). In the optimal growth model $r(x, u) = U(y - k)$ and $\Gamma(x) = \Gamma(y) = [0, y]$. Since U is increasing and nonnegative, we have

$$\sup_{u\in\Gamma(x)} |r(x,u)| = \sup_{0\leq k\leq y} U(y - k) \leq U(y) \leq \kappa(y)$$

Thus (12.10) holds with $R = 1$. Now let's check that (12.11) also holds. Observe that

$$\sup_{0\leq k\leq y} \int \kappa(f(k,z))\phi(dz) \leq \int \kappa(f(y,z))\phi(dz) = \mathbf{N}\kappa(y)$$

where we are using the fact that κ is increasing on S. Now

$$\mathbf{N}\kappa = \mathbf{N}\sum_{t=0}^{\infty} \delta^t \mathbf{N}^t U = \sum_{t=0}^{\infty} \delta^t \mathbf{N}^{t+1}U = (1/\delta)\sum_{t=0}^{\infty} \delta^{t+1}\mathbf{N}^{t+1}U \leq (1/\delta)\kappa$$

$$\therefore \quad \sup_{0\leq k\leq y} \int \kappa(f(k,z))\phi(dz) \leq \beta\kappa(y), \quad \text{where } \beta := 1/\delta$$

Since δ was chosen to satisfy $\rho < \delta < 1$, we have $1 \leq \beta < 1/\rho$ as desired.

Finally, to complete the verification of assumption 12.2.14, we need to check that $(x, u) \mapsto \int \kappa[F(x, u, z)]\phi(dz)$ is continuous, which in the present case amounts to showing that if (y_n, k_n) is a sequence with $0 \leq k_n \leq y_n$ and converging to (y, k), then

$$\int \kappa(f(k_n, z))\phi(dz) \rightarrow \int \kappa(f(k, z))\phi(dz)$$

Evidently $z \mapsto \kappa(f(k_n, z))$ is dominated by $\kappa(f(\bar{y}, z))$, where $\bar{y} := \sup_n k_n$, and an application of the dominated convergence theorem completes the proof.

---

[9]In view of remark 12.2.15 we need not verify that $\kappa \geq 1$.

Returning to the general case, as in chapter 10 we define the function v_σ by

$$v_\sigma(x) := \mathbb{E} \sum_{t=0}^{\infty} \rho^t r_\sigma(X_t) \text{ for } x \in S, \text{ where } X_{t+1} = F(X_t, \sigma(X_t), W_{t+1}) \text{ with } X_0 = x$$

Unlike the situation where r is bounded, this expression is not obviously finite, or even well defined. Indeed it is not clear that $\sum_{t=0}^{\infty} \rho^t r_\sigma(X_t(\omega))$ is convergent at each $\omega \in \Omega$. And even if this random variable is well defined and finite, the expectation may not be.[10]

To start to get a handle on the problem, let's prove that

Lemma 12.2.18 *For all $x \in S$ and all $\sigma \in \Sigma$ we have $\mathbb{E}|r_\sigma(X_t)| \leq R\beta^t \kappa(x)$.*

Proof. That $\mathbb{E}|r_\sigma(X_t)| = \mathbf{M}_\sigma^t |r_\sigma| \leq R\beta^t \kappa$ pointwise on S can be proved by induction. For $t = 0$ we have $|r_\sigma(x)| \leq R\kappa(x)$ by (12.10). Suppose in addition that $\mathbf{M}_\sigma^t |r_\sigma| \leq R\beta^t \kappa$ holds for some arbitrary $t \geq 0$. Then

$$\mathbf{M}_\sigma^{t+1} |r_\sigma| = \mathbf{M}_\sigma \mathbf{M}_\sigma^t |r_\sigma| \leq \mathbf{M}_\sigma R\beta^t \kappa = R\beta^t \mathbf{M}_\sigma \kappa \leq R\beta^t (\beta\kappa) = R\beta^{t+1}\kappa$$

where the second inequality follows from (12.11). $\square$

Lemma 12.2.19 *For each $\sigma \in \Sigma$ and $x \in S$ we have $\mathbb{E} \sum_{t=0}^{\infty} \rho^t |r_\sigma(X_t)| < \infty$.*

Proof. Pick any $\sigma \in \Sigma$ and $x \in S$. Using the monotone convergence theorem followed by lemma 12.2.18, we obtain

$$\mathbb{E} \sum_{t=0}^{\infty} \rho^t |r_\sigma(X_t)| = \sum_{t=0}^{\infty} \rho^t \mathbb{E}|r_\sigma(X_t)| \leq \sum_{t=0}^{\infty} \rho^t R\beta^t \kappa(x)$$

Since $\rho \cdot \beta < 1$ the right-hand side is finite, as was to be shown. $\square$

This lemma implies that $\lim_{T \to \infty} \sum_{t=0}^{T} \rho^t r_\sigma(X_t(\omega))$ exists (and hence the infinite sum is well defined) for $\mathbb{P}$-almost every $\omega \in \Omega$. The reason is that a real-valued series $\sum_t a_t$ converges whenever it converges absolutely, that is, when $\sum_t |a_t| < \infty$. This absolute convergence is true of $\sum_t \rho^t r_\sigma(X_t(\omega))$ for $\mathbb{P}$-almost every ω by lemma 12.2.19 and the fact that a random variable with finite expectation is finite almost everywhere.[11] It also implies via the dominated convergence theorem (why?) that we can pass expectation through infinite sum to obtain our previous expression for v_σ:

$$v_\sigma(x) := \mathbb{E} \left[\sum_{t=0}^{\infty} \rho^t r_\sigma(X_t) \right] = \sum_{t=0}^{\infty} \rho^t \mathbf{M}_\sigma^t r_\sigma(x)$$

We will need the following lemma, which is proved in the appendix to this chapter:

---

[10] It may be infinite or it may involve an expression of the form $\infty - \infty$.

[11] For a proof of this last fact, see Schilling (2005, cor. 10.13).

Lemma 12.2.20 *Let assumptions 12.2.11–12.2.14 all hold. If $w \in b_\kappa cS$, then the mapping $(x, u) \mapsto \int w[F(x, u, z)]\phi(dz)$ is continuous on* gr Γ.

Parallel to definition 10.1.9 on page 233, for $w \in b_\kappa \mathscr{B}(S)$ we say that $\sigma \in \Sigma$ is *w-greedy* if

$$\sigma(x) \in \underset{u \in \Gamma(x)}{\mathrm{argmax}} \left\{ r(x, u) + \rho \int w[F(x, u, z)]\phi(dz) \right\} \qquad (x \in S) \qquad (12.14)$$

Lemma 12.2.21 *Let assumptions 12.2.11–12.2.14 hold. If $w \in b_\kappa cS$, then Σ contains at least one w-greedy policy.*

The proof follows from lemma 12.2.20, and is essentially the same as that of lemma 10.1.10 on page 233. We can now state the main result of this section.

Theorem 12.2.22 *Under assumptions 12.2.11–12.2.14, the value function v^* is the unique function in $b_\kappa cS$ satisfying*

$$v^*(x) = \max_{u \in \Gamma(x)} \left\{ r(x, u) + \rho \int v^*[F(x, u, z)]\phi(dz) \right\} \qquad (x \in S) \qquad (12.15)$$

A feasible policy is optimal if and only if it is v^*-greedy. At least one such policy exists.

12.2.3 Proofs

Let's turn to the proof of theorem 12.2.22. In parallel to §10.1, let $T_\sigma \colon b_\kappa \mathscr{B}(S) \to b_\kappa \mathscr{B}(S)$ be defined for all $\sigma \in \Sigma$ by

$$T_\sigma w(x) = r(x, \sigma(x)) + \rho \int w[F(x, \sigma(x), z)]\phi(dz) = r_\sigma(x) + \rho \mathbf{M}_\sigma w(x)$$

and let the Bellman operator $T \colon b_\kappa cS \to b_\kappa cS$ be defined by

$$Tw(x) = \max_{u \in \Gamma(x)} \left\{ r(x, u) + \rho \int w[F(x, u, z)]\phi(dz) \right\} \qquad (x \in S)$$

Exercise 12.2.23 Confirm that both T_σ and T do in fact send (measurable) κ-bounded functions into κ-bounded functions. Lemma A.2.31 on page 334 might be helpful in the case of T.

Lemma 12.2.24 *Let $\gamma := \rho\beta$. For every $\sigma \in \Sigma$, the operator T_σ is uniformly contracting on the metric space $(b_\kappa \mathscr{B}(S), d_\kappa)$, with*

$$\|T_\sigma w - T_\sigma w'\|_\kappa \leq \gamma \|w - w'\|_\kappa \quad \forall w, w' \in b_\kappa \mathscr{B}(S) \qquad (12.16)$$

and the unique fixed point of T_σ in $b_\kappa \mathscr{B}(S)$ is v_σ. In addition T_σ is monotone on $b_\kappa \mathscr{B}(S)$, in the sense that if $w, w' \in b\mathscr{B}(S)$ and $w \leq w'$, then $T_\sigma w \leq T_\sigma w'$.

Proof. The proof that T_σ is monotone is left to the reader. The proof that $T_\sigma v_\sigma = v_\sigma$ is identical to the proof in §10.1 for bounded r. The proof that T_σ is a uniform contraction goes as follows: Pick any $w, w' \in b_\kappa \mathscr{B}(S)$. Making use of the linearity and monotonicity of $\mathbf{M}_\sigma$, we have

$$|T_\sigma w - T_\sigma w'| = |\rho \mathbf{M}_\sigma w - \rho \mathbf{M}_\sigma w'| = \rho|\mathbf{M}_\sigma(w - w')|$$
$$\leq \rho \mathbf{M}_\sigma |w - w'| \leq \rho \|w - w'\|_\kappa \mathbf{M}_\sigma \kappa \leq \rho \beta \|w - w'\|_\kappa \kappa$$

The rest of the argument is an exercise. $\qquad\square$

Next we turn to the Bellman operator.

Lemma 12.2.25 *The operator T is uniformly contracting on $(b_\kappa cS, d_\kappa)$, with*

$$\|Tw - Tw'\|_\kappa \leq \gamma \|w - w'\|_\kappa \qquad \forall\, w, w' \in b_\kappa cS \qquad (12.17)$$

where $\gamma := \rho\beta$. In addition T is monotone on $b_\kappa cS$, in the sense that if $w, w' \in b_\kappa cS$ and $w \leq w'$, then $Tw \leq Tw'$.

Exercise 12.2.26 Prove lemma 12.2.25. In particular, prove that T is uniformly contracting with modulus γ by applying theorem 12.2.9 (page 307).

The proof of theorem 12.2.22 now follows from lemmas 12.2.24 and 12.2.25 in an almost identical fashion to the bounded case (see §10.1.3). The details are left to the reader.

12.3   Commentary

Monotonicity in parameters is a major topic in mathematical economics and dynamic programming. Useful references include Lovejoy (1987), Puterman (1994), Topkis (1998), Hopenhayn and Prescott (1992), Huggett (2003), Amir (2005), and Mirman et al. (2008). See Amir et al. (1991) for an early analysis of monotone dynamic programming and optimal growth.

Our treatment of concavity and differentiability is standard. The classic reference is Stokey and Lucas (1989). Corollary 12.1.19 is due to Mirman and Zilcha (1975). The connection between lemma 12.1.21 and differentiability of the value function is due to Benveniste and Scheinkman (1979), and is based on earlier results in Rockafellar (1970).

Global stability of the stochastic optimal growth model under certain Inada-type conditions was proved by Brock and Mirman (1972). See also Mirman (1970, 1972, 1973), Mirman and Zilcha (1975), Hopenhayn and Prescott (1992), Stachurski (2002), Nishimura and Stachurski (2004), Zhang (2007), Kamihigashi (2007), or Chatterjee and

Shukayev (2008). The techniques used here closely follow Nishimura and Stachurski (2004).

Our discussion of unbounded dynamic programming in §12.2 closely follows the theory developed in Hernández-Lerma and Lasserre (1999, ch. 8). Boyd (1990) is an early example of the weighted norm approach in economics, with an application to recursive utility. See also Becker and Boyd (1997). Le Van and Vailakis (2005) is a more recent treatment of the same topic. Stokey and Alvarez (1998) use weighted norm techniques for dynamic programs with certain homogeneity properties. See also Rincon-Zapatero and Rodriguez-Palmero (2003).

Part III

Appendixes

Appendix A

Real Analysis

This appendix reviews some bits and pieces from basic real analysis that are used in the book. If you lack background in analysis, then it's probably best to parse the chapter briefly and try some exercises before starting the main body of the text.

A.1 The Nuts and Bolts

We start off our review with fundamental concepts such as sets and functions, and then move on to a short discussion of probability on finite sample spaces.

 **A.1.1 Sets and Logic**

Pure mathematicians might tell you that everything is a set, or that sets are the only primitive (i.e., the only mathematical objects not defined in terms of something else). We won't take such a purist view. For us a set is just a collection of objects viewed as a whole. Functions are rules that associate elements of one set with elements of another.

Examples of sets include $\mathbb{N}$, $\mathbb{Z}$, and $\mathbb{Q}$, which denote the natural numbers (i.e., positive integers), the integers, and the rational numbers respectively. The objects that make up a set are referred to as its elements. If a is an element of A we write $a \in A$. The set that contains no elements is called the *empty set* and denoted by $\varnothing$. Sets A and B are said to be equal if they contain the same elements. Set A is called a *subset* of B (written $A \subset B$) if every element of A is also an element of B.[1] Clearly, $A = B$ if and only if $A \subset B$ and $B \subset A$.

---

[1]Something to ponder: In mathematics any logical statement that cannot be tested is regarded as (vacuously) true. It follows that $\varnothing$ is a subset of every set.

If S is a given set, then the collection of all subsets of S is itself a set. We denote it by $\mathfrak{P}(S)$.

The *intersection* $A \cap B$ of two sets A and B consists of all elements found in both A and B; A and B are called *disjoint* if $A \cap B = \emptyset$. The *union* of A and B is the set $A \cup B$ consisting of all elements in at least one of the two. The *set-theoretic difference* $A \setminus B$ is defined as

$$A \setminus B := \{x : x \in A \text{ and } x \notin B\}$$

In the case where the discussion is one of subsets of some fixed set A, the difference $A \setminus B$ is called the *complement* of B and written B^c.

If A is an arbitrary "index" set so that $\{K_\alpha\}_{\alpha \in A}$ is a collection of sets, then we define

$$\cap_{\alpha \in A} K_\alpha := \{x : x \in K_\alpha \text{ for all } \alpha \in A\}$$

and

$$\cup_{\alpha \in A} K_\alpha := \{x : \text{there exists an } \alpha \in A \text{ such that } x \in K_\alpha\}$$

The same collection $\{K_\alpha\}_{\alpha \in A}$ is called *pairwise disjoint* if any pair K_α, K_β with $\alpha \neq \beta$ is disjoint.

The following two equalities are known as de Morgan's laws:

1. $\left(\cup_{\alpha \in A} K_\alpha\right)^c = \cap_{\alpha \in A} K_\alpha^c$

2. $\left(\cap_{\alpha \in A} K_\alpha\right)^c = \cup_{\alpha \in A} K_\alpha^c$

Let's see how we prove these kinds of set equalities by going through the proof of the first one slowly. Let $A := \left(\cup_{\alpha \in A} K_\alpha\right)^c$ and $B := \cap_{\alpha \in A} K_\alpha^c$. Take some arbitrary element $x \in A$. Since $x \in A$, it must be that x is not in K_α for any α. In other words, $x \in K_\alpha^c$ for every α. But if this is true, then, by the definition of B, we see that $x \in B$. Since x was arbitrary, we have $A \subset B$. Similar reasoning shows that $B \subset A$, and hence $A = B$.

As we go along, fewer of these steps will be spelled out, so read the proof through a few times if the method does not seem obvious. You will get used to this way of thinking.

The *Cartesian product* of sets A and B is the set of ordered pairs

$$A \times B := \{(a,b) : a \in A, b \in B\}$$

For example, if A is the set of outcomes for a random experiment (experiment A), and B is the set of outcomes for a second experiment (experiment B), then $A \times B$ is the set of all outcomes for the experiment C, which consists of first running A and then running B. The pairs (a, b) are ordered, so (a, b) and (b, a) are not in general the same point. In the preceding example this is necessary so that we can distinguish between the outcomes for the first and second experiment.

Infinite Cartesian products are also useful. If (A_n) is a collection of sets, one for each $n \in \mathbb{N}$, then

$$\times_{n \geq 1} A_n := \{(a_1, a_2, \ldots) : a_n \in A_n\}$$

If $A_n = A$ for all n, then $\times_{n \geq 1} A$ is often written as $A^{\mathbb{N}}$.

So much for sets. Now let's very briefly discuss logic and the language of mathematics. We proceed in a "naive" way (rather than axiomatic), with the idea of quickly introducing the notation and its meaning. If you are not very familiar with formal mathematics, I suggest that you skim through, picking up the basic points and coming back if required.

Logic starts with the notion of mathematical statements, which we denote with capital letters such as P or Q. Typical examples are

$$P = \text{the area of a rectangle is the product of its two sides}$$

$$Q = x \text{ is strictly positive}$$

Next we assign *truth values* to these statements, where each statement is labeled either "true" or "false." A starting point for logic is the idea that every sensible mathematical statement is either true or false. The truth value of "maybe" is not permitted.

In general, mathematical statements should not really be thought of as inherently true or false. For example, you might think that P above is always a true statement. However, it is better to regard P as consistent with the natural world in certain ways, and therefore a useful assumption to make when performing geometric calculations. At the same time, let's not rule out the possibility of assuming that P is false in order to discover the resulting implications.

Simply put, a lot of mathematics is about working out the consistency of given truth values assigned to collections of mathematical statements. This is done according to the rules of logic, which are indeed quite logical. For example, if a statement P is labeled as true, then its *negation* $\sim P$ is false, and $\sim (\sim P)$ must have the same truth value as P.

Statements can be combined using the *elementary connectives* "and" and "or." Statement "P and Q" is true if both P and Q are true, and false otherwise. Statement "P or Q" is false if both P and Q are false, and true otherwise. You might try to convince yourself that

$$\sim (A \text{ or } B) \equiv (\sim A) \text{ and } (\sim B) \quad \& \quad \sim (A \text{ and } B) \equiv (\sim A) \text{ or } (\sim B)$$

where the notation $P \equiv Q$ means that P and Q are *logically equivalent* (i.e., always have the same truth value).

Another form of relationship between statements is implication. For example, suppose that we have sets A and B with $A \subset B$. Let P be the statement $x \in A$ and Q be the statement $x \in B$. If P is labeled as true, then Q must also be true, since elements of A are also elements of B. We say that P implies Q (alternatively: if P, then Q), and write $P \implies Q$.

Sometimes it is not so easy to see that $P \implies Q$. Mathematical proofs typically involve creating a chain of statements $R_1, \ldots, R_n$ with

$$P \implies R_1 \implies \cdots \implies R_n \implies Q$$

Often this is done by working forward from P and backward from Q, and hoping that you meet somewhere in the middle. Another strategy for proving that $P \implies Q$ is to show that $\sim Q \implies \sim P$.[2] For if the latter holds then so must $P \implies Q$ be valid, for when P is true Q cannot be false (if it were, then P could not be true).

The universal quantifier $\forall$ (for all) and the existential quantifier $\exists$ (there exists) are used as follows. If $P(\alpha)$ is statement about an object α, then

$$\forall \alpha \in \Lambda, \ P(\alpha)$$

means that for all elements α of the set Λ, the statement $P(\alpha)$ holds.

$$\exists \alpha \in \Lambda \text{ such that } P(\alpha)$$

means that $P(\alpha)$ is true for at least one $\alpha \in \Lambda$. The following equivalences hold:

$$\sim [\forall \alpha \in \Lambda, \ P(\alpha)] \equiv \exists \alpha \in \Lambda \text{ such that } \sim P(\alpha), \qquad \text{and}$$

$$\sim [\exists \alpha \in \Lambda \text{ such that } P(\alpha)] \equiv \forall \alpha \in \Lambda, \ \sim P(\alpha)$$

A.1.2   Functions

A *function* f from set A to set B, written $A \ni x \mapsto f(x) \in B$ or $f \colon A \to B$, is a rule associating to each and every one of the elements a in A one and only one element $b \in B$.[3] The point b is also written as $f(a)$, and called the *image* of a under f. For $C \subset A$, the set $f(C)$ is the set of all images of points in C, and is called the image of C under f. Formally,

$$f(C) := \{b \in B : f(a) = b \text{ for some } a \in C\}$$

Also, for $D \subset B$, the set $f^{-1}(D)$ is all points in A that map into D under f, and is called the *preimage* of D under f. That is,

$$f^{-1}(D) := \{a \in A : f(a) \in D\}$$

---

[2] The latter implication is known as the *contrapositive* of the former.

[3] Some writers refer to a function f by the symbol $f(x)$, as in "the production function $f(x)$ is increasing…," or similar. Try not to follow this notation. The symbol $f(x)$ represents a value, not a function.

When D consists of a single point $b \in B$ we write $f^{-1}(b)$ rather than $f^{-1}(\{b\})$. In general, $f^{-1}(b)$ may contain many elements of A or none.

Let S be any set. For every $A \subset S$, let $S \ni x \mapsto \mathbb{1}_A(x) \in \{0,1\}$ be the function that takes the value 1 when $x \notin A$ and zero otherwise. This function is called the *indicator function* of A.

Exercise A.1.1 Argue that $\mathbb{1}_{A^c} = \mathbb{1}_S - \mathbb{1}_A$ holds pointwise on S (i.e., $\mathbb{1}_{A^c}(x) = \mathbb{1}_S(x) - \mathbb{1}_A(x)$ at each $x \in S$). In what follows we usually write $\mathbb{1}_S$ simply as 1. Argue further that if $A_1, \ldots, A_n$ is a collection of subsets, then $\max_i \mathbb{1}_{A_i} = 1 - \prod_i \mathbb{1}_{A_i^c}$.

A function $f \colon A \to B$ is called *one-to-one* if distinct elements of A are always mapped into distinct elements of B, and *onto* if every element of B is the image under f of at least one point in A. A function that is both one-to-one and onto is called a *bijection*.

You will be able to verify that $f \colon A \to B$ is a bijection if and only if $f^{-1}(b)$ consists of precisely one point in A for each $b \in B$. In this case f^{-1} defines a function from B to A by setting $f^{-1}(b)$ equal to the unique point in A that f maps into b. This function is called the *inverse* of f. Note that $f(f^{-1}(b)) = b$ for all $b \in B$, and that $f^{-1}(f(a)) = a$ for all $a \in A$.

New functions are often defined from old functions by composition: If $f \colon A \to B$ and $g \colon B \to C$, then $g \circ f \colon A \to C$ is defined at $x \in A$ by $(g \circ f)(x) := g(f(x))$. It is easy to check that if f and g are both one-to-one and onto, then so is $g \circ f$.

Preimages and set operations interact nicely. For example, if $f \colon A \to B$, and E and F are subsets of B, then

$$f^{-1}(E \cup F) = f^{-1}(E) \cup f^{-1}(F)$$

To see this, suppose that $x \in f^{-1}(E \cup F)$. Then $f(x) \in E \cup F$, so $f(x) \in E$ or $f(x) \in F$ (or both). Therefore $x \in f^{-1}(E)$ or $x \in f^{-1}(F)$, whence $x \in f^{-1}(E) \cup f^{-1}(F)$. This proves that $f^{-1}(E \cup F) \subset f^{-1}(E) \cup f^{-1}(F)$. A similar argument shows that $f^{-1}(E \cup F) \supset f^{-1}(E) \cup f^{-1}(F)$, from which equality now follows.

More generally, we have the following results. (Check them.)

Lemma A.1.2 *Let $f \colon A \to B$, and let E and $\{E_\gamma\}_{\gamma \in C}$ all be arbitrary subsets of B.[4] We have*

1. $f^{-1}(E^c) = [f^{-1}(E)]^c$,

2. $f^{-1}(\cup_\gamma E_\gamma) = \cup_\gamma f^{-1}(E_\gamma)$, *and*

3. $f^{-1}(\cap_\gamma E_\gamma) = \cap_\gamma f^{-1}(E_\gamma)$.

The forward image is not as well behaved as the preimage.

---

[4]Here C is any "index" set.

Exercise A.1.3 Construct an example of sets A, B, C, D, with $C, D \subset A$, and function $f: A \to B$, where $f(C \cap D) \neq f(C) \cap f(D)$.

Using the concept of bijections, let us now discuss some different notions of infinity. To start, notice that it is not always possible to set up a bijection between two sets. (Consider the case where one set has two elements and the other has one—try to find a bijection.) When a bijection does exist, the two sets are said to be in *one-to-one correspondence*, or have the same *cardinality*. This notion captures the idea that the two sets "have the same number of elements," but in a way that can be applied to infinite sets.

Definition A.1.4 A nonempty set A is called *finite* if it has the same cardinality as the set $\{1, 2, \ldots, n\}$ for some $n \in \mathbb{N}$. Otherwise, A is called *infinite*. If A is either finite or in one-to-one correspondence with $\mathbb{N}$, then A is called *countable*. Otherwise, A is called *uncountable*.[5]

The distinction between countable and uncountable sets is important, particularly for measure theory. In the rest of this section we discuss examples and results for these kinds of properties. The proofs are a little less than completely rigorous—sometimes all the cases are not covered in full generality—but you can find formal treatments in almost all textbooks on real analysis.

An example of a countable set is $E := \{2, 4, \ldots\}$, the even elements of $\mathbb{N}$. We can set up a bijection $f: \mathbb{N} \to E$ by letting $f(n) = 2n$. The set $O := \{1, 3, \ldots\}$ of odd elements of $\mathbb{N}$ is also countable, under $f(n) = 2n - 1$. These examples illustrate that *for infinite sets, a proper subset can have the same cardinality as the original set.*

Theorem A.1.5 *Countable unions of countable sets are countable.*

Proof. Let $A_n := (a_n^1, a_n^2, \ldots)$ be a countable set, and let $A := \cup_{n \geq 1} A_n$. For simplicity we assume that the sets (A_n) are all infinite and pairwise disjoint. Arranging the elements of A into an infinite matrix, we can count them in the following way:

$$
\begin{array}{ccccc}
a_1^1 & \to & a_1^2 & & a_1^3 & \to & \cdots \\
& \swarrow & & \nearrow & & & \\
a_2^1 & & a_2^2 & & \cdots & & \\
\downarrow & \nearrow & & & & & \\
a_3^1 & & \vdots & & & & \\
\vdots & & & & & &
\end{array}
$$

This system of counting provides a bijection with $\mathbb{N}$. $\qquad\qquad\square$

---

[5] Sets we are calling countable some authors refer to as *at most countable*.

Exercise A.1.6 Show that $\mathbb{Z} := \{\ldots, -1, 0, 1, \ldots\}$ is countable.

Theorem A.1.7 *Finite Cartesian products of countable sets are countable.*

Proof. Let's just prove this for a pair A and B, where both A and B are infinite. In this case, the Cartesian product can be written as

$$
\begin{array}{ccccccc}
(a_1, b_1) & \to & (a_1, b_2) & & (a_1, b_3) & \to & \cdots \\
& \swarrow & & \nearrow & & & \\
(a_2, b_1) & & (a_2, b_2) & & \cdots & & \\
\downarrow & \nearrow & & & & & \\
(a_3, b_1) & & \vdots & & & & \\
\vdots & & & & & &
\end{array}
$$

Now count as indicated. $\square$

Theorem A.1.8 *The set of all rational numbers $\mathbb{Q}$ is countable.*

Proof. The set $\mathbb{Q} = \{p/q : p \in \mathbb{Z}, q \in \mathbb{Z}, q \neq 0\}$ can be put in one-to-one correspondence with a subset of $\mathbb{Z} \times \mathbb{Z} = \{(p, q) : p \in \mathbb{Z}, q \in \mathbb{Z}\}$, which is countable by theorem A.1.7. Subsets of countable sets are countable. $\square$

Not all sets are countable. In fact *countable* Cartesian products of countable sets may be uncountable. For example, consider $\{0, 1\}^{\mathbb{N}}$, the set of all binary sequences $(a_1, a_2, \ldots)$, where $a_i \in \{0, 1\}$. If this set were countable, then it could be listed as follows:

$$
\begin{array}{ccc}
1 & \leftrightarrow & a_1, a_2, a_3, \ldots \\
2 & \leftrightarrow & b_1, b_2, b_3, \ldots \\
\vdots & & \vdots
\end{array}
$$

where the sequences on the right-hand side are binary sequences. Actually such a list is never complete: We can always construct a new binary sequence $c_1, c_2, \ldots$ by setting c_1 to be different from a_1 (zero if a_1 is one, and one otherwise), c_2 to be different from b_2, and so on. This differs from every element in our supposedly complete list (in particular, it differs from the n-th sequence in that their n-th elements differ); a contradiction indicating that $\{0, 1\}^{\mathbb{N}}$ is uncountable.[6]

The cardinality of the set of binary sequences is called the *power of the continuum*. The assertion that there are no sets with cardinality greater than countable and less than the continuum is called the Continuum Hypothesis, and is a rather tricky problem to say the least.

---

[6]This is Cantor's famous diagonal argument.

A.1.3   Basic Probability

In this section we briefly recall some elements of probability on finite sets. Consider a finite set Ω, a typical element of which is ω. A *probability* $\mathbb{P}$ on Ω is a function from $\mathfrak{P}(\Omega)$, the set of all subsets of Ω, into $[0,1]$ with properties

1. $\mathbb{P}(\Omega) = 1$, and

2. if $A, B \subset \Omega$ and $A \cap B = \varnothing$, then $\mathbb{P}(A \cup B) = \mathbb{P}(A) + \mathbb{P}(B)$.

The pair $(\Omega, \mathbb{P})$ is sometimes called a *finite probability space*. Subsets of Ω are also called *events*. The elements ω that make up Ω are the *primitive events*, while general $B \subset \Omega$ is a *composite event*, consisting of $M \leq \#\Omega$ primitive events.[7] The number $\mathbb{P}(B)$ is the "probability that event B occurs." In other words, $\mathbb{P}(B)$ represents the probability that when uncertainty is resolved and some $\omega \in \Omega$ is selected by "nature," the statement $\omega \in B$ is true.

Exercise A.1.9 Let $p \colon \Omega \to [0,1]$, where $\sum_{\omega \in \Omega} p(\omega) = 1$, and let

$$\mathbb{P}(B) := \sum_{\omega \in B} p(\omega) \qquad (B \subset \Omega) \tag{A.1}$$

Show that properties (1) and (2) both hold for $\mathbb{P}$ defined in (A.1).

The next few results follow easily from the definition of a probability.

Lemma A.1.10 *If $A \subset \Omega$, then $\mathbb{P}(A^c) = 1 - \mathbb{P}(A)$.*

Proof. Here of course $A^c := \Omega \setminus A$. The proof is immediate from (1) and (2) above because $1 = \mathbb{P}(\Omega) = \mathbb{P}(A \cup A^c) = \mathbb{P}(A) + \mathbb{P}(A^c)$.   □

Exercise A.1.11 Prove that $\mathbb{P}(\varnothing) = 0$. Prove that if $A \subset B$, then $\mathbb{P}(B \setminus A) = \mathbb{P}(B) - \mathbb{P}(A)$. Prove further that if $A \subset B$, then $\mathbb{P}(A) \leq \mathbb{P}(B)$.

The idea that if $A \subset B$, then $\mathbb{P}(A) \leq \mathbb{P}(B)$ is fundamental. Event B occurs whenever A occurs, so the probability of B is larger. Many crucial ideas in probability boil down to this one point.

Exercise A.1.12 Prove that if A and B are (not necessarily disjoint) subsets of Ω, then $\mathbb{P}(A \cup B) \leq \mathbb{P}(A) + \mathbb{P}(B)$. Construct an example of a probability $\mathbb{P}$ and subsets A, B such that this inequality is strict. Show that in general, $\mathbb{P}(A \cup B) = \mathbb{P}(A) + \mathbb{P}(B) - \mathbb{P}(A \cap B)$.

---

[7]If A is a set, then $\#A$ denotes the number of elements in A.

If A and B are two events and $\mathbb{P}(B) > 0$, then the *conditional probability of A given B* is

$$\mathbb{P}(A \mid B) := \frac{\mathbb{P}(A \cap B)}{\mathbb{P}(B)} \tag{A.2}$$

It represents the probability that A will occur, given the information that B has occurred.

What is the justification for the expression (A.2)? Informally, the probability $\mathbb{P}(C)$ of an event C can be thought of as the fraction of times that C occurs in n independent and identical experiments, as $n \to \infty$. Letting ω_n be the outcome of the n-th trial and $\#A$ be the number of elements in set A, we can write this as

$$\mathbb{P}(C) = \lim_{n \to \infty} \frac{\#\{n : \omega_n \in C\}}{n}$$

The conditional $\mathbb{P}(A \mid B)$ is (approximately) the number of times both A and B occur over a large number of observations, expressed as a fraction of the number of occurrences of B:

$$\mathbb{P}(A \mid B) \cong \frac{\#\{n : \omega_n \in A \text{ and } B\}}{\#\{n : \omega_n \in B\}}$$

Dividing through by n and taking the limit gives

$$\mathbb{P}(A \mid B) \cong \frac{\#\{n : \omega_n \in A \text{ and } B\}/n}{\#\{n : \omega_n \in B\}/n} \to \frac{\mathbb{P}(A \cap B)}{\mathbb{P}(B)}$$

Events A and B are called *independent* if $\mathbb{P}(A \cap B) = \mathbb{P}(A)\mathbb{P}(B)$. You will find it easy to confirm that if A and B are independent, then the conditional probability of A given B is just the probability of A.

We will make extensive use of the *law of total probability*, which says that if $A \subset \Omega$ and $B_1, \ldots, B_M$ is a partition of Ω (i.e., $B_m \subset \Omega$ for each m, the B_m's are mutually disjoint in the sense that $B_j \cap B_k$ is empty when $j \neq k$, and $\cup_{m=1}^M B_m = \Omega$) with $\mathbb{P}(B_m) > 0$ for all j, then

$$\mathbb{P}(A) = \sum_{m=1}^M \mathbb{P}(A \mid B_m) \cdot \mathbb{P}(B_m)$$

The proof is quite straightforward, although you should check that the manipulations of intersections and unions work if you have not seen them before:

$$\mathbb{P}(A) = \mathbb{P}(A \cap \cup_{m=1}^M B_m) = \mathbb{P}(\cup_{m=1}^M (A \cap B_m))$$
$$= \sum_{m=1}^M \mathbb{P}(A \cap B_m) = \sum_{m=1}^M \mathbb{P}(A \mid B_m) \cdot \mathbb{P}(B_m)$$

Now consider a random variable taking values in some collection of numbers S. Formally, a random variable X is a function from the sample space Ω into S. The idea is that "nature" picks out an element ω in Ω according to some probability. The random variable now sends this ω into $X(\omega) \in S$. We can think of X as "reporting" the outcome of the draw to us in a format that is more amenable to analysis. For example, Ω might be a collection of binary sequences, and X translates these sequences into (decimal) numbers.

Each probability $\mathbb{P}$ on Ω and $X \colon \Omega \to S$ induces a *distribution*[8] ϕ on S via

$$\phi(x) = \mathbb{P}\{\omega \in \Omega : X(\omega) = x\} \qquad (x \in S) \tag{A.3}$$

Exercise A.1.13 Show that $\phi(x) \geq 0$ and $\sum_{x \in S} \phi(x) = 1$.

In what follows we will often write the right-hand side of (A.3) simply as $\mathbb{P}\{X = x\}$. Please be aware of this convention. We say that X is distributed according to ϕ, and write $X \sim \phi$.

An aside: If you stick to elementary probability, then you may begin to feel that the distinction between the underlying probability $\mathbb{P}$ and the distribution ϕ of X is largely irrelevant. Why can't we just say that X is a random variable with distribution ϕ, and Y is another random variable with distribution ψ? The meaning of these statements seems clear, and there is no need to introduce $\mathbb{P}$ and Ω, or to think about X and Y as functions.

The short answer to this question is that it is often useful to collect different random variables on the one probability space defined by Ω and $\mathbb{P}$. With this construct one can then discuss more complex events, such as convergence of a sequence of random variables on $(\Omega, \mathbb{P})$ to yet another random variable on $(\Omega, \mathbb{P})$.

Next we define expectation. Let $X \colon \Omega \to S$ and let $\mathbb{P}$ be a probability on Ω. The *expectation* $\mathbb{E}X$ of X is given by

$$\mathbb{E}X := \sum_{\omega \in \Omega} X(\omega)\mathbb{P}\{\omega\} \tag{A.4}$$

Exercise A.1.14 Prove that if $X \sim \phi$, then $\mathbb{E}X = \sum_{x \in S} x\phi(x)$.[9] Prove the more general result that if $Y = h(X)$ for some real-valued function h of X (i.e., $h \colon S \to \mathbb{R}$), then

$$\mathbb{E}Y := \sum_{\omega \in \Omega} h(X(\omega))\mathbb{P}\{\omega\} = \sum_{x \in S} h(x)\phi(x) \tag{A.5}$$

---

[8]What we call a distribution here is often referred to as a probability mass function.
[9]Hint: Divide Ω into sets B_x for $x \in S$, where $B_x := \{\omega \in \Omega : X(\omega) = x\}$.

A.2   The Real Numbers

As usual, $\mathbb{R}$ denotes the so-called real numbers, which you can visualize as the "continuous" real line. We will make use of several of its properties. One property worth mentioning before we start is that *the set $\mathbb{R}$ is uncountable*. This can be proved by showing that $\mathbb{R}$ is in one to one correspondence with the set of all binary sequences—which were shown to be uncountable in §A.1.2. (For the correspondence, think of the way that computers represent numbers in binary form.) $\mathbb{R}$ also has certain algebraic, order, and completeness properties, which we now detail.

A.2.1   Real Sequences

In what follows, if $x \in \mathbb{R}$ then $|x|$ denotes its absolute value. For any $x, y \in \mathbb{R}$ the triangle inequality $|x + y| \leq |x| + |y|$ holds.

Exercise A.2.1 Show that if a, b and x are any real numbers, then

$$|a - b| \leq |a - x| + |x - b| \quad \text{and} \quad \big| |x| - |a| \big| \leq |x - a| \tag{A.6}$$

A subset A of $\mathbb{R}$ is called *bounded* if there is an $M \in \mathbb{N}$ such that $|x| \leq M$, all $x \in A$. The ϵ-ball or ϵ-neighborhood around $a \in \mathbb{R}$ is the set of points $x \in \mathbb{R}$ such that $|a - x| < \epsilon$.[10] An X-valued *sequence* is a function from the natural numbers $\mathbb{N} := \{1, 2, \ldots\}$ to nonempty set X, traditionally denoted by notation such as (x_n). It is called a *real sequence* when $X \subset \mathbb{R}$. A real sequence (x_n) called bounded if its range is a bounded set (i.e., $\exists M \in \mathbb{N}$ such that $|x_n| \leq M$ for all $n \in \mathbb{N}$).

A real sequence (x_n) is said to be *convergent* if there is an $x \in \mathbb{R}$ such that, given any $\epsilon > 0$, we can find an $N \in \mathbb{N}$ with the property $|x_n - x| < \epsilon$ whenever $n \geq N$. This property will often be expressed by saying that (x_n) is *eventually* in any ϵ-neighborhood of x. The point x is called the *limit* of the sequence, and we write $\lim_{n \to \infty} x_n = x$ or $x_n \to x$ as $n \to \infty$.

This definition of convergence can be a little hard to grasp at first. One way is to play the "ϵ, N game." If I claim that a sequence is convergent, then, for every ϵ-neighborhood you give me, I commit to providing you with an index N such that all points further along the sequence than the N-th one (i.e., points $x_N, x_{N+1}, \ldots$) are in that ϵ-neighborhood. For example, consider $x_n = 1/n^2$. I claim x_n converges to zero. When you give me $\epsilon = 1/3$, I can give you $N = 2$, because $n \geq 2$ implies $x_n = 1/n^2 \leq 1/4 < \epsilon$. In fact I can give you an "algorithm" for generating such an N: Given $\epsilon > 0$, take any $N \in \mathbb{N}$ greater than $1/\sqrt{\epsilon}$.

Sometimes the "points" ∞ and $-\infty$ can are regarded as limits of sequences. In what follows, we will say that $x_n \to \infty$, or $\lim_n x_n = \infty$, if for each $M \in \mathbb{N}$ there is an

---

[10]This "ball" will look more ball-like once we move to higher dimensional spaces.

$N \in \mathbb{N}$ such that $x_n \geq M$ whenever $n \geq N$. Similarly we say that $x_n \to -\infty$ if, for each $M \in \mathbb{N}$ there is an $N \in \mathbb{N}$ such that $x_n \leq -M$ whenever $n \geq N$. Also, sequence (x_n) is called *monotone increasing* (resp., *decreasing*) if $x_n \leq x_{n+1}$ (resp, $x_{n+1} \leq x_n$) for all $n \in \mathbb{N}$. If (x_n) is monotone increasing (resp., decreasing) and converges to some $x \in \mathbb{R}$, then we write $x_n \uparrow x$ (resp., $x_n \downarrow x$).

Lemma A.2.2 *Let* (x_n) *be a sequence in* $\mathbb{R}$, *and let* $x \in \mathbb{R}$. *Then* $x_n \to x$ *if and only if* $|x_n - x| \to 0$.

Proof. The first statement says that we can make $|x_n - x|$ less than any given $\epsilon > 0$ by choosing n sufficiently large. The second statement says that we can make $||x_n - x| - 0|$ less than any given $\epsilon > 0$ by choosing n sufficiently large. Clearly, these statements are equivalent.   □

Lemma A.2.3 *Each real sequence has at most one limit.*

Proof. Let $x_n \to a$ and $x_n \to b$. Suppose that $a \neq b$. By choosing ϵ small enough, we can take ϵ-balls B_a and B_b around a and b that are disjoint.[11] By the definition of convergence, (x_n) is eventually in B_a and eventually in B_b. In which case there must be an N such that $x_N \in B_a$ and $x_N \in B_b$. But this is impossible. Hence $a = b$.   □

While most of the numbers that we deal with in every day life can be expressed in terms of integers or rational numbers, for more sophisticated mathematics $\mathbb{Q}$ does not suffice. Simple equations using rational numbers may not have rational solutions, and sequences of rational numbers that seem to converge to *something* may not converge to any rational number. The real numbers "complete" the rational numbers, in the sense that sequences of rationals—or reals—that "appear to converge" will have a limit within the set $\mathbb{R}$.

To make this precise, recall that a sequence (x_n) in $\mathbb{R}$ is called *Cauchy* if, for any $\epsilon > 0$, there exists an $N \in \mathbb{N}$ such that for any n and m greater than N, $|x_n - x_m| < \epsilon$. Now Cauchy sequences seem to be converging to something, so we can express the idea of completeness of $\mathbb{R}$—as opposed to $\mathbb{Q}$—by saying that every Cauchy sequence in $\mathbb{R}$ does converge to an element of $\mathbb{R}$.

Axiom A.2.4 (Completeness of $\mathbb{R}$) *Every Cauchy sequence on the real line is convergent.*

There are formal constructions of the real numbers from the rationals by which this statement can be *proved*, but we will take it as axiomatic. This completeness property of $\mathbb{R}$ is one of the most important and fundamental ideas of real analysis. For example, it allows us to define a solution to a particular problem as the limit of a Cauchy sequence of numbers generated by some approximation process, without fearing the

---

[11]Using (A.6), show that if $\epsilon < |a - b|/2$, then $x \in B_a$ and $x \in B_b$ is impossible.

embarrassment that would result should such a limit point fail to exist. This is important because many types of sequences are Cauchy. The following example is extremely useful in both theory and applications:

Theorem A.2.5 *Every bounded monotone sequence in* $\mathbb{R}$ *is convergent.*

Proof. We prove the case where (x_n) is increasing ($x_{n+1} \geq x_n$ for all n). By axiom A.2.4, it suffices to show that (x_n) is Cauchy. Suppose that it is not. Then we can find an $\epsilon_0 > 0$ such that, given any $N \in \mathbb{N}$, there is a pair $n, m \in \mathbb{N}$ with $N \leq n < m$ and $x_m - x_n \geq \epsilon_0$. But then (x_n) cannot be bounded above. (Why?) Contradiction.[12] $\square$

Exercise A.2.6 Prove that if $(x_n) \subset \mathbb{R}$ is convergent, then (x_n) is bounded.[13]

Now we introduce the notion of subsequences. Formally, a sequence (y_n) is called a *subsequence* of another sequence (x_n) if there is a strictly increasing function $f : \mathbb{N} \to \mathbb{N}$ such that $y_n = x_{f(n)}$ for all $n \in \mathbb{N}$. To put it more simply, (y_n) is the original sequence (x_n) but with some points omitted. The function f picks out a strictly increasing sequence of positive integers $n_1 < n_2 < \cdots$ that are to make up the subsequence, in the sense that $y_1 = x_{n_1}, y_2 = x_{n_2}$, and so forth. Often one writes this new sequence as (x_{n_k}).

Exercise A.2.7 Show that if $(x_n) \subset \mathbb{R}$ converges to $x \in \mathbb{R}$, then so does every subsequence.

Exercise A.2.8 Show that (x_n) converges to $x \in \mathbb{R}$ if and only if every subsequence of (x_n) has a subsubsequence that converges to x.

Theorem A.2.9 *Every real sequence has a monotone subsequence.*

Proof. Call an element x_k in (x_n) *dominant* if all the following elements are less than or equal to it. If there are infinitely many such dominant elements, then we can select these to be our monotone subsequence (which is decreasing). If not, let x_m be the last dominant term. Since x_{m+1} is not dominant, there is a $j > m + 1$ such that $x_j > x_{m+1}$. Since x_j is not dominant there is an $i > j$ such that $x_i > x_j$. Continuing in this way, we can pick out a monotone subsequence (which is increasing). $\square$

Now we have the following crucial property of $\mathbb{R}$. Usually called the Bolzano–Weierstrass theorem, it also extends to higher dimensional space (see theorem 3.2.18 on page 47) and forms the foundations of countless results in analysis.

---

[12]How are you going with proof by contradiction? After a while you will become familiar with the style of argument. The assertion that (x_n) is not Cauchy led to a contradiction—in this case of the hypothesis that (x_n) is bounded. We are forced to conclude that this assertion (i.e., that (x_n) is not Cauchy) is false. In other words, (x_n) is Cauchy.

[13]Hint: How many points are there outside a given ϵ-ball around the limit?

Theorem A.2.10 *Every bounded sequence in $\mathbb{R}$ has a convergent subsequence.*

Proof. Take a given sequence in $\mathbb{R}$. By theorem A.2.9, the sequence has a monotone subsequence, which is a sequence in its own right. Evidently this sequence is also bounded. By theorem A.2.5, every bounded monotone sequence converges. □

The next result is important in practice, and the proof is an exercise.

Theorem A.2.11 *Let (x_n) and (y_n) be two sequences in $\mathbb{R}$, with $\lim x_n = x$ and $\lim y_n = y$. If $x_n \leq y_n$ for all $n \in \mathbb{N}$, then $x \leq y$.*[14]

Often when we use this result one sequence will be a constant. For example, if $x_n \leq b$ for all $n \in \mathbb{N}$, then $\lim x_n \leq b$. Note that taking limits does not preserve strict ordering! For example, $1/n > 0$ for all n, but $\lim_n 1/n > 0$ is false.

Theorem A.2.12 *Let (x_n), (y_n) and (z_n) be three sequences in $\mathbb{R}$, with $x_n \leq y_n \leq z_n$ for all $n \in \mathbb{N}$. If $x_n \to a$ and $z_n \to a$ both hold, then $y_n \to a$.*

Proof. Fix $\epsilon > 0$. We can choose an $N \in \mathbb{N}$ such that if $n \geq N$, then $x_n > a - \epsilon$ and $z_n < a + \epsilon$. (Why?) For such n we must have $|y_n - a| < \epsilon$. □

You might have thought it would be simpler to argue that, since $x_n \leq y_n \leq z_n$ for all n, we have $\lim x_n \leq \lim y_n \leq \lim z_n$ from theorem A.2.11. But this is not permissible because we did not know at the start of the proof that $\lim y_n$ exists. Theorem A.2.11 expressly requires that the limits exist. (This is an easy mistake to make.)

Theorem A.2.13 *Let (x_n) and (y_n) be real sequences. If $x_n \to x$ and $y_n \to y$, then $x_n + y_n \to x + y$.*

Proof. Fix $\epsilon > 0$. By the triangle inequality,

$$|(x_n + y_n) - (x + y)| \leq |x_n - x| + |y_n - y| \tag{A.7}$$

Choose $N \in \mathbb{N}$ such that $|x_n - x| < \epsilon/2$ whenever $n \geq N$, and $N' \in \mathbb{N}$ such that $|y_n - y| < \epsilon/2$ whenever $n \geq N'$. For $n \geq \max\{N, N'\}$, the right-hand side of (A.7) is less than ϵ. □

Exercise A.2.14 Show that if $a \in \mathbb{R}$ and $x_n \to x$, then $ax_n \to ax$.

Theorem A.2.15 *Let (x_n) and (y_n) be real sequences. If $x_n \to x$ and $y_n \to y$, then $x_n y_n \to xy$.*

---

[14]Hint for the proof: Suppose that $x > y$. Take ϵ-balls around each point that do not intersect. (Convince yourself that this is possible.) Now try to contradict $x_n \leq y_n$ for all n.

Proof. In view of exercise A.2.6, there is a positive integer M such that $|x_n| \leq M$ for all $n \in \mathbb{N}$. By the triangle inequality,

$$|x_n y_n - xy| = |x_n y_n - x_n y + x_n y - xy| \leq |x_n y_n - x_n y| + |x_n y - xy|$$
$$= |x_n||y_n - y| + |y||x_n - x| \leq M|y_n - y| + |y||x_n - x| < \epsilon$$

The result now follows from exercise A.2.14 and theorem A.2.13. $\square$

If (x_n) is a sequence in $\mathbb{R}$, the term $\sum_{n \geq 1} x_n$ or $\sum_n x_n$ is defined, when it exists, as the limit of the sequence (s_k), where $s_k := \sum_{n=1}^{k} x_n$. If $s_k \to \infty$, then we write $\sum_n x_n = \infty$. Of course the limit may fail to exist entirely, as for $x_n = (-1)^n$.

Lemma A.2.16 *Let $(x_n) \subset \mathbb{R}_+$. If $\sum_n x_n < \infty$, then $x_n \to 0$.*

Proof. Suppose instead that $x_n \to 0$ fails. Then $\exists \, \epsilon > 0$ such that $x_n > \epsilon$ infinitely often. (Why?) Hence $\sum_n x_n = \infty$. (Why?) Contradiction. $\square$

A.2.2   Max, Min, Sup, and Inf

Let x and y be any two real numbers. We will use the notation $x \vee y$ for the maximum of x and y, while $x \wedge y$ is their minimum. The following equalities are bread and butter:

Lemma A.2.17 *For any $x, y \in \mathbb{R}$ and any $a \geq 0$ we have the following identities:*

1. $x + y = x \vee y + x \wedge y$.

2. $|x - y| = x \vee y - x \wedge y$.

3. $|x - y| = x + y - 2(x \wedge y)$.

4. $|x - y| = 2(x \vee y) - x - y$.

5. $a(x \vee y) = (ax) \vee (ay)$.

6. $a(x \wedge y) = (ax) \wedge (ay)$.

To see that $x + y = x \vee y + x \wedge y$, pick any $x, y \in \mathbb{R}$. Suppose without loss of generality that $x \leq y$. Then $x \vee y + x \wedge y = y + x$, as was to be shown. The remaining equalities are left as exercises.

Exercise A.2.18 Show that if $x_n \to x$ in $\mathbb{R}$, then $|x_n| \to |x|$. (Hint: Use (A.6) on page 327.) Using this result and identities 3 and 4 in lemma A.2.17, argue that if $x_n \to x$ and $y_n \to y$, then $x_n \wedge y_n \to x \wedge y$ and $x_n \vee y_n \to x \vee y$.

If $A \subset \mathbb{R}$, the maximum of A, when it exists, is a number $m \in A$ with $a \leq m$ for all $a \in A$. The minimum is defined analogously. For any finite collection of real numbers, the maximum and minimum always exist. For infinite collections this is not the case. To deal with infinite sets we introduce the notion of suprema and infima.

Given a set $A \subset \mathbb{R}$, an *upper bound* of A is any number u such that $a \leq u$ for all $a \in A$. If $s \in \mathbb{R}$ is an upper bound for A and also satisfies $s \leq u$ for every upper bound u of A, then s is called the *supremum* of A. You will be able to verify that at most one such s exists. We write $s = \sup A$.

Lemma A.2.19 *Suppose that s is an upper bound of A. The following statements are then equivalent:*

1. *$s = \sup A$.*

2. *$s \leq u$ for all upper bounds u of A.*

3. *$\forall \epsilon > 0, \exists a \in A$ with $a > s - \epsilon$.*

4. *There exists a sequence $(a_n) \subset A$ with $a_n \uparrow s$.*

Exercise A.2.20 Prove lemma A.2.19.

Exercise A.2.21 Show that $\sup(0, 1) = 1$ and $\sup(0, 1] = 1$. Show that if a set A contains one of its upper bounds u, then $u = \sup A$.

Theorem A.2.22 *Every nonempty subset of $\mathbb{R}$ that is bounded above has a supremum in $\mathbb{R}$.*

The proof is omitted, but this is in fact *equivalent* to axiom A.2.4. Either one can be treated as the axiom. They assert the "completeness" of the real numbers.

If A is not bounded above, then it is conventional to set $\sup A := \infty$. With this convention, the following statement is true:

Lemma A.2.23 *If $A, B \subset \mathbb{R}$ with $A \subset B$, then $\sup A \leq \sup B$.*

Proof. If $\sup B = \infty$ the result is trivial. Suppose instead that B is bounded above, and let $\bar{b} := \sup B$, $\bar{a} = \sup A$. By lemma A.2.19, there is a sequence $(a_n) \subset A$ with $a_n \uparrow \bar{a}$. But $\bar{b}$ is an upper bound for A (why?), so $a_n \leq \bar{b}$ for all n. It now follows from theorem A.2.11 that $\bar{a} = \lim a_n \leq \bar{b}$.   □

For $A \subset \mathbb{R}$ a *lower bound* of A is any number l such that $a \geq l$ for all $a \in A$. If $i \in \mathbb{R}$ is an lower bound for A and also satisfies $i \geq l$ for every lower bound l of A, then i is called the *infimum* of A. At most one such i exists. We write $i = \inf A$. Every nonempty subset of $\mathbb{R}$ bounded from below has an infimum.

Exercise A.2.24 Let A be bounded below. Show that $i = \inf A$ if and only if i is a lower bound of A and, for each $\epsilon > 0$, there is an $a \in A$ with $a < i + \epsilon$.

Lemma A.2.25 *If $A, B \subset \mathbb{R}$ with $A \subset B$, then $\inf A \geq \inf B$.*

Proof. The proof is an exercise.   □

For $(x_n) \subset \mathbb{R}$ we set

$$\liminf x_n := \lim_{n \to \infty} \inf_{k \geq n} x_k \quad \text{and} \quad \limsup x_n := \lim_{n \to \infty} \sup_{k \geq n} x_k$$

If (x_n) is bounded, then both $\liminf x_n$ and $\limsup x_n$ always exist in $\mathbb{R}$. (Why?)

Exercise A.2.26 For A a bounded subset of $\mathbb{R}$, let $-A$ be all $b \in \mathbb{R}$ such that $b = -a$ for some $a \in A$. Show that $-\sup A = \inf(-A)$. Let (x_n) be a bounded sequence of real numbers. Show that $-\limsup x_n = \liminf -x_n$.

Exercise A.2.27 Let (x_n) be a sequence of real numbers, and let $x \in \mathbb{R}$. Show that $\lim_n x_n = x$ if and only if $\limsup_n x_n = \liminf_n x_n = x$.

Exercise A.2.28 Let (x_n), (y_n), and (z_n) be sequences of real numbers with $x_n \leq y_n + z_n$ for all $n \in \mathbb{N}$. Show that the following inequality always holds:

$$\limsup x_n \leq \limsup y_n + \limsup z_n$$

Exercise A.2.29 Show that $(x_n) \subset \mathbb{R}_+$ and $\limsup x_n = 0$ implies $\lim x_n = 0$.[15]

Let $f \colon A \to \mathbb{R}$, where A is any nonempty set. We will use the notation

$$\sup f :=: \sup_{x \in A} f(x) := \sup\{f(x) : x \in A\}$$

Also, if $g \colon A \to \mathbb{R}$, then $f + g$ is defined by $(f + g)(x) = f(x) + g(x)$, while $|f|$ is defined by $|f|(x) = |f(x)|$.

Lemma A.2.30 *Let $f, g \colon A \to \mathbb{R}$, where A is any nonempty set. Then*

$$\sup(f + g) \leq \sup f + \sup g$$

Proof. We can and do suppose that $\sup f$ and $\sup g$ are finite. (Otherwise the result is trivial.) For any $x \in A$, $f(x) \leq \sup f$ and $g(x) \leq \sup g$.

$$\therefore \quad f(x) + g(x) \leq \sup f + \sup g$$

$$\therefore \quad \sup(f + g) \leq \sup f + \sup g$$

                                                                              □

---

[15]Hint: A neat argument follows from theorem A.2.12.

Lemma A.2.31 *If $f \colon A \to \mathbb{R}$, then $|\sup f| \leq \sup |f|$.*

Proof. We can and do suppose that $\sup |f| < \infty$. Evidently $\sup f \leq \sup |f|$.[16] To complete the proof, we need only show that $-\sup f \leq \sup |f|$ also holds. This is the case because

$$0 = \sup(-f + f) \leq \sup(-f) + \sup f \leq \sup |f| + \sup f$$

$\square$

Exercise A.2.32 Show via counterexample that the statement $|\sup f| = \sup |f|$ does not hold in general.

Let $A \subset \mathbb{R}$. A function $f \colon A \to \mathbb{R}$ is called *monotone increasing* on A if, whenever $x, y \in A$ and $x \leq y$, we have $f(x) \leq f(y)$. It is called *monotone decreasing* if, whenever $x \leq y$, we have $f(x) \geq f(y)$. We say strictly monotone increasing or strictly monotone decreasing if the previous inequalities can be replaced with strict inequalities.

Exercise A.2.33 Let S be any set, let $g \colon S \to \mathbb{R}$, and let $\bar{x}$ be a maximizer of g on S, in the sense that $g(\bar{x}) \geq g(x)$ for all $x \in S$. Prove that if $f \colon \mathbb{R} \to \mathbb{R}$ is monotone increasing, then $\bar{x}$ is a maximizer of $f \circ g$ on S.

A.2.3   Functions of a Real Variable

Let's recall some basics about functions when send subsets of $\mathbb{R}$ into $\mathbb{R}$. Below we define such concepts as continuity, differentiability, convexity, and concavity. If you are rusty on these definitions, then it is probably worth skim-reading this section and completing a few of the exercises.

Let $A \subset \mathbb{R}$ and let $f, g \colon A \to \mathbb{R}$. As usual, the sum of f and g is the function $f + g$ defined by $(f + g)(x) := f(x) + g(x)$. Similarly, the product fg is defined by $(fg)(x) := f(x)g(x)$. The product of real number α and f is the function $(\alpha f)(x) := \alpha f(x)$. Recall that f is called *bounded* if its range is a bounded set (i.e., $\exists M \in \mathbb{N}$ such that $|f(a)| \leq M$ for all $a \in A$).

Exercise A.2.34 Show that if f and g are bounded and $\alpha \in \mathbb{R}$, then $f + g$, fg, and αf are also bounded functions.

Function $f \colon A \to \mathbb{R}$ is said to be *continuous* at $a \in A$ if for every sequence (x_n) in A converging to a we have $f(x_n) \to f(a)$. (Sketch it.) It is called continuous on A (or just continuous) whenever it is continuous at every $a \in A$. Continuity of functions captures the idea that small changes to the input do not lead to sudden jumps in the

---

[16]Pick any $x \in A$. Then $f(x) \leq |f(x)| \leq \sup |f|$. Since x is arbitrary, $\sup f \leq \sup |f|$.

output. Notice that in requiring that $f(x_n) \to f(a)$ for each $x_n \to a$, we require that not only does $f(x_n)$ actually converge for each choice of $x_n \to a$, but all these sequences converge to the same limit, and moreover that limit is $f(a)$.

Exercise A.2.35 Prove carefully that the functions $f(x) = x + 1$ and $g(x) = x^2$ are continuous. Give an example of a function that is not continuous, showing how it fails the definition.

More generally, for the same $f \colon A \to \mathbb{R}$ and for $a \in A$, we say that $y = \lim_{x \to a} f(x)$ if $f(x_n) \to y$ for every sequence $(x_n) \subset A$ with $x_n \to a$. Note that $\lim_{x \to a} f(x)$ may not exist. It may be the case that different sequences converging to a yield different limits for the sequence $f(x_n)$, or indeed that $f(x_n)$ does not converge at all. But this new notation is useful because we can now say that f is continuous at a if and only if $\lim_{x \to a} f(x)$ exists and is equal to $f(a)$.

Exercise A.2.36 Show that if f and g are continuous functions and $\alpha \in \mathbb{R}$, then $f + g$, fg and αf are also continuous.

Exercise A.2.37 A function $f \colon A \to \mathbb{R}$ is said to be *continuous from the left* at $x \in A$ if $f(x_n) \to f(x)$ for every sequence $x_n \uparrow x$; and *continuous from the right* at $x \in A$ if $f(x_n) \to f(x)$ for every sequence $x_n \downarrow x$. Clearly, a function continuous at x is both continuous from the left at x and continuous from the right at x. Show that the converse also holds.[17]

One of the many delightful results concerning continuous functions is the intermediate value theorem:

Theorem A.2.38 *Let $f \colon [a, b] \to \mathbb{R}$, where $a < b$. If f is continuous on $[a, b]$ and $f(a) < 0 < f(b)$, then there exists an $s \in (a, b)$ with $f(s) = 0$.*

Proof. Let $A := \{x \in [a, b] : f(x) < 0\}$, and let s be the supremum of this set. (Why can we be sure that such a supremum exists?) We claim that $f(s) = 0$. To see why this must be the case, observe that since $s = \sup A$ there exists a sequence (x_n) with $f(x_n) < 0$ and $x_n \uparrow s$. (Why?) By continuity of f, we have $\lim f(x_n) = f(s)$. But $f(x_n) < 0$ for all n, so $\lim f(x_n) \leq 0$. Hence $f(s) \leq 0$. On the other hand, since s is an upper bound of A, we know that $x > s$ implies $x \notin A$, in which case $f(x) \geq 0$. Take a strictly decreasing sequence (x_n) in $(s, b]$ with $x_n \downarrow s$. (Convince yourself that such a sequence does exist.) As $f(x_n) \geq 0$ for all n it follows that $\lim f(x_n) = f(s) \geq 0$. Therefore $f(s) = 0$. $\qquad\square$

Exercise A.2.39 Using theorem A.2.38 (the result, not the proof), show that the same result holds when $f(b) < 0 < f(a)$.

---

[17]Hint: You might like to make use of exercise A.2.8 and theorem A.2.9.

Let's briefly review differentiability. Let $f\colon (a,b) \to \mathbb{R}$, and let $x \in (a,b)$. The function f is said to be *differentiable at x* if, for every sequence (h_n) converging to zero and satisfying $h_n \neq 0$ and $x + h_n \in (a,b)$ for each n, the sequence

$$\frac{f(x + h_n) - f(x)}{h_n}$$

converges, and the limit is independent of the choice of (h_n). If such a limit exists, it is denoted by $f'(x)$. The function f is called *differentiable* if it is differentiable at each point in its domain, and *continuously differentiable* if, in addition to being differentiable, $x \mapsto f'(x)$ is continuous everywhere on the domain of f.

Exercise A.2.40 Let $f\colon \mathbb{R} \to \mathbb{R}$ be defined by $f(x) = x^2$. Prove that $f'(x) = 2x$ for any $x \in \mathbb{R}$.

A function f from an interval I to $\mathbb{R}$ is called *convex* (resp., *strictly convex*) if

$$\lambda f(x) + (1 - \lambda)f(y) \geq f(\lambda x + (1 - \lambda)y)$$

for all $\lambda \in [0,1]$ and $x, y \in I$ (resp., for all $x \neq y$ and all $\lambda \in (0,1)$)), and *concave* (resp., *strictly concave*) if

$$\lambda f(x) + (1 - \lambda)f(y) \leq f(\lambda x + (1 - \lambda)y)$$

for all $\lambda \in [0,1]$ and $x, y \in I$ (resp., for all $x \neq y$ and all $\lambda \in (0,1)$). Since f is concave if and only if $-f$ is convex, we can think of convexity is the fundamental property; concavity is merely a shorthand way of referring to convexity of $-f$.

There are numerous connections between continuity, differentiability, and convexity. For example, if $f\colon [a,b] \to \mathbb{R}$ is convex, then it is continuous everywhere on (a,b). Also you are no doubt aware that if f is twice differentiable, then nonnegativity of f'' on (a,b) is equivalent to convexity on (a,b). These facts can be proved from the definitions above.

Finally, let's consider right and left derivatives. Let $f\colon (a,b) \to \mathbb{R}$. For fixed $x \in (a,b)$ we define

$$D(x,h) := \frac{f(x+h) - f(x)}{h} \qquad (h \neq 0 \text{ and } x + h \in (a,b))$$

If for each sequence $h_n \downarrow 0$ the limit $\lim_{n \to \infty} D(x, h_n)$ exists and is equal to the same number, we call that number the *right-hand derivative* of f at x, and denote it by $f'_+(x)$. If for each sequence $h_n \uparrow 0$ the limit $\lim_{n \to \infty} D(x, h_n)$ exists and is equal to the same number, we call that number the *left-hand derivative* of f at x, and denote it by $f'_-(x)$. It turns out that f is differentiable at x if and only if both the left- and right-hand derivatives exist at x and are equal. The proof is not too difficult if you feel like doing it as an exercise.

The following lemma collects some useful facts:

Lemma A.2.41 *If f is concave on (a, b), then f'_+ and f'_- exist everywhere on (a, b). For each $x \in (a, b)$,*

$$f'_+(x) = \sup_{h>0} D(x, h) \text{ and } f'_-(x) = \inf_{h<0} D(x, h)$$

Moreover $f'_+ \leq f'_-$ everywhere on (a, b). If $f'_+(x) = f'_-(x)$ at some point $x \in (a, b)$, then f is differentiable at x, and $f'(x) = f'_+(x) = f'_-(x)$.

Exercise A.2.42 Prove lemma A.2.41. First show that when f is concave, $D(x, h)$ is decreasing in h. Next apply existence results for limits of monotone bounded sequences.

Appendix B

Chapter Appendixes

B.1 Appendix to Chapter 3

Let us briefly discuss the topic of parametric continuity. The question we address is whether or not the solution to a given optimization problem varies continuously with the parameters that define the problem. The classic theorem in this area is Berge's theorem of the maximum. You should familiarize yourself at least with the statement of the theorem.

To begin, let A and B be two sets. A function Γ from A into $\mathfrak{P}(B)$ (i.e., into the subsets of B) is called a *correspondence* from A to B. Correspondences are often used to define constraint sets. For example, $a \in A$ might be the price of a commodity, or a level of wealth, and $\Gamma(a) \subset B$ is the budget set associated with that value of the parameter.

Now suppose that A and B are metric spaces, and let Γ be a correspondence from A to B. We say that Γ is *compact-valued* if $\Gamma(a)$ is a compact subset of B for every $a \in A$, and nonempty if $\Gamma(a) \neq \emptyset$ for every $a \in A$. A nonempty compact-valued correspondence Γ from A to B is called *upper-hemicontinuous* at $a \in A$ if, for each sequence $(a_n) \subset A$ with $a_n \to a$, and each sequence $(b_n) \subset B$ with $b_n \in \Gamma(a_n)$ for all $n \in \mathbb{N}$, the sequence (b_n) has a convergent subsequence whose limit is in $\Gamma(a)$. It is called *lower-hemicontinuous* at a if, for each $(a_n) \subset A$ with $a_n \to a$ and each $b \in \Gamma(a)$, there is a sequence $(b_n) \subset B$ with $b_n \in \Gamma(a_n)$ for all $n \in \mathbb{N}$, and $b_n \to b$ as $n \to \infty$. Finally, Γ is called *continuous* at a if it is both upper-hemicontinuous and lower-hemicontinuous at a. It is called continuous if it is continuous at a for each $a \in A$.

The following lemma treats an important special case:

Lemma B.1.1 *Let $A \subset \mathbb{R}$, let g and h be continuous functions from A to $\mathbb{R}$, and let $\Gamma \colon A \to \mathfrak{P}(\mathbb{R})$ be defined by*

$$\Gamma(x) = \{y \in \mathbb{R} : g(x) \le y \le h(x)\} \qquad (x \in A)$$

If g and h are continuous functions, then the correspondence Γ is also continuous.

Proof. Pick any $a \in A$. First let's check upper-hemicontinuity at a. Let $(a_n) \subset A$, $a_n \to a$, and let $(b_n) \subset B$, $b_n \in \Gamma(a_n)$ for all n. We claim the existence of a subsequence (b_{n_j}) and a $b \in \Gamma(a)$ with $b_{n_j} \to b$ as $j \to \infty$.

To see why—fill in any gaps in the argument to your own satisfaction—note that (a_n) is bounded and $C := \{a\} \cup \{a_n\}_{n \in \mathbb{N}}$ is closed, from which it follows that $G := \inf_{x \in C} g(x)$ and $H = \sup_{x \in C} h(x)$ exist (see theorem 3.2.20 on page 48). But as $G \le b_n \le H$ for all n, the sequence b_n is bounded and hence contains a convergent subsequence $b_{n_j} \to b$. Observing that $g(a_{n_j}) \le b_{n_j} \le h(a_{n_j})$ for all $j \in \mathbb{N}$, we can take the limit to obtain $g(a) \le b \le h(a)$. In other words, $b \in \Gamma(a)$, as was to be proved.

Regarding lower-hemicontinuity at a, given (a_n) with $a_n \to a$ and $b \in \Gamma(a)$, we claim there is a sequence $(b_n) \subset B$ with $b_n \in \Gamma(a_n)$ for all $n \in \mathbb{N}$ and $b_n \to b$. To see this, suppose first that $b = g(a)$. Setting $b_n = g(a_n)$ gives the desired convergence. The case of $b = h(a)$ is treated similarly. Suppose instead that $g(a) < b < h(a)$. It follows that for N sufficiently large we have $g(a_n) < b < h(a_n)$ whenever $n \ge N$. Taking $b_1, \ldots, b_{N-1}$ arbitrary and $b_n = b$ for all $n \ge N$ gives a suitable sequence $b_n \to b$. $\square$

Exercise B.1.2 Let $\Gamma \colon A \to B$ be a correspondence such that $\Gamma(a)$ is a singleton $\{b_a\}$ for each $a \in A$. Show that if Γ is a continuous correspondence, then $a \mapsto b_a$ is a continuous function.

Now we can state Berge's theorem.

Theorem B.1.3 *Let Θ and U be two metric spaces, let Γ be a correspondence from Θ to U, and let*

$$\operatorname{gr}\Gamma := \{(\theta, u) \in \Theta \times U : u \in \Gamma(\theta)\}$$

If $f \colon \operatorname{gr}\Gamma \to \mathbb{R}$ is continuous, and Γ is nonempty, compact-valued and continuous, then the function

$$g \colon \Theta \ni \theta \mapsto \max_{u \in \Gamma(\theta)} f(\theta, u) \in \mathbb{R}$$

is continuous on Θ. The correspondence of maximizers

$$M \colon \Theta \ni \theta \mapsto \operatorname*{argmax}_{u \in \Gamma(\theta)} f(\theta, u) \subset U$$

is compact-valued and upper-hemicontinuous on Θ. In particular, if $M(\theta)$ is single-valued, then it is continuous.

In the theorem, continuity of f on $\operatorname{gr}\Gamma$ means that if $(\theta, u) \in \operatorname{gr}\Gamma$ and (θ_n, u_n) is a sequence in $\operatorname{gr}\Gamma$ with $(\theta_n, u_n) \to (\theta, u)$, then $f(\theta_n, u_n) \to f(\theta, u)$. (This is a stronger requirement than assuming f is continuous in each individual argument while the other is held fixed.) The theorem is well-known to economists, and we omit the proof. See Aliprantis and Border (1999, thm. 16.31), or Stokey and Lucas (1989, thm. 3.6).

The next result is a direct implication of Berge's theorem B.1.3 but pertains to parametric continuity of fixed points.

Theorem B.1.4 *Let* Θ, U *and* Γ *be as in theorem B.1.3. Let* $g \colon \operatorname{gr}\Gamma \to U$, *and let*

$$F(\theta) := \{u \in U : u = g(\theta, u)\} \qquad (\theta \in \Theta)$$

If $F(\theta)$ *is nonempty for each* $\theta \in \Theta$, g *is continuous on* $\operatorname{gr}\Gamma$, *and* Γ *is nonempty, compact-valued and continuous, then* $\theta \mapsto F(\theta)$ *is compact-valued and upper-hemicontinuous on* Θ. *In particular, if* $F(\theta)$ *is single-valued, then it is continuous.*

Proof. Continuity of g on $\operatorname{gr}\Gamma$ means that if $(\theta, u) \in \operatorname{gr}\Gamma$ and (θ_n, u_n) is a sequence in $\operatorname{gr}\Gamma$ with $(\theta_n, u_n) \to (\theta, u)$, then $g(\theta_n, u_n) \to g(\theta, u)$. Let $f \colon \operatorname{gr}\Gamma \to \mathbb{R}$ be defined by

$$f(\theta, u) = -\rho(u, g(\theta, u))$$

where ρ is the metric on U. You will be able to show that f is also continuous on $\operatorname{gr}\Gamma$. Theorem B.1.3 then implies that $\theta \mapsto M(\theta)$ is compact-valued and upper-hemicontinuous, where $M(\theta)$ is the set of maximizers $\operatorname{argmax}_{u \in \Gamma(\theta)} f(\theta, u)$.

Now pick any $\theta \in \Theta$. As $F(\theta)$ is assumed to be nonempty, the set of maximizers of f and fixed-points of g coincide. That is, $M(\theta) = F(\theta)$. Since θ is arbitrary, M and F are the same correspondence on Θ, and $\theta \mapsto F(\theta)$ is also compact-valued and upper-hemicontinuous. $\qquad\square$

Next we turn to the

Proof of theorem 3.2.38. Uniqueness is by exercise 3.2.34. To prove existence, define $r \colon S \to \mathbb{R}$ by $r(x) = \rho(Tx, x)$. It is not too difficult to show that r is continuous (with respect to ρ). Since S is compact, r has a minimizer x^*. But then $Tx^* = x^*$ must hold, because otherwise

$$r(Tx^*) = \rho(TTx^*, Tx^*) < \rho(Tx^*, x^*) = r(x^*)$$

contradicting the definition of x^*.

Next we show that $T^n x \to x^*$ as $n \to \infty$ for all $x \in S$. To see this, pick any $x \in S$ and consider the real sequence (α_n) defined by $\alpha_n := \rho(T^n x, x^*)$. Since T is contracting, the sequence (α_n) is monotone decreasing, and therefore converges (why?) to some limit $\alpha \geq 0$. I claim that $\alpha = 0$.

To see this, we can argue as follows: By compactness of S the sequence $(T^n x)$ has a subsequence $(T^{n(k)}x)$ with $T^{n(k)}x \to x'$ for some $x' \in S$. It must be the case that $\rho(x', x^*) = \alpha$. The reason is that $y \mapsto \rho(y, x^*)$ is continuous as a map from S to $\mathbb{R}$ (example 3.1.12 on page 40). Hence $\rho(T^{n(k)}x, x^*) \to \rho(x', x^*)$. But $\rho(T^{n(k)}x, x^*) \to \alpha$ and sequences have at most one limit, so $\rho(x', x^*) = \alpha$.

It is also the case that $\rho(Tx', x^*) = \alpha$. To see this, note that by continuity of T we have

$$T(T^{n(k)}x) = T^{n(k)+1}x \to Tx'$$

Since $y \mapsto \rho(y, x^*)$ is continuous, we have $\rho(T^{n(k)+1}x, x^*) \to \rho(Tx', x^*)$. At the same time, $\rho(T^{n(k)+1}x, x^*) \to \alpha$ is also true, so $\rho(Tx', x^*) = \alpha$.

We have established the existence of a point $x' \in S$ such that both $\rho(x', x^*)$ and $\rho(Tx', x^*)$ are equal to α. If $\alpha > 0$ the points x' and x^* are distinct, implying

$$\alpha = \rho(x', x^*) > \rho(Tx', Tx^*) = \rho(Tx', x^*) = \alpha$$

Contradiction.   □

B.2   Appendix to Chapter 4

We now provide the proof of theorem 4.3.17 on page 89. To begin our proof, consider the following result:

Lemma B.2.1 *If ϕ and ψ are elements of $\mathscr{P}(S)$ and $h \colon S \to \mathbb{R}_+$, then*

$$\left| \sum_{x \in S} h(x)\phi(x) - \sum_{x \in S} h(x)\psi(x) \right| \leq \frac{1}{2} \sup_{x,x'} |h(x) - h(x')| \cdot \|\phi - \psi\|_1$$

Proof. Let $\rho(x) := \phi(x) - \psi(x)$, $\rho^+(x) := \rho(x) \vee 0$, $\rho^-(x) := (-\rho(x)) \vee 0$. It is left to the reader to show that $\rho(x) = \rho^+(x) - \rho^-(x)$, that $|\rho(x)| = \rho^+(x) + \rho^-(x)$, and that $\sum_{x \in S} \rho^+(x) = \sum_{x \in S} \rho^-(x) = (1/2)\|\rho\|_1$.

In view of the equality $|a - b| = a \vee b - a \wedge b$ (lemma A.2.17, page 331), we have

$$\left| \sum h\phi - \sum h\psi \right| = \left| \sum h\rho \right| = \left| \sum h\rho^+ - \sum h\rho^- \right|$$
$$= \left(\sum h\rho^+ \right) \bigvee \left(\sum h\rho^- \right) - \left(\sum h\rho^+ \right) \bigwedge \left(\sum h\rho^- \right)$$

Consider the two terms to the right of the last equality. If $\sup h := \sup_{x \in S} h(x)$ and $\inf h := \inf_{x \in S} h(x)$, then the first term satisfies

$$\left(\sum h\rho^+ \right) \bigvee \left(\sum h\rho^- \right) \leq \left(\sup h \sum \rho^+ \right) \bigvee \left(\sup h \sum \rho^- \right)$$
$$= \sup h \left(\sum \rho^+ \right) \bigvee \left(\sum \rho^- \right) = \sup h \frac{\|\rho\|_1}{2}$$

while the second satisfies

$$\left(\sum h\rho^+\right) \bigwedge \left(\sum h\rho^-\right) \ge \left(\inf h \sum \rho^+\right) \bigwedge \left(\inf h \sum \rho^-\right)$$
$$= \inf h \left(\sum \rho^+\right) \bigwedge \left(\sum \rho^-\right) = \inf h \frac{\|\rho\|_1}{2}$$

Combining these two bounds, we get

$$\left|\sum h\phi - \sum h\psi\right| \le (\sup h - \inf h)\frac{\|\rho\|_1}{2}$$

This is the same bound as given in the statement of the lemma. $\square$

We need two more results to prove theorem 4.3.17, both of which are straightforward.

Lemma B.2.2 *Let p, $\mathbf{M}$, ϕ, and ψ be as in theorem 4.3.17. Then*[1]

$$\|\phi\mathbf{M} - \psi\mathbf{M}\|_1 \le \frac{1}{2}\sup_{x,x'} \|p(x,dy) - p(x',dy)\|_1 \cdot \|\phi - \psi\|_1$$

Proof. In the proof of this lemma, if $\phi \in \mathscr{P}(S)$ and $A \subset S$, we will write $\phi(A)$ as a shorthand for $\sum_{y\in A} \phi(x)$. So pick any $A \subset S$. In view of lemma B.2.1, we have

$$\left|\sum_{x\in S} P(x,A)\phi(x) - \sum_{x\in S} P(x,A)\psi(x)\right| \le \frac{1}{2}\sup_{x,x'} |P(x,A) - P(x',A)| \cdot \|\phi - \psi\|_1$$

Applying the result of exercise 4.3.2, we obtain

$$\left|\sum_{x\in S} P(x,A)\phi(x) - \sum_{x\in S} P(x,A)\psi(x)\right| \le \frac{1}{4}\sup_{x,x'} \|p(x,dy) - p(x',dy)\|_1 \cdot \|\phi - \psi\|_1$$

which is another way of writing

$$|\phi\mathbf{M}(A) - \psi\mathbf{M}(A)| \le \frac{1}{4}\sup_{x,x'} \|p(x,dy) - p(x',dy)\|_1 \cdot \|\phi - \psi\|_1$$

$$\therefore \quad \sup_{A\subset S} |\phi\mathbf{M}(A) - \psi\mathbf{M}(A)| \le \frac{1}{4}\sup_{x,x'} \|p(x,dy) - p(x',dy)\|_1 \cdot \|\phi - \psi\|_1$$

Using exercise 4.3.2 again, we obtain the bound we are seeking. $\square$

---

[1]Here $\|p(x,dy) - p(x',dy)\|_1$ is to be interpreted as $\sum_{y\in S} |p(x,y) - p(x'y)|$.

To prove the first claim in theorem 4.3.17, it remains only to show that

$$\frac{1}{2}\sup_{x,x'}\|p(x,dy) - p(x',dy)\|_1 = 1 - \inf_{x,x'}\sum_{y \in S}p(x,y) \wedge p(x',y)$$

It is sufficient (why?) to show that $\|p(x,dy) - p(x',dy)\|_1/2 = 1 - \sum_{y \in S}p(x,y) \wedge p(x',y)$ for any pair x, x'. Actually this is true for any pair of distributions, as shown in the next and final lemma.

Lemma B.2.3 *For any pair $\mu, \nu \in \mathscr{P}(S)$, we have $\|\mu - \nu\|_1/2 = 1 - \sum_{y \in S}\mu(y) \wedge \nu(y)$.*

Proof. From lemma A.2.17 (page 331) one can show that given any pair of real numbers a and b, we have $|a - b| = a + b - 2a \wedge b$. Hence for each $x \in S$ we obtain

$$|\mu(x) - \nu(x)| = \mu(x) + \nu(x) - 2\mu(x) \wedge \nu(x)$$

Summing over x gives the identity we are seeking. $\qquad\square$

The first claim in theorem 4.3.17 is now established. Regarding the second claim, we have

$$1 - \alpha(p) = \frac{1}{2}\sup_{x,x'}\|p(x,dy) - p(x',dy)\|_1$$

$$= \sup_{x \neq x'}\frac{\|p(x,dy) - p(x',dy)\|_1}{\|\delta_x - \delta_{x'}\|_1} \leq \sup_{\mu \neq \nu}\frac{\|\mu\mathbf{M} - \nu\mathbf{M}\|_1}{\|\mu - \nu\|_1}$$

The claim now follows from the definition of the supremum.

B.3 Appendix to Chapter 6

Proof of theorem 6.3.8. Pick any u and v in $b\mathcal{U}$. Observe that

$$u = u + v - v \leq v + |u - v| \leq v + \|u - v\|_\infty$$

where (in)equalities are pointwise on $\mathcal{U}$. By the monotonicity property of T, we have $Tu \leq T(v + \|u - v\|_\infty)$. Applying (6.29), we have $Tu - Tv \leq \lambda\|u - v\|_\infty$. Reversing the roles of u and v gives $Tv - Tu \leq \lambda\|u - v\|_\infty$. These two inequalities are sufficient for the proof. (Why?) $\qquad\square$

B.4   Appendix to Chapter 8

First let us prove lemma 8.2.3, beginning with some preliminary discussion: We can extend $\mathbf{M}$ to act on all functions f in $L_1(S)$ by setting $f\mathbf{M}(y) := \int p(x,y)f(x)dx$. The inequality $\|f\mathbf{M}\|_1 \leq \|f\|_1$ always holds. Under the following condition it is strict:

Lemma B.4.1 *Let $f \in L_1(S)$. Then $\|f\mathbf{M}\|_1 < \|f\|_1$ if and only if*

$$\lambda[(f^+\mathbf{M}) \wedge (f^-\mathbf{M})] > 0$$

Proof. In view of lemma A.2.17 (page 331) the pointwise inequality

$$|f\mathbf{M}| = |f^+\mathbf{M} - f^-\mathbf{M}| = f^+\mathbf{M} + f^-\mathbf{M} - 2(f^+\mathbf{M}) \wedge (f^-\mathbf{M})$$

holds. Integrating over S and making some simple manipulations, we have

$$\|f\mathbf{M}\|_1 = \lambda(f^+) + \lambda(f^-) - 2\lambda[(f^+\mathbf{M}) \wedge (f^-\mathbf{M})]$$

$$\therefore \quad \|f\mathbf{M}\|_1 = \|f\|_1 - 2\lambda[(f^+\mathbf{M}) \wedge (f^-\mathbf{M})]$$

The proof is done.   □

Proof of lemma 8.2.3. Note that it is sufficient to prove the stated result for the case $t = 1$ because if it holds at $t = 1$ for an arbitrary kernel q and its associated Markov operator $\mathbf{N}$, then it holds for $q := p^t$, and the Markov operator associated with p^t is $\mathbf{M}^t$ (lemma 8.1.13, page 196).

So choose any distinct $\phi, \psi \in D(S)$, and let $f := \phi - \psi$. In view of lemma B.4.1 we will have $\|\phi\mathbf{M} - \psi\mathbf{M}\|_1 < \|\phi - \psi\|$ whenever

$$\int \left[\left(\int p(x,y)f^+(x)dx \right) \bigwedge \left(\int p(x,y)f^-(x)dx \right) \right] dy > 0 \qquad \text{(B.1)}$$

With a little bit of effort one can show that, for each $y \in S$, we have

$$\left(\int p(x,y)f^+(x)dx \right) \bigwedge \left(\int p(x',y)f^-(x')dx' \right)$$
$$\geq \int \int p(x,y) \wedge p(x',y)f^+(x)f^-(x')dxdx'$$

Integrating over y shows that (B.1) dominates

$$\int \int \left[\int p(x,y) \wedge p(x',y)dy \right] f^+(x)f^-(x')dxdx'$$

Since the inner integral is always positive by hypothesis, and both f^+ and f^- are nontrivial (due to distinctness of ϕ and ψ), this term is strictly positive. Lemma 8.2.3 is now established.   □

Proof of proposition 8.2.12. In lemma 11.2.9 it is shown that if $\lambda(w \cdot \phi) < \infty$ and the geometric drift condition holds then $(\phi \mathbf{M}^t)_{t \geq 0}$ is tight. It remains to establish this result for general $\psi \in D(S)$. We establish it via the following two claims:

1. The set D_0 of all $\phi \in D(S)$ with $\lambda(w \cdot \phi) < \infty$ is dense in $D(S)$.[2]

2. If there exists a sequence $(\phi_n) \subset D(S)$ such that $(\phi_n \mathbf{M}^t)_{t \geq 0}$ is tight for each $n \in \mathbb{N}$ and $d_1(\phi_n, \psi) \to 0$, then $(\psi \mathbf{M}^t)_{t \geq 0}$ is tight.

The two claims are sufficient because if claim 1 holds, then there is a dense subset D_0 of $D(S)$ such that trajectories starting from D_0 are all tight. Since D_0 is dense, the existence of a sequence with the properties in claim 2 is assured.

Regarding the first claim, let $C_n := \{x : w(x) \leq n\}$ and pick any $\phi \in D(S)$. Define $\phi_n := a_n \mathbb{1}_{C_n} \phi$, where a_n is the normalizing constant $1/\lambda(\mathbb{1}_{C_n} \phi)$. It can be verified that the sequence (ϕ_n) lies in D_0 and converges pointwise to ϕ. Scheffè's lemma (see Taylor, 1997, p. 186) implies that for densities pointwise convergence implies d_1 convergence. Since ϕ is an arbitrary density, D_0 is dense.

Regarding claim 2, pick any $\epsilon > 0$, and choose n such that $d_1(\phi_n, \psi) \leq \epsilon/2$. Non-expansiveness of $\mathbf{M}$ implies that $d_1(\phi_n \mathbf{M}^t, \psi \mathbf{M}^t) \leq \epsilon/2$ for all t. Since $(\phi_n \mathbf{M}^t)$ is tight, there exists a compact set K such that $\lambda(\mathbb{1}_{K^c} \phi_n \mathbf{M}^t) \leq \epsilon/2$ for all t. But then

$$\lambda(\mathbb{1}_{K^c} \psi \mathbf{M}^t) = \lambda(\mathbb{1}_{K^c} |\psi \mathbf{M}^t - \phi_n \mathbf{M}^t + \phi_n \mathbf{M}^t|) \leq d_1(\psi \mathbf{M}^t, \phi_n \mathbf{M}^t) + \lambda(\mathbb{1}_{K^c} \phi_n \mathbf{M}^t) \leq \epsilon$$

for all $t \in \mathbb{N}$. Hence $(\psi \mathbf{M}^t)_{t \geq 0}$ is tight as claimed.   □

Proof of proposition 8.2.13. Fix $\epsilon > 0$. We claim the existence of a $\delta > 0$ such that $\int_A \psi \mathbf{M}^t(x) dx < \epsilon$ whenever $\lambda(A) < \delta$. Since $(\psi \mathbf{M}^t)_{t \geq 0}$ is tight, there exists a compact set K such that

$$\lambda(\mathbb{1}_{K^c} \psi \mathbf{M}^t) :=: \int_{K^c} \psi \mathbf{M}^t \, d\lambda < \frac{\epsilon}{2} \qquad \forall \, t \in \mathbb{N} \tag{B.2}$$

For any Borel set $A \subset S$ the decomposition

$$\int_A \psi \mathbf{M}^t \, d\lambda = \int_{A \cap K} \psi \mathbf{M}^t \, d\lambda + \int_{A \cap K^c} \psi \mathbf{M}^t \, d\lambda \tag{B.3}$$

holds. Consider the first term in the sum. We have

$$\int_{A \cap K} (\psi \mathbf{M}^t)(x) \lambda(dx) = \int_{A \cap K} \left[\int p(x, y)(\psi \mathbf{M}^{t-1})(x) \lambda(dx) \right] \lambda(dy)$$

$$= \int \left[\int_{A \cap K} p(x, y) \lambda(dy) \right] (\psi \mathbf{M}^{t-1})(x) \lambda(dx)$$

---

[2] A subset A of a metric space U is called *dense* in U if every element of U is the limit of a sequence in A.

But by the hypothesis $p \leq m$ and the fact that the image of the continuous function m is bounded on K by some constant $N < \infty$,

$$\int_{A \cap K} p(x,y)\lambda(dy) \leq \int_{A \cap K} m(y)\lambda(dy) \leq N \cdot \lambda(A)$$

$$\therefore \quad \int_{A \cap K} \psi \mathbf{M}^t \, d\lambda = \int \left[\int_{A \cap K} p(x,y)\lambda(dy) \right] \psi \mathbf{M}^{t-1} \, d\lambda \leq N\lambda(A) \qquad \text{(B.4)}$$

Combining (B.2), (B.3), and (B.4), we obtain the bound

$$\int_A (\psi \mathbf{M}^t)(x)\lambda(dx) \leq N \cdot \lambda(A) + \frac{\epsilon}{2}$$

for any t and any $A \in \mathscr{B}(S)$. Setting $\delta := \epsilon / (2N)$ now gives the desired result. $\qquad \square$

B.5 Appendix to Chapter 10

Next we give the proof of theorem 10.2.11. To simplify notation, we write $\| \cdot \|_\infty$ as $\| \cdot \|$. The theorem claims that if σ is the policy generated by the approximate value iteration algorithm, then

$$\|v^* - v_\sigma\| \leq \frac{2}{(1-\rho)^2} \left(\rho \|v_n - v_{n-1}\| + \|v^* - Lv^*\| \right)$$

This result follows immediately from the next two lemmas.

Lemma B.5.1 *The v_n-greedy policy σ satisfies*

$$\|v^* - v_\sigma\| \leq \frac{2}{1-\rho} \|v_n - v^*\| \qquad \text{(B.5)}$$

Proof. We have

$$\|v^* - v_\sigma\| \leq \|v^* - v_n\| + \|v_n - v_\sigma\| \qquad \text{(B.6)}$$

The second term on the right-hand side of (B.6) satisfies

$$\|v_n - v_\sigma\| \leq \|v_n - Tv_n\| + \|Tv_n - v_\sigma\| \qquad \text{(B.7)}$$

Consider the first term on the right-hand side of (B.7). Observe that for any $w \in b\mathscr{B}(S)$, we have

$$\|w - Tw\| \leq \|w - v^*\| + \|v^* - Tw\| \leq \|w - v^*\| + \rho\|v^* - w\| = (1 + \rho)\|w - v^*\|$$

Substituting in v_n for w, we obtain

$$\|v_n - Tv_n\| \leq (1 + \rho)\|v_n - v^*\| \qquad \text{(B.8)}$$

Now consider the second term on the right-hand side of (B.7). Since σ is v_n-greedy, we have $Tv_n = T_\sigma v_n$, and

$$\|Tv_n - v_\sigma\| = \|T_\sigma v_n - v_\sigma\| = \|T_\sigma v_n - T_\sigma v_\sigma\| \leq \rho \|v_n - v_\sigma\|$$

Substituting this bound and (B.8) into (B.7), we obtain

$$\|v_n - v_\sigma\| \leq (1 + \rho)\|v_n - v^*\| + \rho\|v_n - v_\sigma\|$$

$$\therefore \quad \|v_n - v_\sigma\| \leq \frac{1 + \rho}{1 - \rho}\|v_n - v^*\|.$$

This inequality and (B.6) together give

$$\|v^* - v_\sigma\| \leq \|v^* - v_n\| + \frac{1 + \rho}{1 - \rho}\|v_n - v^*\|$$

Simple algebra now gives (B.5). $\qquad\square$

Lemma B.5.2 *For every $n \in \mathbb{N}$, we have*

$$(1 - \rho)\|v^* - v_n\| \leq \|v^* - Lv^*\| + \rho\|v_n - v_{n-1}\|$$

Proof. Let $\hat{v}$ be the fixed point of $\hat{T}$. By the triangle inequality,

$$\|v^* - v_n\| \leq \|v^* - \hat{v}\| + \|\hat{v} - v_n\| \qquad (B.9)$$

Regarding the first term on the right-hand side of (B.9), we have

$$\|v^* - \hat{v}\| \leq \|v^* - \hat{T}v^*\| + \|\hat{T}v^* - \hat{v}\|$$
$$= \|v^* - Lv^*\| + \|\hat{T}v^* - \hat{T}\hat{v}\| \leq \|v^* - Lv^*\| + \rho\|v^* - \hat{v}\|$$

$$\therefore \quad (1 - \rho)\|v^* - \hat{v}\| \leq \|v^* - Lv^*\| \qquad (B.10)$$

Regarding the second term in the sum (B.9), we have

$$\|\hat{v} - v_n\| \leq \|\hat{v} - \hat{T}^{n+1}v_0\| + \|\hat{T}^{n+1}v_0 - \hat{T}^n v_0\| \leq \rho\|\hat{v} - v_n\| + \rho\|v_n - v_{n-1}\|$$

$$\therefore \quad (1 - \rho)\|\hat{v} - v_n\| \leq \rho\|v_n - v_{n-1}\| \qquad (B.11)$$

Combining (B.9), (B.10), and (B.11) gives the bound we are seeking. $\qquad\square$

B.6 Appendix to Chapter 11

Proof of lemma 11.1.29. It is an exercise to show that if $\phi, \psi \in \mathscr{P}(S)$, then

$$\|\phi - \psi\|_{TV} = 2(\phi - \psi)^+(S) = 2(\phi - \psi)^-(S) = 2(\phi - \psi)(S^+) \qquad (B.12)$$

where S^+ is a positive set for the signed measure $\phi - \psi$. Now suppose that S^+ is a maximizer of $|\phi(B) - \psi(B)|$ over $\mathscr{B}(S)$. In this case, we have

$$\sup_{B \in \mathscr{B}(S)} |\phi(B) - \psi(B)| = |\phi(S^+) - \psi(S^+)| = (\phi - \psi)(S^+)$$

and the claim in lemma 11.1.29 follows from (B.12). Hence we need only show that S^+ is indeed a maximizer. To do so, pick any $B \in \mathscr{B}(S)$, and note that

$$
\begin{aligned}
|\phi(B) - \psi(B)| &= |(\phi - \psi)^+(B) - (\phi - \psi)^-(B)| \\
&= (\phi - \psi)^+(B) \vee (\phi - \psi)^-(B) - (\phi - \psi)^+(B) \wedge (\phi - \psi)^-(B)
\end{aligned}
$$

where the second equality follows from lemma A.2.17 on page 331.

$$\therefore \quad |\phi(B) - \psi(B)| \leq (\phi - \psi)^+(B) \vee (\phi - \psi)^-(B) \leq (\phi - \psi)^+(S)$$

But $(\phi - \psi)^+(S) = (\phi - \psi)(S^+)$ by definition, so S^+ is a maximizer as claimed. $\qquad \square$

Next let's prove theorem 11.2.5. In the proof, P is a stochastic kernel and $\mathbf{M}$ is the Markov operator. By assumption, $\mathbf{M}h \in bcS$ whenever $h \in bcS$.

Proof of theorem 11.2.5. Let ψ be as in the statement of the theorem, so $(\psi \mathbf{M}^t)_{t \geq 1}$ is tight. Let $\nu_n := \frac{1}{n} \sum_{t=1}^{n} \psi \mathbf{M}^t$. The sequence $(\nu_n)_{n \geq 1}$ is also tight (proof?), from which it follows (see Prohorov's theorem, on page 257) that there exists a subsequence (ν_{n_k}) of (ν_n) and a $\nu \in \mathscr{P}(S)$ such that $d_{FM}(\nu_{n_k}, \nu) \to 0$ as $k \to \infty$. It is not hard to check that, for all $n \in \mathbb{N}$, we have

$$\nu_n \mathbf{M} - \nu_n = \frac{\psi \mathbf{M}^{n+1} - \psi \mathbf{M}}{n}$$

We aim to show that $d_{FM}(\nu \mathbf{M}, \nu) = 0$, from which it follows that ν is stationary for $\mathbf{M}$. From the definition of the Fortet–Mourier distance (see page 256), it is sufficient to show that for any bounded Lipschitz function $h \in b\ell S$ with $\|h\|_{b\ell} \leq 1$ we have $|\nu \mathbf{M}(h) - \nu(h)| = 0$.

So pick any such h. Observe that

$$|\nu \mathbf{M}(h) - \nu(h)| \leq |\nu \mathbf{M}(h) - \nu_n \mathbf{M}(h)| + |\nu_n \mathbf{M}(h) - \nu_n(h)| + |\nu_n(h) - \nu(h)| \qquad (B.13)$$

for all $n \in \mathbb{N}$. All three terms on the right-hand side of (B.13) converge to zero along the subsequence (n_k), which implies $|\nu\mathbf{M}(h) - \nu(h)| = 0$. To see that this is the case, consider the first term. We have

$$|\nu\mathbf{M}(h) - \nu_{n_k}\mathbf{M}(h)| = |\nu(\mathbf{M}h) - \nu_{n_k}(\mathbf{M}h)| \to 0$$

where the equality is from the duality property in theorem 9.2.15 (page 226), and convergence is due to the fact that $\mathbf{M}h$ is bounded and continuous, and $d_{FM}(\nu_{n_k}, \nu) \to 0$ as $k \to \infty$.

Consider next the second term in (B.13). That $|\nu_{n_k}\mathbf{M}(h) - \nu_{n_k}(h)|$ converges to zero as $k \to \infty$ follows from the bound

$$|\nu_{n_k}\mathbf{M}(h) - \nu_{n_k}(h)| = \frac{1}{n_k}|\psi\mathbf{M}^{n_k+1}(h) - \psi\mathbf{M}(h)| \leq \frac{2}{n_k}$$

That the final term in the sum (B.13) converges to zero along the subsequence (n_k) is trivial, and this completes the proof of theorem 11.2.5. $\qquad\square$

Proof of lemma 11.3.3. The x_b that solves $P(x_b) = \alpha \int p^*(z)\phi(dz)$ satisfies $D(\alpha P(0)) \leq x_b$ because $P(x_b) = \alpha \int p^*(z)\phi(dz) \leq \alpha p^*(0) = \alpha P(0)$. Here we are using the fact that $p^*(0) = P(0)$.[3] Also $D(\alpha P(0)) > 0$ because $D(P(0)) = 0$ and D is strictly decreasing. Since $x_b \geq D(\alpha P(0))$, we have shown that $x_b > 0$.

We claim in addition that if $x \leq x_b$, then $p^*(x) = P(x)$ and $I(x) = 0$. That $p^*(x) = P(x)$ implies $I(x) = 0$ is immediate from the definition: $I(x) = x - D(p^*(x))$ (see page 146). Hence we need only prove that when $x \leq x_b$ we have $p^*(x) = 0$. But if $x \leq x_b$, then $P(x_b) \leq P(x)$, and hence

$$P(x) \geq \alpha \int p^*(z)\phi(dz) \geq \alpha \int p^*(\alpha I(x) + z)\phi(dz)$$

That $p^*(x) = P(x)$ is now clear from the definition of p^*.[4] $\qquad\square$

B.7   Appendix to Chapter 12

Proof of proposition 12.1.17. Pick any $y > 0$ and define the $h: [0, y] \to \mathbb{R}$ by

$$h(k) := U(y - k) + W(k), \quad W(k) := \rho \int w[f(k, z)]\phi(dz)$$

---

[3]To prove this, one can show via (6.28) on page 147 that if $p \leq P(0)$, then $Tp \leq P(0)$, from which it follows that $p^* = \lim_n T^n P \leq P(0)$. Therefore $p^*(0) \leq P(0)$. On the other hand, $p^* \geq P$, so $p^*(0) \geq P(0)$. Hence $p^*(0) = P(0)$.

[4]For the definition refer to (6.26) on page 146.

Given any $\epsilon > 0$, we have

$$\frac{h(y) - h(y - \epsilon)}{\epsilon} = -\frac{U(\epsilon)}{\epsilon} + \frac{W(y) - W(y - \epsilon)}{\epsilon} \qquad \text{(B.14)}$$

If $\sigma(y) = y$, then $h(y) \geq h(y - \epsilon)$, and (B.14) is nonnegative for all $\epsilon > 0$. But this is impossible: On one hand, $W(k)$ is concave and its left-hand derivative exists at y (lemma A.2.41, page 337), implying that the second term on the right-hand side converges to a finite number as $\epsilon \downarrow 0$. On the other hand, the assumption $U'(0) = \infty$ implies that the first term converges to $-\infty$. □

Proof of proposition 12.1.18. Fix $w \in \mathscr{C}ibcS$ and $y > 0$. Define

$$W(x, k) := U(x - k) + \rho \int w(f(k, z)) \phi(dz) \qquad (x > 0 \text{ and } k \leq x)$$

From proposition 12.1.17 we have $\sigma(y) < y$. From this inequality one can establish the existence of an open neighborhood G of zero with $0 \leq \sigma(y) \leq y + h$ for all $h \in G$.

$$\therefore \quad W(y + h, \sigma(y)) \leq Tw(y + h) = W(y + h, \sigma(y + h)) \qquad \forall h \in G$$

It then follows that for all $h \in G$,

$$Tw(y + h) - Tw(y) \geq W(y + h, \sigma(y)) - W(y, \sigma(y)) = U(y - \sigma(y) + h) - U(y - \sigma(y))$$

Take $h_n \in G, h_n > 0, h_n \downarrow 0$. Since $h_n > 0$, we have

$$\frac{Tw(y + h_n) - Tw(y)}{h_n} \geq \frac{U(y - \sigma(y) + h_n) - U(y - \sigma(y))}{h_n} \qquad \forall n \in \mathbb{N}$$

Let DTw_+ denote the right derivative of Tw, which exists by concavity of Tw. Taking limits gives $DTw_+(y) \geq U'(y - \sigma(y))$.

Now take $h_n \in G, h_n < 0, h_n \uparrow 0$. Since $h_n < 0$, we get the reverse inequality

$$\frac{Tw(y + h_n) - Tw(y)}{h_n} \leq \frac{U(y - \sigma(y) + h_n) - U(y - \sigma(y))}{h_n} \qquad \forall n \in \mathbb{N}$$

and taking limits gives $DTw_-(y) \leq U'(y - \sigma(y))$. Thus

$$DTw_-(y) \leq U'(y - \sigma(y)) \leq DTw_+(y)$$

But concavity of Tw and lemma A.2.41 imply that $DTw_+(y) \leq DTw_-(y)$ also holds. In which case the left and right derivatives are equal (implying differentiability of Tw at y), and their value is $U'(y - \sigma(y))$. □

Proof of proposition 12.1.23. Fix $y > 0$. Let $k^* := \sigma(y)$ be optimal investment, and let $v := v^*$ be the value function. In light of proposition 12.1.17 we have $k^* < y$. Let's assume for now that

$$h(k) := U(y - k) + \rho \int v(f(k, z))\phi(dz)$$

is differentiable on $[0, y)$, and that

$$h'(k) = -U'(y - k) + \rho \int U' \circ c(f(k, z))f'(k, z)\phi(dz) \qquad (B.15)$$

(If $k = 0$, then by differentiability we mean that the right-hand derivative $h'_+(0)$ exists, although it is permitted to be $+\infty$.)

The inequality (12.3) is then equivalent to $h'(k^*) \leq 0$. This must hold because k^* is a maximizer and $k^* < y$, in which case $h'(k^*) > 0$ is impossible. Thus it remains only to show that h is differentiable on $[0, y)$, and that h' is given by (B.15). In view of corollary 12.1.19, it suffices to show that

$$h'(k) = -U'(y - k) + \rho \int \frac{\partial}{\partial k}v(f(k, z))\phi(dz) \qquad (0 < k < y)$$

The only difficultly in the preceding set of arguments is in showing that

$$\frac{d}{dk}\int v(f(k, z))\phi(dz) = \int \frac{\partial}{\partial k}v(f(k, z))\phi(dz) \qquad (B.16)$$

To see this, define

$$g(k) := \int v(f(k, z))\phi(dz) \qquad (k > 0)$$

and consider the derivative at fixed $k > 0$. Let $h_0 < 0$ be such that $k + h_0 > 0$. For all $h > h_0$ it is not hard to show that

$$\frac{g(k + h) - g(k)}{h} = \int \left[\frac{v(f(k + h, z)) - v(f(k, z))}{h}\right]\phi(dz)$$

Since $k \mapsto v(f(k, z))$ is concave for each z, the inequality

$$\frac{v(f(k + h, z)) - v(f(k, z))}{h} \leq \frac{v(f(k + h_0, z)) - v(f(k, z))}{h_0} := M(z)$$

holds for all z (see exercise A.2.42 on page 337). The function M is bounded and therefore ϕ-integrable. As a result the dominated convergence theorem implies that for $h_n \to 0$ with $h_n > h_0$ and $h_n \neq 0$ we have

$$
\begin{aligned}
g(k) &= \lim_{n \to \infty} \int \left[\frac{v(f(k + h_n, z)) - v(f(k, z))}{h_n} \right] \phi(dz) \\
&= \int \lim_{n \to \infty} \left[\frac{v(f(k + h_n, z)) - v(f(k, z))}{h_n} \right] \phi(dz) \\
&= \int \frac{\partial}{\partial k} v(f(k, z)) \phi(dz)
\end{aligned}
$$

The last equality is due to the fact that $k \mapsto v(f(k, z))$ is differentiable in k for each z. $\qquad \square$

Proof of proposition 12.1.24. Starting with the second claim, in the notation of the proof of proposition 12.1.23, we are claiming that if $h'(k^*) < 0$, then $k^* = 0$. This follows because if $k^* > 0$, then k^* is interior, in which case $h'(k^*) = 0$.

The first claim will be established if, whenever $f(0, z) = 0$ for each $z \in Z$, we have $0 < \sigma(y)$ for all $y > 0$. (Why?) Suppose instead that $\sigma(y) = 0$ at some $y > 0$, so

$$
v(y) = U(y) + \rho \int v[f(0, z)] \phi(dz) = U(y) \tag{B.17}
$$

where we have used $U(0) = 0$. Define also

$$
v_\xi := U(y - \xi) + \rho \int v[f(\xi, z)] \phi(dz) \tag{B.18}
$$

where ξ is a positive number less than y. Using optimality and the fact that $U(y) = v(y)$, we get

$$
0 \leq \frac{v(y) - v_\xi}{\xi} = \frac{U(y) - U(y - \xi)}{\xi} - \rho \int \frac{v[f(\xi, z)]}{\xi} \phi(dz) \qquad \forall \xi < y
$$

Note that the first term on the right-hand side of the equal sign converges to the finite constant $U'(y)$ as $\xi \downarrow 0$. We will therefore induce a contradiction if the second term (i.e., the integral term) converges to plus infinity. Although our simple version of the monotone convergence theorem does not include this case, it is sufficient to show that the integrand converges to infinity as $\xi \downarrow 0$ for each fixed z; interested readers should consult, for example, Dudley (2002, thm. 4.3.2).[5] To see that this is so, observe that for any $z \in Z$,

$$
\lim_{\xi \downarrow 0} \frac{v[f(\xi, z)]}{\xi} = \lim_{\xi \downarrow 0} \frac{v[f(\xi, z)]}{f(\xi, z)} \frac{f(\xi, z)}{\xi} \geq \lim_{\xi \downarrow 0} \frac{U[f(\xi, z)]}{f(\xi, z)} \frac{f(\xi, z)}{\xi} \to \infty
$$

We have used here the fact that $v \geq U$ pointwise on S. $\qquad \square$

---

[5]We are using the fact that the integrand increases monotonically as $\xi \downarrow 0$ for each fixed z, as follows from concavity of f in its first argument, and the fact that the value function is increasing.

Proof of lemma 12.1.28. Set $w_1 := (U' \circ c)^{1/2}$, as in the statement of the proposition. We have

$$\int w_1[f(\sigma(y),z)]\phi(dz) = \int \left[U' \circ c[f(\sigma(y),z)] \frac{f'(\sigma(y),z)}{f'(\sigma(y),z)} \right]^{1/2} \phi(dz)$$

To break up this expression, we make use of the fact that if g and h are positive real functions on S, then by the Cauchy–Schwartz inequality (Dudley 2002, cor. 5.1.4),

$$\int (gh)^{1/2} d\phi \leq \left(\int g\, d\phi \cdot \int h\, d\phi \right)^{1/2} \tag{B.19}$$

It follows that

$$\int w_1[f(\sigma(y),z)]\phi(dz)$$

$$\leq \left[\int U' \circ c[f(\sigma(y),z)]f'(\sigma(y),z)\phi(dz) \right]^{1/2} \left[\int \frac{1}{f'(\sigma(y),z)}\phi(dz) \right]^{1/2}$$

Substituting in the Euler equation, we obtain

$$\int w_1[f(\sigma(y))z]\phi(dz) \leq \left[\frac{U' \circ c(y)}{\rho} \right]^{1/2} \left[\int \frac{1}{f'(\sigma(y),z)}\phi(dz) \right]^{1/2}$$

Using the definition of w_1, this expression can be rewritten as

$$\int w_1[f(\sigma(y))z]\phi(dz) \leq \left[\int \frac{1}{\rho f'(\sigma(y),z)}\phi(dz) \right]^{1/2} w_1(y)$$

From assumption 12.1.27 one can deduce the existence of a $\delta > 0$ and an $\alpha_1 \in (0,1)$ such that

$$\left[\int \frac{1}{\rho f'(\sigma(y),z)}\phi(dz) \right]^{1/2} < \alpha_1 < 1 \text{ for all } y < \delta$$

$$\therefore \quad \int w_1[f(\sigma(y),z)]\phi(dz) \leq \alpha_1 w_1(y) \qquad (y < \delta)$$

On the other hand, if $y \geq \delta$, then

$$\int w_1[f(\sigma(y),z)]\phi(dz) \leq \beta_1 := \int w_1[f(\sigma(\delta),z)]\phi(dz)$$

The last two inequalities together give the bound

$$\int w_1[f(\sigma(y),z)]\phi(dz) \leq \alpha_1 w_1(y) + \beta_1 \qquad (y \in S)$$

This completes the proof of lemma 12.1.28. $\square$

Proof of lemma 12.2.17. Our first observation is that if w is any continuous and increasing function on $\mathbb{R}_+$ with $\mathbf{N}w(y) = \int w(f(y,z))\phi(dz) < \infty$ for every $y \in \mathbb{R}_+$, then $\mathbf{N}w$ is also continuous and increasing on $\mathbb{R}_+$. Monotonicity of $\mathbf{N}w$ is obvious. Continuity holds because if $y_n \to y$, then (y_n) is bounded by some $\bar{y}$, and $w(f(y_n, \cdot)) \le w(f(\bar{y}, \cdot))$. As $\int w(f(\bar{y}, z))\phi(dz) < \infty$, the dominated convergence theorem gives us $\lim_{n \to \infty} \mathbf{N}w(y_n) \to \mathbf{N}w(y)$.

A simple induction argument now shows that $\mathbf{N}^t U$ is increasing and continuous on $\mathbb{R}_+$ for every $t \ge 0$. The fact that κ is monotone increasing on $\mathbb{R}_+$ is immediate from this result, given that $\kappa = \sum_t \delta^t \mathbf{N}^t U$. (Why?) Continuity of κ on $\mathbb{R}_+$ can be established by corollary 7.3.15 on page 182. Take $y_n \to y$ and note again the existence of a $\bar{y}$ such that $y_n \le \bar{y}$ for every $n \in \mathbb{N}$. Thus $\delta^t \mathbf{N}^t U(y_n) \le \delta^t \mathbf{N}^t U(\bar{y})$ for every n and every t. Moreover $\delta^t \mathbf{N}^t U(y_n) \to \delta^t \mathbf{N}^t U(y)$ as $n \to \infty$ for each t. Corollary 7.3.15 now gives

$$\lim_{n \to \infty} \kappa(y_n) = \sum_{t=0}^{\infty} \lim_{n \to \infty} \delta^t \mathbf{N}^t U(y_n) = \sum_{t=0}^{\infty} \delta^t \mathbf{N}^t U(y) = \kappa(y)$$

$\square$

Proof of lemma 12.2.20. Pick $v \in b_\kappa cS$, $(x, u) \in \mathrm{gr}\,\Gamma$ and $(x_n, u_n) \to (x, u)$. Let $\hat{v} := v + \|v\|_\kappa \kappa$. Observe that $\hat{v}$ is both continuous and nonnegative. Let $\hat{v}_k$ be a sequence of bounded continuous nonnegative functions on S with $\hat{v}_k \uparrow \hat{v}$ pointwise. (Can you give an explicit example of such a sequence?) By the dominated convergence theorem, for each $k \in \mathbb{N}$ we have

$$\liminf_n \int \hat{v}[F(x_n, u_n, z)]\phi(dz) \ge \liminf_n \int \hat{v}_k[F(x_n, u_n, z)]\phi(dz) = \int \hat{v}_k[F(x, u, z)]\phi(dz)$$

Taking limits with respect to k gives

$$\liminf_n \int \hat{v}[F(x_n, u_n, z)]\phi(dz) \ge \int \hat{v}[F(x, u, z)]\phi(dz)$$

It follows that

$$\hat{g}(x, u) := \int \hat{v}[F(x, u, z)]\phi(dz)$$

is lower-semicontinuous (lsc) on $\mathrm{gr}\,\Gamma$. And if $\hat{g}$ is lsc, then so is

$$g(x, u) := \int v[F(x, u, z)]\phi(dz)$$

as $g(x, u) = \hat{g}(x, u) - \|v\|_\kappa \int \kappa[F(x, u, z)]\phi(dz)$.

Since v was an arbitrary element of $b_\kappa cS$, and since $-v$ is also in $b_\kappa cS$, we can also conclude that $-g$ is lsc on $\mathrm{gr}\,\Gamma$—equivalently, g is usc on $\mathrm{gr}\,\Gamma$ (recall exercise 3.1.15 on page 40). But if g is both lsc and usc on $\mathrm{gr}\,\Gamma$, then g is continuous on $\mathrm{gr}\,\Gamma$, as was to be shown.

$\square$

Bibliography

Adda, J., and R. Cooper. 2003. *Dynamic Economics: Quantitative Methods and Applications.* Cambridge: MIT Press.

Aiyagari, S. R. 1994. Uninsured idiosyncratic risk and aggregate saving. *Quarterly Journal of Economics* 109: 659–684.

Aliprantis, C. D., and K. C. Border. 1999. *Infinite Dimensional Analysis.* New York: Springer.

Aliprantis, C. D., and O. Burkinshaw. 1998. *Principles of Real Analysis.* London: Academic Press.

Alvarez, F., and N. L. Stokey. 1998. Dynamic programming with homogeneous functions. *Journal of Economic Theory* 82: 167–89.

Amir, R. 1997. A new look at optimal growth under uncertainty. *Journal of Economic Dynamics and Control* 22: 67–86.

Amir, R. 2005. Supermodularity and complementarity in economics: An elementary survey. *Southern Economic Journal* 71: 636–60.

Amir, R., L. J. Mirman, and W. R. Perkins. 1991. One-sector nonclassical optimal growth: Optimality conditions and comparative dynamics. *International Economic Review* 32: 625–44.

Amman, H. M., D. A. Kendrick, and J. Rust, eds. 1990. *Handbook of Computational Economics.* Burlington, MA: Elsevier.

Angeletos, G-M. 2007. Uninsured idiosyncratic investment risk and aggregate saving. *Review of Economic Dynamics* 10: 1–30.

Aruoba, S. B., J. Fernàndez-Villaverde, and J. Rubio-Ramírez. 2006. Comparing solution methods for dynamic equilibrium economies. *Journal of Economic Dynamics and Control* 30: 2447–508.

Azariadis, C. 1993. *Intertemporal Macroeconomics.* New York: Blackwell.

Azariadis, C., and A. Drazen. 1990. Threshold externalities in economic development. *Quarterly Journal of Economics* 105: 501–26.

Barnett, W. A., and A. Serletis. 2000. Martingales, nonlinearity, and chaos. *Journal of Economic Dynamics and Control* 24: 703–24.

Bartle, R., and D. Sherbet. 1992. *Introduction to Real Analysis.* New York: Wiley.

357

Becker, R. A., and J. H. Boyd. 1997. *Capital Theory, Equilibrium Analysis and Recursive Utility*. New York: Blackwell.

Bellman, R. E. 1957. *Dynamic Programming*. Princeton: Princeton University Press.

Benhabib, J., and K. Nishimura. 1985. Competitive equilibrium cycles. *Journal of Economic Theory* 35: 284–306.

Benveniste, L. M. , and J. A. Scheinkman. 1979. On the differentiability of the value function in dynamic models of economics. *Econometrica* 47: 727–32.

Bertsekas, D. P. 1995. *Dynamic Programming and Optimal Control*. New York: Athena Scientific.

Bewley, T. 2007. *General Equilibrium, Overlapping Generations Models, and Optimal Growth Theory*. Cambridge: Harvard University Press.

Bhattacharya, R. N., and O. Lee. 1988. Asymptotics of a class of Markov processes which are not in general irreducible. *Annals of Probability* 16: 1333–47.

Bhattacharya, R., and M. Majumdar. 2007. *Random Dynamical Systems: Theory and Applications*. Cambridge: Cambridge University Press.

Böhm, V., and L. Kaas. 2000. Differential savings, factor shares, and endogenous growth cycles. *Journal of Economic Dynamics and Control* 24: 965–80.

Boldrin, M., and L. Montrucchio. 1986. On the indeterminacy of capital accumulation paths. *Journal of Economic Theory* 40: 26–39.

Boyd, J. H. 1990. Recursive utility and the Ramsey problem. *Journal of Economic Theory* 50: 326–45.

Breiman, L. 1992. *Probability*. SIAM Classics in Applied Mathematics, Philadelphia: SIAM.

Bremaud, P. 1999. *Markov Chains*. New York: Springer.

Brock, W. A., and L. J. Mirman. 1972. Optimal economic growth and uncertainty: The discounted case. *Journal of Economic Theory* 4: 479–513.

Brock, W. A. 1982. Asset prices in a production economy. In *Economics of Information and Uncertainty*. J. J. McCall, ed. Chicago: University of Chicago Press, pp. 1–43.

Brock, W. A., and C. H. Hommes. 1998. Heterogeneous beliefs and routes to chaos in a simple asset pricing model. *Journal of Economic Dynamics and Control* 22: 1235–74.

Canova, F. 2007. *Methods for Applied Macroeconomic Research*. Princeton: Princeton University Press.

Caputo, M. R. 2005. *Foundations of Dynamic Economic Analysis: Optimal Control Theory and Applications*. Cambridge: Cambridge University Press.

Chan, K. S., and H. Tong. 1986. On estimating thresholds in autoregressive models. *Journal of Time Series Analysis* 7: 179–90.

Chatterjee, P., and M. Shukayev. 2008. Note on a positive lower bound of capital in the stochastic growth model. *Journal of Economic Dynamics and Control* 32: 2137–47.

Chiarella, C. 1988. The cobweb model its instability and the onset of chaos. *Economic Modelling* 5: 377–84.

Christiano, L. J., and J. D. M. Fisher. 2000. Algorithms for solving dynamic models with occasionally binding constraints. *Journal of Economic Dynamics and Control* 24: 1179–232.

Coleman, W. J. 1990. Solving the stochastic growth model by policy-function iteration. *Journal of Business and Economic Statistics* 8: 27–29.

Datta, M., L. J. Mirman, O. F. Morand, and K. Reffett. 2005. Markovian equilibrium in infinite horizon economies with incomplete markets and public policy. *Journal of Mathematical Economics* 41: 505–44.

Deaton, A., and G. Laroque. 1992. On the behavior of commodity prices. *Review of Economic Studies* 59: 1–23.

Dechert, W. D., and S. I. O'Donnell. 2006. The stochastic lake game: A numerical solution. *Journal of Economic Dynamics and Control* 30: 1569–87.

Den Haan, W. J., and A. Marcet. 1994. Accuracy in simulations. *Review of Economic Studies* 61: 3–17.

Dobrushin, R. L. 1956. Central limit theorem for nonstationary Markov chains. *Theory of Probability and its Applications* 1: 65–80.

Doeblin, W. 1938. Exposé de la theorie des chaîns simples constantes de Markov à un nombre fini d'états. *Revue Mathematique de l'Union Interbalkanique* 2: 77–105.

Donaldson, J. B., and R. Mehra. 1983. Stochastic growth with correlated production shocks. *Journal of Economic Theory* 29: 282–312.

Dudley, R. M. 2002. *Real Analysis and Probability.* Cambridge: Cambridge University Press.

Duffie, D. 2001. *Dynamic Asset Pricing Theory.* Princeton: Princeton University Press.

Durlauf, S. 1993. Nonergodic economic growth. *Review of Economic Studies* 60: 349–66.

Durrett, R. 1996. *Probability: Theory and Examples.* New York: Duxbury Press.

Ericson, R., and A. Pakes. 1995. Markov-perfect industry dynamics: A framework for empirical work. *Review of Economic Studies* 62: 53–82.

Farmer, R. E. A. 1999. *The Macroeconomics of Self-Fulfilling Prophecies.* Cambridge: MIT Press.

de la Fuente, A. 2000. *Mathematical Methods and Models for Economists.* Cambridge: Cambridge University Press.

Galor, O. 1994. A two-sector overlapping generations model: A global characterization of the dynamical system. *Econometrica* 60: 1351–86.

Gandolfo, G. 2005. *Economic Dynamics: Study Edition.* New York: Springer.

Glynn, P. W., and S. G. Henderson. 2001. Computing densities for Markov chains via simulation. *Mathematics of Operations Research* 26: 375–400.

Gordon, G. J. 1995. Stable function approximation in dynamic programming. Mimeo. Carnegie–Mellon University.

Grandmont, J-M. 1985. On endogenous competitive business cycles. *Econometrica* 53: 995–1046.

Green, E. J., and R. H. Porter. 1984. Noncooperative collusion under imperfect price information. *Econometrica* 52: 87–100.

Greenwood, J., and G. W. Huffman. 1995. On the existence of nonoptimal equilibria. *Journal of Economic Theory* 65: 611–23.

Grüne, L., and W. Semmler. 2004. Using dynamic programming with adaptive grid scheme for optimal control. *Journal of Economic Dynamics and Control* 28: 2427–56.

Häggström, O. 2002. *Finite Markov Chains and Algorithmic Applications.* Cambridge: Cambridge University Press.

Hall, R. E. 1978. Stochastic implications of the life cycle-permanent income hypothesis: Theory and evidence. *Journal of Political Economy* 86: 971–87.

Hamilton, J. D. 2005. What's real about the business cycle? *Federal Reserve Bank of St. Louis Review.* July–August: 435–52.

Heer, B., and A. Maussner. 2005. *Dynamic General Equilibrium Modelling.* New York: Springer.

Hernández-Lerma, O., and J. B. Lasserre. 1996. *Discrete Time Markov Control Processes: Basic Optimality Criteria.* New York: Springer.

Hernández-Lerma, O., and J. B. Lasserre. 1999. *Further Topics on Discrete Time Markov Control Processes.* New York: Springer.

Hernández-Lerma, O., and J. B. Lasserre. 2003. *Markov Chains and Invariant Probabilities.* Boston: Birkhäuser.

Holmgren, R. A. 1996. *A First Course in Discrete Dynamical Systems.* New York: Springer.

Hopenhayn, H. A. 1992. Entry, exit, and firm dynamics in long run equilibrium. *Econometrica* 60: 1127–50.

Hopenhayn, H. A., and E. C. Prescott. 1992. Stochastic monotonicity and stationary distributions for dynamic economies. *Econometrica* 60: 1387–1406.

Huggett, M. 1993. The risk-free rate in heterogeneous-agent incomplete-insurance economies. *Journal of Economic Dynamics and Control* 17: 953–969.

Huggett, M. 2003. When are comparative dynamics monotone? *Review of Economic Dynamics* 6: 1–11.

Jones, G. L. 2004. On the Markov chain central limit theorem. *Probability Surveys* 1: 299–320.

Judd, K. L. 1992. Projection methods for solving aggregate growth models. *Journal of Economic Theory* 58: 410–52.

Judd, K. L. 1998. *Numerical Methods in Economics.* Cambridge: MIT Press.

Kamihigashi, T. 2007. Stochastic optimal growth with bounded or unbounded utility and with bounded or unbounded shocks. *Journal of Mathematical Economics* 43: 477–500.

Kamihigashi, T., and J. Stachurski. 2008. Asymptotics of stochastic recursive economies under monotonicity. Mimeo. Kyoto University.

Kandori, M., G. J. Mailath, and R. Rob. 1993. Learning, mutation and long run equilibria in games. *Econometrica* 61: 29–56.

Kendrick, D. A., P. R. Mercado, and H. M. Amman. 2005. *Computational Economics*. Princeton: Princeton University Press.

Kikuchi, T. 2008. International asset market, nonconvergence, and endogenous fluctuations. *Journal of Economic Theory* 139: 310–34.

Kolmogorov, A. N. 1955. *Foundations of the Theory of Probability*. Chelsea, NY: Nathan Morrison.

Kolmogorov, A. N., and S. V. Fomin. 1970. *Introductory Real Analysis*. New York: Dover Publications.

Krebs, T. 2004. Non-existence of recursive equilibria on compact state spaces when markets are incomplete. *Journal of Economic Theory* 115: 134–50.

Kristensen, D. 2007. Geometric ergodicity of a class of Markov chains with applications to time series models. Mimeo. University of Wisconsin.

Krusell, P., and A. Smith. 1998. Income and wealth heterogeneity in the macroeconomy. *Journal of Political Economy* 106: 867–96.

Krylov, N., and N. Bogolubov. 1937. Sur les properties en chaine. *Comptes Rendus Mathematique* 204: 1386–88.

Kubler, F., and K. Schmedders. 2002. Recursive equilibria in economies with incomplete markets. *Macroeconomic Dynamics* 6: 284–306.

Kydland, F., and E. C. Prescott. 1982. Time to build and aggregate fluctuations. *Econometrica* 50: 1345–71.

Langtangen, H. P. 2008. *Python Scripting for Computational Science*. New York: Springer.

Lasota, A. 1994. Invariant principle for discrete time dynamical systems. *Universitatis Iagellonicae Acta Mathematica* 31: 111–27.

Lasota, A., and M. C. Mackey. 1994. *Chaos, Fractals and Noise: Stochastic Aspects of Dynamics*. New York: Springer.

Le Van, C., and R-A. Dana. 2003. *Dynamic Programming in Economics*. New York: Springer.

Le Van, C., and Y. Vailakis. 2005. Recursive utility and optimal growth with bounded or unbounded returns. *Journal of Economic Theory* 123: 187–209.

Light, W. 1990. *Introduction to Abstract Analysis*. Oxford, UK: Chapman and Hall.

Lindvall, T. 1992. *Lectures on the Coupling Method*. New York: Dover Publications.

Ljungqvist, L., and T. Sargent. 2004. *Recursive Macroeconomic Theory*. Cambridge: MIT Press

Long, J., and C. Plosser. 1983. Real business cycles. *Journal of Political Economy* 91: 39–69.

Lovejoy, W. 1987. Ordered solutions for dynamic programs. *Mathematics of Operations Research* 12: 269–76.

Lucas, R. E., Jr., and E. C. Prescott. 1971. Investment under uncertainty. *Econometrica* 39: 659–81.

Lucas, R. E., Jr. 1978. Asset prices in an exchange economy. *Econometrica* 46: 1429–45.

Maliar, L., and S. Maliar. 2005. Solving nonlinear dynamic stochastic models: An algorithm computing value function by simulations. *Economics Letters* 87: 135–40.

Marcet, A. 1988. Solving nonlinear models by parameterizing expectations. Mimeo. Carnegie Mellon University.

Marimon, R., and A. Scott, eds. 2001. *Computational Methods for the Study of Dynamic Economies.* Oxford: Oxford University Press.

Matsuyama, K. 2004. Financial market globalization, symmetry-breaking, and endogenous inequality of nations. *Econometrica* 72: 853–84.

McCall, J. J. 1970. Economics of information and job search. *Quarterly Journal of Economics* 84: 113–26.

McGrattan, E. R. 2001. Application of weighted residual methods to dynamic economic models. In *Computational Methods for the Study of Dynamic Economies.* R. Marimon and A. Scott, eds. Oxford: Oxford University Press, pp. 114–43.

McLennan, A., and R. Tourky. 2005. From imitation games to Kakutani. Mimeo. University of Melbourne.

Medio, A. 1995. *Chaotic Dynamics: Theory and Applications to Economics.* Cambridge: Cambridge University Press.

Mehra, R., and E. C. Prescott. 1985. The equity premium: A puzzle. *Journal of Monetary Economics* 15: 145–61.

Meyn, S. P., and R. L. Tweedie. 1993. *Markov Chains and Stochastic Stability.* London: Springer.

Miao, J. 2006. Competitive equilibria of economies with a continuum of consumers and aggregate shocks. *Journal of Economic Theory* 128: 274–98.

Miranda, M., and P. L. Fackler. 2002. *Applied Computational Economics and Finance.* Cambridge: MIT Press.

Mirman, L. J. 1970. Two essays on uncertainty and economics. PhD Thesis, University of Rochester.

Mirman, L. J., 1972. On the existence of steady state measures for one sector growth models with uncertain technology. *International Economic Review* 13: 271–86.

Mirman, L. J. 1973. The steady state behavior of a class of one sector growth models with uncertain technology. *Journal of Economic Theory* 6: 219–42.

Mirman, L. J., and I. Zilcha. 1975. On optimal growth under uncertainty. *Journal of Economic Theory* 11: 329–39.

Mirman, L. J., O. F. Morand, and K. L. Reffett. 2008. A qualitative approach to Markovian equilibrium in infinite horizon economies with capital. *Journal of Economic Theory* 139: 75–98.

Mirman, L. J., K. Reffett, and J. Stachurski. 2005. Some stability results for Markovian economic semigroups. *International Journal of Economic Theory* 1: 57–72.

Mitra, T., and G. Sorger. 1999. Rationalizing policy functions by dynamic optimization. *Econometrica* 67: 375–92.

Nirei, M. 2008. Aggregate fluctuations of discrete investments. Mimeo. Carleton University.

Nishimura, K., G. Sorger, and M. Yano. 1994. Ergodic chaos in optimal growth models with low discount rates. *Economic Theory* 4: 705–17.

Nishimura, K., and J. Stachurski. 2005. Stability of stochastic optimal growth models: A new approach. *Journal of Economic Theory* 122: 100–18.

Norris, J. R. 1997. *Markov Chains.* Cambridge: Cambridge University Press.

Nummelin, E. 1984. *General Irreducible Markov Chains and Nonnegative Operators.* Cambridge: Cambridge University Press.

Ok, E. A. 2007. *Real Analysis with Economic Applications.* Princeton: Princeton University Press.

Olsen, L., and S. Roy. 2006. Theory of stochastic optimal growth. In *Handbook of Optimal Growth,* vol. 1. C. Le Van, R-A. Dana, T. Mitra and K. Nishimura, eds. New York: Springer, pp. 297–335.

Pakes, A., and P. McGuire. 2001. Stochastic algorithms, symmetric Markov perfect equilibria and the curse of dimensionality. *Econometrica* 69: 1261–81.

Pollard, D. 2002. *A User's Guide to Measure Theoretic Probability.* Cambridge: Cambridge University Press.

Prescott, E. C., and R. Mehra. 1980. Recursive competitive equilibrium: The case of homogeneous households. *Econometrica* 48: 1365–79.

Puterman, M. 1994. *Markov Decision Processes: Discrete Stochastic Dynamic Programming.* New York: Wiley.

Quah, D. T. 1993. Empirical cross-section dynamics in economic growth. *European Economic Review* 37: 426–34.

Razin, A., and J. A. Yahav. 1979. On stochastic models of economic growth. *International Economic Review* 20: 599–604.

Reffett, K., and O. F. Morand. 2003. Existence and uniqueness of equilibrium in nonoptimal unbounded infinite horizon economies. *Journal of Monetary Economics* 50: 1351–73.

Rios-Rull, V. 1996. Life-cycle economies with aggregate fluctuations. *Review of Economic Studies* 63: 465–90.

Rincon-Zapatero, J. P., and C. Rodriguez-Palmero. 2003. Existence and uniqueness of solutions to the Bellman equation in the unbounded case. *Econometrica* 71: 1519–55.

Roberts, G. O., and J. S. Rosenthal. 2004. General state space Markov chains and MCMC algorithms. *Probability Surveys* 1: 20–71.

Rockafellar, R. T. 1970. *Convex Analysis.* Princeton: Princeton University Press.

Rogerson, R., R. Shimer, and R. Wright. 2005. Search-theoretic models of the labor market: A survey. *Journal of Economic Literature* 43: 959–88.

Rosenthal, J. S. 2002. Quantitative convergence rates of Markov chains: A simple account. *Electronic Communications in Probability* 7: 123–28.

Rust, J. 1996. Numerical dynamic programming in economics. In *Handbook of Computational Economics*. H. Amman, D. Kendrick, and J. Rust, eds. Burlington, MA: Elsevier, pp. 619–729.

Sargent, T. J. 1987. *Dynamic Macroeconomic Theory*. Cambridge: Harvard University Press.

Santos, M. S., and J. Vigo-Aguiar. 1998. Analysis of a numerical dynamic programming algorithm applied to economic models. *Econometrica* 66: 409–26.

Santos, M. 1999. Numerical solutions of dynamic economic models. In *Handbook of Macroeconomics*, vol. 1A. J. B. Taylor and M. Woodford, eds. Burlington, MA: Elsevier, pp. 311–86.

Scheinkman, J. A., and J. Schectman. 1983. A simple competitive model with production and storage. *Review of Economic Studies* 50: 427–41.

Schilling, R. L. 2005. *Measures, Integrals and Martingales*. Cambridge: Cambridge University Press.

Shiryaev, A. N. 1996. *Probability*. New York: Springer.

Shone, R. 2003. *Economic Dynamics: Phase Diagrams and their Economic Application*. Cambridge: Cambridge University Press.

Stachurski, J. 2002. Stochastic optimal growth with unbounded shock. *Journal of Economic Theory* 106: 40–65.

Stachurski, J. 2003. Economic dynamical systems with multiplicative noise. *Journal of Mathematical Economics* 39: 135–52.

Stachurski, J., and V. Martin. 2008. Computing the distributions of economics models via simulation. *Econometrica* 76: 443–50.

Stachurski, J. 2008. Continuous state dynamic programming via nonexpansive approximation. *Computational Economics* 31: 141–60.

Samuelson, P. A. 1971. Stochastic speculative price. *Proceedings of the National Academy of Science* 68: 335–7.

Sorger, G. 1992. On the minimum rate of impatience for complicated optimal growth paths. *Journal of Economic Theory* 56: 160–79.

Stokey, N. L. 2008. *The Economics of Inaction*. Princeton: Princeton University Press.

Stokey, N. L., and R. E. Lucas, with E. C. Prescott. 1989. *Recursive Methods in Economic Dynamics*. Cambridge: Harvard University Press.

Sundaram, R. K. 1996. *A First Course in Optimization Theory*. Cambridge: Cambridge University Press.

Tauchen, G., and R. Hussey. 1991. Quadrature-based methods for obtaining approximate solutions to nonlinear asset pricing models. *Econometrica* 59: 371–96.

Taylor, J. C. 1997. *An Introduction to Measure and Probability*. New York: Springer.

Tesfatsion, L., and K. L. Judd, eds. 2006. *Handbook of Computational Economics, Volume 2: Agent-Based Computational Economics.* Burlington, MA: Elsevier.

Topkis, D. 1998. *Supermodularity and Complementarity.* Princeton: Princeton University Press.

Torres, R. 1990. Stochastic dominance. Mimeo. Northwestern University.

Turnovsky, S. 2000. *Methods of Macroeconomic Dynamics.* Cambridge: MIT Press.

Uhlig, H. 2001. A toolkit for analysing nonlinear dynamic stochastic models easily. In *Computational Methods for the Study of Dynamic Economies.* R. Marimon and A. Scott, eds. Oxford: Oxford University Press, pp. 30–62.

Venditti, A. 1998. Indeterminacy and endogenous fluctuations in a two-sector optimal growth model with externalities. *Journal of Economic Behavior and Organization* 33: 521–42.

Williams, D. 1991. *Probability with Martingales.* Cambridge, UK: Cambridge University Press.

Williams, N. 2004. Small noise asymptotics for a stochastic growth model. *Journal of Economic Theory* 119: 271–98.

Williams, J. C., and B. C. Wright. 1991. *Storage and Commodity Markets.* Cambridge: Cambridge University Press.

Zelle, J. M. 2003. *Python Programming: An Introduction to Computer Science.* Wilsonville, OR: Franklin Beedle and Associates.

Zhang, Y. 2007. Stochastic optimal growth with a non-compact state space. *Journal of Mathematical Economics* 43: 115–29.

Index

$A \setminus B$, 318
$B(\epsilon; x)$, 39
$D(S)$, 185
$D(\epsilon; x)$, 42
L_1, 184
$P_x \wedge P_{x'}$, 265
S^+, S^-, etc., 252
T, Bellman op., 235
T_σ, 235
$\mathbb{1}_A$, $\mathbb{1}_B$, etc., 321
$\mathbb{N}$, 317
$\mathbb{Q}$, 317
Σ, set of policies, 232
$\mathbb{Z}$, 317
$\mathscr{B}(S)$, $\mathscr{B}(\mathbb{R})$, etc., 165
$\mathscr{C}ibcS$, 299
δ_x, δ_z, etc., 78, 167
ϵ-ball, 39
$\exists$, 320
$\forall$, 320
$\gamma(x, x')$, 277
gr Γ, 230
$\implies$, 319
κ-bounded, 306
$\mathscr{L}$, Lebesgue measurable sets, 162
$\mathscr{L}_1$, 178
μ-a.e., 181
$\mu'(x, x', dy)$, 277
$\mu(f)$, $\lambda(f)$, etc., 177
$\mu(x, x', dy)$, 277
μ^+, μ^-, etc., 252

$\nu(x, x', dy)$, 277
$\mathscr{P}(S)$, 68, 167
$\mathfrak{P}(U)$, $\mathfrak{P}(S)$, etc., 318
σ-algebra, 164
$\sigma(\mathscr{C})$, $\sigma(\mathscr{O})$, etc., 164
bU, bS, etc., 37
$b\ell S$, 256
$b\mathscr{M}(S)$, 252
$b\mathscr{S}$, $b\mathscr{B}(S)$, etc., 174
$b_\kappa S$, 306
$b_\kappa \mathscr{B}(S)$, 306
$b_\kappa cS$, 306
bcU, bcS, etc., 46
d_{FM}, 256
d_∞, 38
d_κ, 306
$f \circ g$, 321
f^+, f^-, etc., 178
$ibcS$, 255
$m\mathscr{S}$, $m\mathscr{B}(S)$, etc., 174
$m\mathscr{S}^+$, 174
r_σ, 232
$s\mathscr{S}$, $s\mathscr{B}(S)$, etc., 172
$s\mathscr{S}^+$, $s\mathscr{B}(S)^+$, etc., 172
w-greedy, 102, 134, 233, 311
$x \vee y$, 331
$x \wedge y$, 331
45 degree diagram
    deterministic model, 57
    stochastic model, 188

Additivity, 161

Adherence, 41
Affinity
    between densities, 201
    between measures, 264
Almost everywhere, 181
Aperiodicity, 96, 292
AR(p) model, 221
Atom, 276
Attractor, 56, 62
Attribute, 19

Banach's fixed point theorem, 53
Banach, S., 51
Bellman equation, 102, 134, 234, 311
Bellman operator, 102, 134, 235, 311
Bellman, R., 99
Berge's theorem, 340
Bifurcation diagram, 65
Bijection, 321
Binding, 17
Blackwell's condition, 149, 307
Bolzano–Weierstrass thm., 47, 329
Borel measurable, 174
Borel sets, 165
Boundedness
    in $\mathbb{R}$, 327
    in metric space, 39
break, 23
Brouwer's fixed point thm., 52
Brouwer, L.E.J., 51

C language, 15
__call__, 33
Carathéodory's condition, 162
Cardinality, 322
Cartesian product, 318
Cauchy sequence, 44, 328
Central limit theorem
    IID, 122, 256
    Markov, 292, 294

Change-of-variable, 194
Chaotic dynamics, 62
Chapman–Kolmogorov eq., 79
Chebychev's inequality, 214
class, 30
Class, Python, 29
Closed ball, 42
Closed set, 41
Closure, 43
Compact-valued, 339
Compactness
    and fixed points, 52, 54
    and optima, 49
    and small sets, 293
    and the Borel sets, 165
    definition, 46
    in $(\mathbb{R}^k, d_2)$, 48
    of state space, 259
Complement, 318
Completeness
    definition, 44
    of $(\mathscr{C}, d_\infty)$, 147
    of $(\mathscr{P}(S), d_{TV})$, 254
    of $(\mathbb{R}^k, d_2)$, 45
    of $(bc\mathcal{U}, d_\infty)$, 46
    of $(b\mathcal{U}, d_\infty)$, 45
    of $(D(S), d_1)$, 185
    of $(L_1(\lambda), d_1)$, 185
    of $\mathbb{R}$, 328
Concatenation, 27
Concavity, 295, 336
Condition M, 266
Conditional probability, 325
Constructor, 30
Continuity
    in metric space, 40
    of correspondences, 339
    of real functions, 334
    open sets definition, 43
Continuum, 323

Contracting, 52
Contrapositive, 320
Convergence
    almost sure, 247
    in $\mathbb{R}$, 327
    in metric space, 38
    in probability, 248
    pointwise, 46, 307
    setwise, 255
    uniform, 46
    weak, 255
    with prob. one, 247
Convex
    function, 336
    set, 52
Coordination game, 110
Correlated shocks, 221, 231, 262
Correspondence, 339
Countable, 322
Countable additivity, 163
Countable cover, 161, 169
Counting measure, 167, 177
Coupling, 272
    finite case, 112
    inequality, 114, 273
    time, 114
Covariance, 213
Credit constraints, 266
Cython, 16

de Morgan's laws, 318
Decreasing
    function, 295
    set, 295
def, 25
Density, 185
    existence of, 217
    expectation w.r.t., 217
    kernel, *see* Stochastic kernel
Dictionary, Python, 27

Differentiability, 336
Diminishing returns, 4, 304
Dirac probability, 167
Distance, 36
Distribution
    finite, 68
    function, 170
    joint, 80
    marginal, 72, 76, 118, 223
    of a random variable, 212
    stationary, *see* Stationary
    unconditional, 72
Dobrushin coefficient
    and coupling, 276
    density case, 201
    finite case, 89
    measure case, 265
Doc string, 26
Doeblin, W., 112, 294
Dominated convergence thm., 182
Drift condition
    density case, 204
    for monotone systems, 283
    measure case, 260
    to a small set, 291
DS (class name), 62
Dynamical system, 55

Empirical distribution, 122, 131, 270
Equivalence class, 184
Equivalence of metrics, 49
Euclidean distance, 35
Euler (in)equality, 302
Expectation
    finite, 326
    general, 212
Expression, Python, 21

Feasible policy, *see* Policy
Feller property, 258

Financial markets, 266
First passage time, 125
Fitted value iteration, 135, 244
Fixed point, 51, 55
for, 23
Fortet–Mourier distance, 256
Fréchet, M., 54
Fubini's theorem, 215
Function
    mathematical, 320
    Python, 25

Generating class, 164, 175
Geometric drift, *see* Drift cond.
Greedy policy, *see* w-greedy

Hahn–Jordan theorem, 252
Hamilton's kernel, 70
Hartman–Grobman theorem, 67
Hausdorff, F., 36, 54
Heine–Borel thm., 48
Hemicontinuity, 339
Homeomorphism, 67

Identifier, 17
IDLE, 17
if/else, 21
IID, 216
Image (of a set), 320
Image measure, 182, 212
Immutable, 27
Imperfect markets, 268
Inada condition, 284, 304
Increasing
    correspondence, 295
    function, 58, 295
    set, 280, 295
Increasing differences, 297
Independence, 215
Indicator function, 321
Infimum

of a subset of $\mathbb{R}$, 332
of two measures, 264
`__init__`, 32
Instance, 29
Integrable function, 178
Interior, 43
Interior Point, 41
Intermediate value thm., 335
Intersection, 318
Invariant
    point, 55
    set, 55
Inventory dynamics, 92
Inverse function, 321
IPython, 17
Irreducibility, 96, 292
Iterate, 55
Iterated function system, 109

Jordan decomposition, 252

Krylov–Bogolubov thm., 259

Lagrange stability, *see* Stability
lambda, 26
Law of large numbers
    correlated, 250
    IID, 94, 249
    Markov, density, 206
    Markov, finite, 94, 251
    Markov, general, 265, 292
Law of Total Probability, 325
Lebesgue
    integral, 177, 178
    measurable set, 162
    measure, 162, 169
    outer measure, 161
Left derivative, 336
Liminf, 333
Limit
    in $\mathbb{R}$, 327

in metric space, 38
Limsup, 333
Linear systems, 59
Linearization, 66
LinInterp (class name), 138
Lipschitz function, 256
List comprehension, 23
List, Python, 18
LLN, *see* Law of large numbers
Logic, 319
Look-ahead estimator
    marginal distribution, 128, 141
    stationary distribution, 131
Loop
    for, 13, 23
    while, 12, 22
Lower bound, 332
Lower-semicontinuity, 40

Markov chain
    density case, 188
    finite, 71
    general case, 219
    periodic, 88
Markov operator
    continuity w.r.t. d_{FM}, 258
    contraction properties, 89, 201, 265
    density case, 195
    finite state, 75, 78
    for SRS, 226
    general case, 224, 225
math, Python module, 19
MATLAB, 16
Matplotlib, 21, 62
Matsuyama, K., 266
Maximizer, 49
Maximum, 49
MC (class name), 72
Measurable
    function, 174, 176

selection, 234
set, 164
space, 164
Measure, 166
    probability, 167
    signed, 252
Measure space, 166
Method, 29
Metric, 36
Metric space, 36
Minimizer, 49
Minimum, 49
Mixing, 200, 202
Module, 19
Monotone convergence thm., 182
Monotone sequence, 327
Monotone subseq., 329
Monotonicity
    and stability, 279
    in parameters, 297
    of an SRS, 280
    of functions, 58
    of measures, 166
    of optimal policies, 299
    of the Bellman operator, 236
Mutable, 27

Namespace, 20
Negative set, 252
Nonconvex growth model
    density kernel, 194
    deterministic, 57
    simulation, 124
    stability, 209
Nonexpansiveness, 52
    and approximation, 136, 244
    of **M** w.r.t. d_1, 85, 196
    of **M** w.r.t. d_{TV}, 253
Norm, 37
Norm-like function, 204, 259

NumPy, 21

Object, Python, 17, 29
Object-oriented programming, 29
One-to-one, 321
Onto, 321
Open ball, 39
Open cover, 47
Open set, 41
Optimal growth
    concavity of v^*, 300
    drift condition, 303
    monotonicity, 297, 299
    optimization, 230, 234
    outline, 133
    stability, 302, 305
    uniqueness of policy, 301
Optimal policy
    definition, 102, 134, 233
    monotonicity of, 299
Order inducing, 282
Order mixing, 281, 283
Order norm-like, 282
Outer measure, 169

Parametric continuity, 339
Parametric monotonicity, 298
Path dependence, 58, 111
Policy
    feasible, 232
    interiority, 301
    optimal, *see* Optimal policy
    savings, 1, 133
    stationary Markov, 231
Policy iteration, 105, 142, 241
Positive set, 252
Pre-measure, 169
Precompactness
    and Lagrange stability, 58
    and norm-like functions, 204

and Prohorov's theorem, 257
    definition, 46
    of densities, 204
Preimage
    definition, 320
    rules concerning, 321
Probability
    finite, 324
    measure, 167
Probability space, 167
Prohorov's theorem, 256
Pseudocode, 11
Pseudometric space, 184
Pylab, 62
PyX, 21

Quadratic map, 62
Quah's kernel, 71

random, Python module, 20
Random variable, 212
Real
    numbers, 327
    sequence, 327
Regeneration, 272
return, 26
Return time, 91
Reward function, 230
Riemann integral, 171, 178
Right derivative, 336

Sage, 16
Scheffés identity, 186
SciPy, 21
Script, 19
SDP, 229
self, 32
Semi-ring, 169
Sequence, 38, 327
Signed measure, *see* Measure
Simple function, 171, 172

Small set, 290
Solow–Swan
    density kernel, 193
    deterministic, 57
    simulation, 118
    stationary distribution, 261
    stochastic kernel, 220
    stochastic stability, 283
SRS, *see* Stochastic recursive seq.
SRS (class name), 118
Stability
    global, 56, 88, 90, 198
    Lagrange, 58, 203
    local, 56
Stationary
    density, 129, 197
    distribution, 85, 257
    point, 55
Stochastic kernel
    density, 126, 188
    finite, 68
    for SRS, 220
    general case, 218
    higher order, 76, 196, 225
Stochastic process, 216, 247
Stochastic recursive sequence
    and density kernel, 189
    canonical, 219
    finite case, 107
    on $\mathbb{R}$, 117
    simulation of, 118
Strict concavity, 295, 301
Strictly increasing diff., 297
String, 18
Sub-additivity, 162, 166
Subsequence
    in $\mathbb{R}$, 329
    in metric space, 39
Supremum, 332

Taylor expansion, 66
Threshold autoregression
    density kernel, 193
    stability, 208
    STAR, 118, 208, 290, 293
Tightness
    of densities, 200, 203
    of measures, 257, 259
Topological conjugacy, 67
Total variation norm, 253
Trajectory, 55
Transition function, 230
Triangle inequality, 36, 183
Tuple, Python, 27

Unbounded rewards, 306
Uncountable, 322
Uniform integrability, 203
Uniformly contracting, 52
Union, 318
Unit simplex, 68
Unpacking, 18
Upper bound, 332
Upper-semicontinuity, 40

Value function
    concavity of, 299
    definition, 102, 134, 233
    differentiability of, 301
    monotonicity of, 296
Value iteration, 103, 238
Variable, Python, 17
Variance, 213
Vector, 35
Vectorized function, 21

Weak convergence, *see* Convergence
Weierstrass theorem, 49
Weighted supremum norm, 306
while, 22